T0202752

# Monetizing Machine Learning

## Quickly Turn Python ML Ideas into Web Applications on the Serverless Cloud

Manuel Amunategui

Mehdi Roopaei

Apress®

*Monetizing Machine Learning: Quickly Turn Python ML Ideas into Web Applications on the Serverless Cloud*

Manuel Amunategui
Portland, Oregon, USA

Mehdi Roopaei
Platteville, Wisconsin, USA

ISBN-13 (pbk): 978-1-4842-3872-1
https://doi.org/10.1007/978-1-4842-3873-8

ISBN-13 (electronic): 978-1-4842-3873-8

Library of Congress Control Number: 2018956745

Managing Director, Apress Media LLC: Welmoed Spahr
Acquisitions Editor: Susan McDermott
Development Editor: Laura Berendson
Coordinating Editor: Rita Fernando

Cover designed by eStudioCalamar

Distributed to the book trade worldwide by Springer Science+Business Media New York, 233 Spring Street, 6th Floor, New York, NY 10013. Phone 1-800-SPRINGER, fax (201) 348-4505, e-mail orders-ny@springer-sbm.com, or visit www.springeronline.com. Apress Media, LLC is a California LLC and the sole member (owner) is Springer Science + Business Media Finance Inc (SSBM Finance Inc). SSBM Finance Inc is a **Delaware** corporation.

For information on translations, please e-mail rights@apress.com, or visit www.apress.com/rights-permissions.

Apress titles may be purchased in bulk for academic, corporate, or promotional use. eBook versions and licenses are also available for most titles. For more information, reference our Print and eBook Bulk Sales web page at www.apress.com/bulk-sales.

Any source code or other supplementary material referenced by the author in this book is available to readers on GitHub via the book's product page, located at www.apress.com/9781484238721. For more detailed information, please visit www.apress.com/source-code.

Printed on acid-free paper

# Table of Contents

# About the Authors

**Manuel Amunategui** is VP of Data Science at SpringML, a Google Cloud and Salesforce preferred partner, and holds Masters in Predictive Analytics and International Administration. Over the past 20 years, he has implemented hundreds of end-to-end customer solutions in the tech industry. The experience from consulting in machine learning, healthcare modeling, six years on Wall Street in the financial industry, and four years at Microsoft, has opened his eyes to the lack of applied data science educational and training material available. To help alleviate this gap, he has been advocating for applied data science through blogs, vlogs, and educational material. He has grown and curated various highly focused and niche social media channels including a YouTube channel (`www.youtube.com/user/mamunate/videos`) and a popular applied data science blog: amunategui.github.io (`http://amunategui.github.io`).

**Mehdi Roopaei (M'02–SM'12)** is a Senior Member of IEEE, AIAA, and ISA. He received a Ph.D. degree in Computer Engineering from Shiraz University on Intelligent Control of Dynamic Systems in 2011. He was a Postdoctoral Fellow at the University of Texas at San Antonio, 2012-Summer 2018, and holds the title of Assistant Professor at the University of Wisconsin-Platteville, Fall 2018. His research interests include AI-Driven Control Systems, Data-Driven Decision Making, Machine Learning and Internet of Things (IoT), and Immersive Analytics. He is Associate Editor of IEEE Access and sits on the Editorial Board of the IoT Elsevier journal. He was guest editor for the special issue: "IoT Analytics for Data Streams" at IoT Elsevier and published a book *Applied Cloud Deep Semantic Recognition: Advanced Anomaly Detection* (CRC Press, 2018). He was IEEE chapter chair officer for joint communication and signal processing communities at San Antonio, Jan-July 2018. He has more than 60 peer-reviewed technical publications, serves on the program committee at several conferences, and is a technical reviewer in many journals.

# About the Technical Reviewers

**Rafal Buch** is a technologist living and working in New York as a financial systems architect. He's been doing software engineering for two decades and spends most of his free time hacking in coffee shops and exploring new technologies. Blog: rafalbuch.com

**Matt Katz** has been working in financial technology since 2001 and still gets excited about new stuff all the time. He lives online at www.morelightmorelight.com and he lives offline in New York with his two strange children and one amazing, patient wife.

# Acknowledgments

To the friends, family, editors, and all those involved in one way or another in helping make this project a reality–a huge thanks! Without your help, this book would have never seen the light of day.

# Introduction

A few decades ago, as a kid learning to program, I had an ASCII gaming book for my Apple II (of which the name eludes me) that started each chapter with a picture of the finished game. This was the teaser and the motivator in a book that was otherwise made up of pages and pages of nothing else but computer code. This was years before GitHub and the Internet. As if it were only yesterday, I remember the excitement of racing through the code, copying it line-by-line, fixing typos and wiping tears just to play the game. Today, a lot has changed, but even though the code is downloadable, we put a screenshot of the final product at the beginning of each chapter, so you too can feel the motivation and excitement of working through the concepts.

## Low-Barrier-To-Entry and Fast-To-Market

This book will guide you through a variety of projects that explore different Python machine learning ideas and different ways of transforming them into web applications. Each chapter ends with a serverless web application accessible by anyone around the world with an Internet connection. These projects are based on classic and popular Python data science problems that increase in difficulty as you progress. A modeling solution is studied, designed, and an interesting aspect of the approach is finally extended into an interactive and inviting web application.

Being a data scientist is a wonderful profession, but there is a troubling gap in the teaching material when trying to become one. Data science isn't about statistics and modeling; it is about fulfilling human needs and solving real problems. Not enough material tackles the big picture. it seems that whenever you start talking about the big picture, you have to sign a non-disclosure agreement (NDA). This is an essential area of study and if you are like me, you need to understand why you are doing something in order to do it right. These aren't difficult topics, especially when you use the right tools to tackle them.

We won't focus on **"becoming a data scientist"** as an end goal; there are plenty of books on that topic already. Instead, we'll focus on getting machine learning products to market quickly, simply, and with the user/customer in mind at all times! That's what is missing in this profession's educational syllabus. If you build first and then talk to your customer, your pipelines will be flawed and your solutions will miss their target. I have redrawn Drew Conway's Data Science Venn Diagram with the customer as top priority (Figure 1).

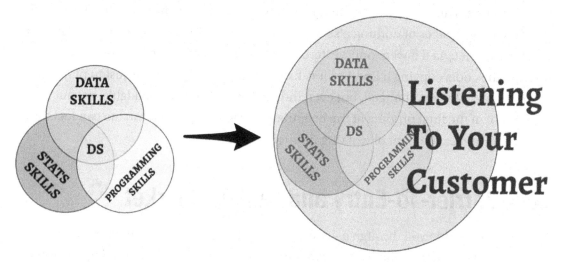

***Figure 1.*** *The classic data science Venn diagram next to my updated version*

[1]Mehdi and I worked hard on the content of this book. We took our time to develop the concepts, making sure they were of practical use to our reader (i.e., our customer–always keep the customer in mind at all times). I built the material and Mehdi edited it. This is an ambitious book in terms of scope and technologies covered. Choices and compromises had to be made to focus on the quickest ways of getting practical use out of the material. The tools are constantly changing. Some things in this book are going to be stale by the time you read them, and that is OK (you can go to the GitHub repo for updates). Here, everything changes all the time, but things tend to change for the better! So, learning new tricks often means learning better, faster, and more powerful ways to do things. This book will not only show you how to build web applications but also point you in the right direction to deepen your knowledge in those areas of particular interest.

---

[1]http://drewconway.com/zia/2013/3/26/the-data-science-venn-diagram

If this was a class, I'd have you sign a "**compete agreement**": yes, the opposite of a non-compete. I would have you go through this book, understand the tools, and then copy them and make them your own. These are meant to be templates to quickly get your platforms up and running to focus on the bigger things, to build impactful tools for your customers and friends. When you understand this, that's the day you graduate with all the entitlements and privileges of being called a "**data science professional.**"

## What is the Serverless Cloud?

Cloud providers have gone to great efforts to improve web hosting solutions and bring costs down. The recent advent of the "**serverless**" option, which abstracts a large swath of the configuring process, is available on three of the four cloud providers covered in this book. This means you can get your projects up and running on fully managed platforms, with automatic load-balancing, throughput scaling, fast deployments, etc., without having to select, configure, or worry about any of it. The level of disengagement with these architectural and monitoring options is up to you. You can choose what you want to control and what you want to delegate to the provider. One thing is guaranteed: the site will automatically adjust with traffic and offer unparalleled uptime.

This allows us to focus on what is important without getting bogged down by the trappings and support needed to get there. These so-called "**trappings**" are critical and need to be taken very seriously. This is why we are looking at four reputable cloud providers that will give us the peace of mind required to fully focus on our web applications and not worry about the site crashing or the provider going dark. Let's focus on what is important and get to work!

## Critical Path in Web Application Development

So many machine learning models stagnate in their original coded state: hard to understand, with no easy way to invite others to benefit from its insights. These models are doomed to be forgotten. Even those that manage to escape the confines of an integrated development interface fall short of their potential when reduced to a static chart or a cryptic modeling score. This book aims to avoid this trap by breaking down the process of extending a machine learning model into a universally accessible web application. Each chapter follows these three critical steps:

1. **Modeling the right way**. We start at the end, by understanding what users want to see, and by investing time and thought on the final goal and user experience. We ensure that the appropriate modeling approach is used in order to reach a web-application state rapidly and without surprise (Figure 2).

**Figure 2.** *Always check that there is an audience for your idea before building (source Lucas Amunategui)*

2.  **Designing and developing a local web application**. This step requires leveraging various web front-end technologies to offer the needed interactivity and dynamism to highlight a model's insight and sustain a user's interest. The final product at this stage is indistinguishable from the next one except that it is hosted on your local machine, not the cloud.

3.  **Deploying onto a popular and reliable serverless cloud provider**. Each provider has unique requirements, advantages, and disadvantages that need to be understood and addressed. This is the final stage where the world gets to enjoy and learn from your work.

We start by tackling easy ways of offering intelligent interactivity, like leveraging a model's coefficients or saving a trained model, then move to the complex, like using a database to track engagement or relying on open-source pretrained models for image recognition. A fictional case study around stock market predictions is started in the first section, then revisited in subsequent ones. New features are added to it until it culminates into a complex dashboard with a paywall to offer customized intelligence to paying subscribers.

By focusing on classic, data science problems, coupled with popular, open-source technologies like Python, Flask, Ajax, Bootstrap, JavaScript, and HTML, you should find yourself on familiar ground and if you don't, you'll have a drastically reduced learning curve. We focus on simple tools and simple techniques to reliably and rapidly get machine learning ideas out into the wild. The tools and approaches are revisited in each chapter, so don't worry if some aspects aren't clear from the start; keep going and things will keep getting clearer.

We also rotate cloud providers in each chapter, so you will get exposed to the most popular providers. This will give you plenty of insights into which provider to select for your future project. I recommend going through all chapters, as each will show different ways of doing things, highlighting a provider's strengths along with showing unique tips and tricks to get things done quickly.

# You, the Reader

This book is for those interested in extending statistical models, machine learning pipelines, data-driven projects, or any stand-alone Python script into interactive dashboards accessible by anyone with a web browser. The Internet is the most powerful medium with an extremely low barrier to entry–anybody can access it, and this book is geared to those who want to leverage that.

This book assumes you have Python and programming experience. You should have a code editor and interpreter in working order at your disposal. You should have the ability to test and troubleshoot issues, install Python libraries, and be familiar with popular packages such as NumPy, Pandas, and Scikit-learn. An introduction to these basic concepts isn't covered in this book. The code presented here uses Python 3.x only and hasn't been tested for lower versions. Also, a basic knowledge of web-scripting languages will come in handy.

This book is geared towards those with an entrepreneurial bent who want to get their ideas onto the Web without breaking the bank, small companies without an IT staff, students wanting exposure and real-world training, and for any data science professional ready to take things to the next level.

# How to Use This Book

Each chapter starts with a picture of the final web application along with a description of what it will do. This approach serves multiple purposes:

- It works as a motivator to entice you to put in the work.

- It visually explains what the project is going to be about.

- And more importantly, it teaches how critical it is to have a clear customer-centric understanding of the end game whenever tackling a project.

The book will only show highlights of the source code, but complete versions are attached in the corresponding repositories. This includes a Jupyter notebook when covering data exploration and zipped folders for web applications.

The practical projects presented here are simple, clear, and can be used as templates to jump-start many other types of applications. Whether you want to understand how to create a web application around numerical or categorical predictions, the analysis of text, the creation of powerful and interactive presentations, to offer access to restricted data, or to leverage web plugins to accept subscription payments and donations, this book will help you get your projects into the hands of the world quickly.

---

**Note** For edits/bugs, please report back at www.apress.com/9781484238721.

---

# Tools of the Trade and Miscellaneous Tips

Here is a brief look at the tools that will transform our machine learning ideas into web applications quickly, simply, and beautifully. This isn't meant to be a comprehensive or complete list, just a taste of the different technologies used and pointers for you to follow if you want to learn more (and I hope you will).

## Jupyter Notebooks

The book only shows code snippets, so you need to download and run the Jupyter notebook for each chapter in order to run things for yourself and experiment with the various features and different parameters. If you aren't familiar with Jupyter notebooks, they are web-based interactive Python interpreters great for building, tweaking, and publishing anything that makes use of Python scripting. It attaches to a fully functioning Python kernel (make it a Python 3.x one) and can load and run libraries and scripts just like any other interpreter. To install Jupyter notebooks, follow the official docs at http://jupyter.readthedocs.io/en/latest/install.html.

There are various ways to install it, including the "**pip3**" command; check official documentation for the different ways of doing it if this approach doesn't work for you (Listing 1).

*Listing 1.* Install Jupyter

```
sudo pip3 install jupyter
```

To use a Jupyter notebook is both easy and powerful at the same time. You simply need to download the notebook to your local machine (it will be the file with a *.ipynb extension), open a command/terminal shell window, navigate to that folder, and run the "**notebook**" command (Listing 2).

***Listing 2.*** Run a Notebook (check official docs for alternative ways of starting notebooks)

```
jupyter notebook
```

This command will open a web page showing the content of the folder from where it was launched (Figure 3). You can navigate down a folder structure by clicking the folder icon right above the file listings.

***Figure 3.*** *Jupyter notebook landing page*

To open a Jupyter notebook, simply click any file with the "*.**ipynb**" extension and you are good to go! If you want to create a brand-new notebook, click the "**new**" button at the right of the dashboard next to the refresh button.

---

**Note**    For additional information, issues with Jupyter notebooks, and attaching kernels, see: http://jupyter-notebook-beginner-guide.readthedocs. io/en/latest/execute.html.

---

# Flask

Flask is a lightweight but very powerful server-side web framework. It is the brains behind all the applications presented in this book. It is also the glue between our Python data producing functions and web pages. That is one of the reasons I love working with Flask, as it allows us to link stand-alone Python scripts to server-side web frameworks without leaving the Python language; it makes passing data between objects a whole lot easier!

Flask comes with the bare minimum to get a web page published. If you need additional support, like a database, form controls, etc., you will have to install additional libraries and that is why it is called a lightweight, microframework. This is also what makes it so easy to use, as you only have to learn a few more tricks; everything else uses the tried-and-true Python libraries that we're already familiar with.

Unfortunately, we can only work in Python for so long and eventually you will need to step into front-end web scripting. But don't let that bother you; there are so many great examples on the Web (Stackoverflow.com, w3schools.com) and the incredible looking GetBootstrap.com templates to get you there as quickly as possible.

---

**Note** For more information on Flask, check the official Flask documentation at `http://flask.pocoo.org/`.

---

# HTML

HTML, which means Hypertext Markup Language, needs no introduction. This is the lowest common denominator in terms of web technologies. It has been around for years and is used to create practically all web pages and web applications in existence.

For those wanting to brush up on this topic, the amount of free material on the Web can be overwhelming. I recommend anything from w3schools.com. Their material is well organized, comprehensive, and often interactive.

# CSS

Cascading Style Sheets (CSS) is what makes most websites out there look great! We use two types of CSS files here: the CSS links loaded in the "**<HEAD>**" section of most web pages (the most common) and custom CSS as shown in code snippet in Listing 3.

***Listing 3.*** Custom CSS Script Block

```
<STYLE>
.btn-circle.btn-xl {
    width: 70px;
    height: 70px;
    padding: 10px 2px;
    border-radius: 35px;
    font-size: 17px;
    line-height: 1.33;
}
</STYLE>
```

The CSS files that are hosted on outside servers cannot be customized but are usually best-in-class. But there are times you simply need to customize a feature on your page, and that is when you create a local CSS file or a style tag directly in the HTML page. It is then applied to a particular tag or area using the "**class**" parameter (Listing 4).

***Listing 4.*** Applying CSS Tag to an HTML Tag

```
<button type="button" onclick="calculateBikeDemand(this)"
id="season_spring" class="btn btn-info btn-circle btn-xl">
<i class="fa fa-check">Spring</i></button>
```

CSS defines in great detail what size, color, font, everything and anything under the sun, should look like. It also allows the generalization of your look-and-feel through your web portal. You create it once and can have all your pages invoke it to inherit that particular style.

---

**Note**   For additional information and training on CSS, check out the w3schools.com.

---

# Jinja2

Jinja2 is used to generate markup and HTML code, and works tightly with Flask variables. It is created by Armin Ronacher, and is widely used to handle Flask-generated data and if/then logic directly in HTML templates.

In this HTML template example, a Flask-generated value called "**previous_slider_value**" is injected into the slider's "**value**" parameter using Jinja2. Note the use of double curly brackets (Listing 5).

***Listing 5.*** Jinja2 Passing Data to HTML Input Control

```
<input type="range" min="1" max="100" value="{{previous_slider_value}}"
id="my_slider">
```

> **Note**   For additional information on Jinja2, check out the docs at
> `http://jinja.pocoo.org/docs/2.10/`.

# JavaScript

JavaScript is a real programming language in and of itself. It can add extremely powerful behavior to any of your front-end controls. JavaScript brings a great level of interactivity to a web page, and we will leverage it throughout each chapter.

Here is an interesting example where we capture the mouse-up event of an HTML slider control to submit the form to the Flask server. The idea is that whenever a user changes the slider value, Flask needs to do some server-side processing using the new slider value and regenerate the web page (Listing 6).

***Listing 6.*** JavaScript Capturing Slider "**onmouseup**" Event

```
slider1.onmouseup = function ()
{
    document.getElementById("submit_params").submit();
}
```

> **Note**   For additional information and training on JavaScript, check out
> w3schools.com.

# jQuery

JQeury is a customized JavaScript library to facilitate the handling of complex front-end and behavior events, and insures compatibility between different browser versions.

jQuery will facilitate things like button, drop-down dynamic behavior, even Ajax (a critical technology used heavily in many of this book's projects).

---

**Note** For more information on JQuery, check out the official documents at JQuery.com.

---

# Ajax

Ajax is a great front-end scripting technique that can add dynamic server-side behavior to a web page. It allows sending and receiving data without rebuilding or reloading the entire page as you would do with form submits. One area where it is commonly used is on map web pages, such as Google Maps, which allows dragging and sliding the map without reloading the entire page after every move.

---

**Note** For additional information and training on Ajax, check out w3schools.com.

---

# Bootstrap

Bootstrap is a very powerful, almost magical tool for front-end web work. It is used by almost 13% of the Web according to BuiltWith Trends.[2] It contains all sorts of great looking styles and behavior for most web tags and controls. By simply linking your web page to the latest Bootstrap, CSS will give any boring HTML page an instant and professional looking makeover!

If you look at any of the HTML files in this book, the first thing you will notice are the links wrapped in "**LINK**" and "**SCRIPT**" tags at the top of the page. These represent an enormous shortcut in building a web page (Listing 7).

---

[2]https://trends.builtwith.com/docinfo/Twitter-Bootstrap

***Listing 7.*** Link Tag to Inherit Bootstrap CSS Styles

```
<LINK rel="stylesheet" href="https://maxcdn.bootstrapcdn.com/
bootstrap/4.0.0/css/bootstrap.min.css">
```

All the HTML files in this book (and more than likely any web page you will create in the future) will use these links to download premade bootstrap and JavaScript scripts and automatically inherit the beautiful fonts, colors, styles, and behaviors that are prevalent all over the Internet. This allows you to immediately inherit best-in-class looks and behavior with minimal effort.

---

**Note**   For additional information and training on Bootstrap, check out the official documents on GetBootstrap.com.

---

# Web Plugins

Web plugins have a huge advantage: they push a large swath of hardware, data, and/or security management onto someone else, preferably someone specialized in that area. There is no reason to reinvent the wheel, waste valuable time, or introduce security risks. Let others take care of that and focus on what you do best; this is what this book is all about!

Unfortunately, we only have the bandwidth to explore a few of these, but here is a list of good ones that I've either used in the past or heard good things from others (and there are hundreds more out there that are probably as good–look for those that offer good terms for small businesses along with demo or test accounts to experiment before committing).

## Membership Platforms

There are several platforms to note.

**Memberful** (`www.memberful.com`)

Memberful is the plugin we will work with and implement in this book. I personally really like Memberful.com and think it is a great choice for anybody looking for an easy way to manage a paywall section of a website. It offers credit card payment through Stripe.com, offers user-management features, and is tightly integrated within your own web application.

**Patreon** (`www.patreon.com`)

Patreon is a membership platform and plugin for artists and content creators.

**Wild Apricot** (`www.wildapricot.com`)

Wild Apricot is a membership platform for small and nonprofit organizations.

**Subhub** (`www.subhub.com`)

Subhub is a membership platform designed for entrepreneurs, experts, and organizations.

**Membergate** (`www.membergate.com`)

Membergate is a platform for corporate communications, newsletters, associations, and restricted access sites.

# Payment Platforms

There are several platforms available.

**Paypal Donations** (`www.paypal.com/us/webapps/mpp/donation`)

I've used Paypal plugins in the past and have been delighted with the ease of installation and use. All you need is a Paypal account in good standing and the rest is a cinch.

Paypal Express (`www.paypal.com/us/webapps/mpp/express-checkout`)

Paypal Express is still Paypal but for quick and easy checkouts.

**Stripe** (`http://stripe.com/`)

Stripe is a payment options that easily allow websites to accept online credit card payments. It is the payment engine behind Memberful.com that we will see in the last chapter of this book.

# Analytics

Building your own web-usage tracker requires a lot of Flask custom code on every page, along with a database to save those interactions and an analytical engine to make sense of it. That's a lot of work! Instead, with Google Analytics, all we have to do is add a JavaScript snippet of code at the top of each page. It is free for basic analysis, which is fine for our needs.

# Message Boards

I have used `https://disqus.com` in the past to add message boards to static websites. It creates the appearance of professional-looking message boards directly on your site, all the while being managed elsewhere.

# Mailing Lists

I have used formspree.io for many years and love it! It is trivial to add to any static web page along with a text box and submit button. Users can add their email address on your web page and `https://formspress.io` will email you the submitted information. This is a great option if you are hosting a static site or don't want to deal with managing your own database.

# Git

Git is a great version control tool; it allows you to store your code's creation, changes, updates, and any deletions happening in a repository. It is tightly integrated with GitHub, which is critical for code safeguard and collaboration. It is also integrated on most of the cloud providers out there. In some chapters we will use it and in others we won't. If you end up working on larger applications or collaborate with others, I highly recommend you start using it.

Most cloud providers support online code repositories like GitHub, BitBucket, and others. These online repos work with Git, so learning the basics will give you a big leg up. The process of deploying web applications on Microsoft Azure is tightly integrated with Git, so please take a look at this basic primer or go online for some great tutorials such as try.github.io[3]:

- **git init**: creates a local repository

- **git clone `https://github.com/`**... clones a GitHub repository to your local drive

- **git status**: list files that are changed and awaiting commit + push to repo

- **git add .**: add all files (note period)

- **git add '*.txt'**: add all text files

- **git commit**: commit waiting files

- **git log**: see commit history

---

[3]`https://try.github.io/`

- **git push (or git push azure master)**: push branches to remote master

- **git pull**: get remote changes to local repo

- **git reset \***: to undo git

- **git rm --cached <file>**: stop tracking a file

# Virtual Environments

Using a virtual environment offers many advantages:

- Creates an environment with no installed Python libraries

- Knows exactly which Python libraries are required for your application to run

- Keeps the rest of your computer system safe from any Python libraries you install in this environment

- Encourages experimentation

To start a virtual environment, use the "**venv**" command. If it isn't installed on your computer, it is recommended you do so (it is available via common installers such as pip, conda, brew, etc). For more information on installing virtual environments for your OS, see the "**venv - Creation of virtual environments**" user guide: `https://docs.python.org/3/library/venv.html`.

Open a command window and call the Python 3 "**venv**" function on the command line to create a sandbox area (Listings 8 and 9).

***Listing 8.*** Creating a Python Virtual Environment

```
$ python3 -m venv some_name
```

***Listing 9.*** Activating the Environment

```
$ source some_name/bin/activate
```

Once you are done, you can deactivate your virtual environment with the command in Listing 10.

*Listing 10.* Deactivating Virtual Environment

```
$ deactivate
```

# Creating a "requirements.txt" File

The "**requirements.txt**" file is used by most cloud providers to list any Python libraries needed for a hosted web application. In most cases, it is packaged alongside the web files and sent to its "**serverless**" destination for setup.

You can create your own "**requirements.txt**" and house it in the same folder as you main Flask Python script. Let's see how we can use virtual environments to create a complete "**requirements.txt**" file. When using a virtual environment, you are creating a safe sandbox free of any Python libraries. This allows you to install only what you need and run a "**pip freeze**" command to get a snapshot of the libraries and current version numbers. Mind you, you don't need to do this if you already know what libraries, dependencies, and version numbers you need. As a matter of fact, you can use the "**requirements.txt**" files packaged with this book content just as well.

## Step 1

Start with a clean slate by creating a virtual environment in Python as shown in Listing 11.

*Listing 11.* Starting Virtual Environment

```
$ python3 -m venv some_env_name
$ source some_env_name/bin/activate
```

## Step 2

Use "**pip3**" to install libraries needed to run a local web application, as shown in listing 12.

*Listing 12.* Installing Some Libraries as an Example

```
$ pip3 install flask
$ pip3 install pandas
$ pip3 install sklearn
```

# Step 3

Freeze the environment and pipe all installed Python libraries, including version numbers in the "**requirements.txt**" file, as shown in Listing 13.

***Listing 13.*** Installed Required Libraries

```
$ pip3 freeze > requirements.txt
```

# Step 4

Finally, deactivate your virtual environment, as shown in Listing 14.

***Listing 14.*** Deactivate out of venv

```
$ deactivate
```

There you go: you've just created a "**requirements.txt**" file. Check out its content by calling "**vi**" (click Escape and q to exit). The contents of your "**requirements.txt**" may look very different, and that's OK (Listing 15).

***Listing 15.*** Checking the Content of "**requirements.txt**" File

**Input:**

```
$ vi requirements.txt
```

**Ouput:**

```
click==6.7
Flask==0.12.2
itsdangerous==0.24
Jinja2==2.10
MarkupSafe==1.0
numpy==1.14.2
scikit-learn
scipy
python-dateutil==2.7.2
pytz==2018.4
six==1.11.0
```

```
Werkzeug==0.14.1
Pillow>=1.0
matplotlib
gunicorn>=19.7.1
wtforms>=2.1
```

Inside the requirements.txt file, you can require a specific version by using the "=="
sign (Listing 16)

***Listing 16.*** Exact Assignment

```
Flask==0.12.2
```

You can also require a version equal to and larger, or equal to and smaller (Listing 17)

***Listing 17.*** Directional Assignment

```
Flask >= 0.12
```

Or you can simply state the latest version that the installer can find (Listing 18)

***Listing 18.*** Use Latest Version Available

```
Flask
```

# Conclusion

This was only meant to be a brief introduction to the tools used in this book. Use these as
jumping off points to explore further and to deepen your knowledge in the areas that are
of particular interest to you.

## CHAPTER 1

# Introduction to Serverless Technologies

We're going to create a very simple Flask web application (Figure 1-1) that we will reuse in the next four sections when we explore cloud-based services from Amazon AWS, Google Cloud, Microsoft Azure, and Python Anywhere.

*Figure 1-1. Flask*

It is a good idea to start with a local version of your website working on your local machine before venturing out onto the cloud.

---

**Note**  Download the files for Chapter 1 by going to `www.apress.com/9781484238721` and clicking the source code button. Open Jupyter notebook "**chapter1.ipynb**" to follow along with this chapter's content.

---

1

© Manuel Amunategui, Mehdi Roopaei 2018
M. Amunategui and M. Roopaei, *Monetizing Machine Learning*, https://doi.org/10.1007/978-1-4842-3873-8_1

# A Simple Local Flask Application

The code in this section is very simple, and you can either write it from scratch or use the files in folder "**simple-local-flask-application**."

## Step 1: Basic "Hello World!" Example

Let's create a very simple Flask script straight from the official Flask help docs (Listing 1-1).

***Listing 1-1.*** Simple Flask Script

```
from flask import Flask
app = Flask(__name__)

@app.route("/")
def hello():
    return "Hello World!"
```

That's it! Even-though this doesn't do much, it represents the minimum of what is needed to get a Flask website up and running; it is a real web application. Save the script as "**main.py**" anywhere on your local machine (and you can name it anything you want).

## Step 2: Start a Virtual Environment

It's always a good idea to segregate your development work using virtual environments from the rest of your machine (and this also comes in handy when building "**requirements.txt**" files–see the section on "**Creating a 'requirements.txt' File**" in the introduction). Let's start a virtual environment (Listing 1-2).

***Listing 1-2.*** Starting a Virtual Environment

```
$ python3 -m venv simple_flask
$ source simple_flask/bin/activate
```

# Step 3: Install Flask

Install Flask. This assumes you can run "**pip3**" to install libraries; otherwise use whatever install tools you would normally use for Python 3.x or check the official Flask docs.[1] See Listing 1-3.

***Listing 1-3.*** Install Flask

```
$ pip3 install Flask
```

# Step 4: Run Web Application

Open a command/terminal window and enter the following command on the Mac or Windows (Listings 1-4 and 1-5).

***Listing 1-4.*** On the Mac

```
$ export FLASK_APP=main.py
$ flask run
```

***Listing 1-5.*** On Windows

```
$ export FLASK_APP= main.py
$ python -m flask run
```

# Step 5: View in Browser

You should see the following message in the command window offering a local "**HTTP**" address to follow. Copy it and drop it into the address bar of your browser (Listing 1-6).

---

[1]http://flask.pocoo.org/

***Listing 1-6.*** Flask Application Successfully Running on Local Machine

```
manuel$ export FLASK_APP=main.py
manuel$ flask run
 * Serving Flask app "main.py"
 * Environment: production
   WARNING: Do not use the development server in a production environment.
   Use a production WSGI server instead.
 * Debug mode: off
 * Running on http://127.0.0.1:5000/ (Press CTRL+C to quit)
```

Then open a browser and copy/paste (or type in) the local address listed (Figure 1-2).

***Figure 1-2.*** *Local Flask application running as expected*

## Step 6: A Slightly Faster Way

There you have it, a real server-generated web page. There is an even easier way you can get your Flask app up and running locally by adding the following two lines to the end of your "**main.py**" script. This only works in local mode but allows the script itself to run the instantiated Flask application and allows you to skip the exporting step (Listing 1-7).

***Listing 1-7.*** Automatically Starting Flask Scripts in Local Mode

```
if __name__ == '__main__':
    app.run(debug=True)
```

Save the amended "**main.py**" script, which should look like Listing 1-8.

***Listing 1-8.*** Full Flask Script

```
from flask import Flask
app = Flask(__name__)

@app.route("/")
def hello():
    return "Hello World!"

if __name__=='__main__':
        app.run(debug=True)
```

Go back to your command/terminal window pointing to the amended script and enter the shorter version (Listing 1-9).

***Listing 1-9.*** Easier Command to Start Local Flask Script

```
$ python3 main.py
```

## Step 7: Closing It All Down

To stop the web application from serving the "**Hello World!**" page, hit "**ctrl-c**" in your terminal window.

We turned on the "**debug**" flag to true in the last line of the script. This will print any Flask errors with the script directly into the browser. This is a great feature for quickly developing and debugging scripts, but remember to turn it to false before moving it to the cloud.

Finally, terminate your virtual environment (Listing 1-10).

***Listing 1-10.*** Closing the Virtual Environment

```
$ deactivate
```

## Introducing Serverless Hosting on Microsoft Azure

The Azure Cloud offers an easy-to-use, serverless and fully managed platform for web applications, with plenty of customizable options ranging from storage to databases, monitoring, and analytics (Figure 1-3).

*Figure 1-3.* *Microsoft Azure*

Let's see how we can run our basic Flask application on Microsoft Azure's serverless web apps. Here, we'll keep the steps as simple as possible, as we'll drill down deeper into this provider in subsequent chapters.

---

**Note**   Download the files for Chapter 1 by going to www.apress.com/ 9781484238721 and clicking the source code button, and open the "**serverless-hosting-on-microsoft-azure**" folder.

---

## Step 1: Get an Account on Microsoft Azure

You will need an account on Microsoft Azure and at the time of this writing, Microsoft offers a convenient $200 30-day trial on all services and 12 months access. For more information, see: https://azure.microsoft.com/en-us/free/.

## Step 2: Download Source Files

Download the files for this chapter onto your local machine and navigate to the folder named "**serverless-hosting-on-microsoft-azure**." Your local folder structure should look like the following (notice the name of the Flask script "**main.py**," the default on Azure; Listing 1-11).

***Listing 1-11.*** All Files Needed for Our Web Application on Azure

```
serverless-hosting-on-microsoft-azure/
├── main.py
├── ptvs_virtualenv_proxy.py
├── requirements.txt
└── web.3.4.config
```

## Supporting Files

The "**requirements.txt**" file holds the Python library names the web application needs and is used by the serverless cloud during the application's deployment. You can create your own "**requirements.txt**" and house it in the same folder as the Flask script "**main. py**." In this case it contains only one library and a version requirement (Listing 1-12).

***Listing 1-12.*** All Files for our Web Application

```
Flask>=0.12
```

The "**web.3.4.config**" is the web server's configuration file. We will use the Python 3.4 version and go with the defaults. If you decide to explore this cloud-provider further, then definitely crack it open and take a look inside.

## Step 3: Install Git

For this project you will need to have Git installed on your local machine (you can find the install binaries at www.git-scm.com/downloads). As stated earlier, Git is a source code versioning tool and it is a fully prepared Git package that we will push out to Microsoft Azure (see the brief primer on Git in the introduction section). In most chapters we would create a virtual environment to run the following steps, but as this is already a big project, we'll keep it simple and skip it.

Open your terminal/command-line window and point it to this chapter's "**serverless-hosting-on-microsoft-azure**" folder and initialize a Git session (Listing 1-13).

***Listing 1-13.*** Initialize a Git Session

```
$ git init
```

Next, add all the web-application files from the "**serverless-hosting-on-microsoft-azure**" folder and check its status (Listing 1-14 and Figure 1-4).

*Listing 1-14.* Add All Files in Folder to Git and Check Status

```
$ git add .
$ git status
```

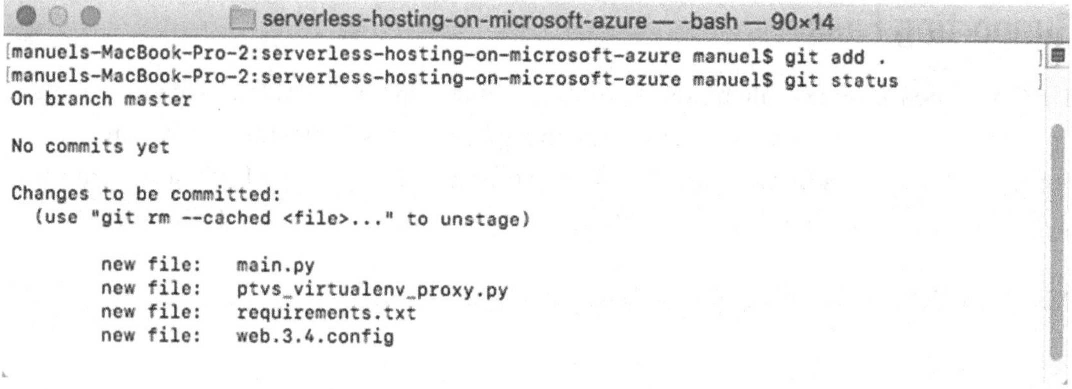

```
● ○ ●          serverless-hosting-on-microsoft-azure — -bash — 90×14
[manuels-MacBook-Pro-2:serverless-hosting-on-microsoft-azure manuel$ git add .
[manuels-MacBook-Pro-2:serverless-hosting-on-microsoft-azure manuel$ git status
On branch master

No commits yet

Changes to be committed:
  (use "git rm --cached <file>..." to unstage)

        new file:   main.py
        new file:   ptvs_virtualenv_proxy.py
        new file:   requirements.txt
        new file:   web.3.4.config
```

*Figure 1-4. Shows the web application files ready for commit*

Do a local Git commit and add a comment that makes sense, in case you need to revisit your past actions in the future (Listing 1-15).

*Listing 1-15.* Committing Files to Git

```
$ git commit -am "Intro to Cloud Azure commit"
```

All the needed files are in the local repository. For more information on the Git Deployment to Azure App Service, go to https://docs.microsoft.com/en-us/azure/app-service/app-service-deploy-local-git.

# Step 4: Open Azure Cloud Shell

Log into your Microsoft Azure dashboard and open the Azure Cloud Shell by clicking the caret-underscore (Figure 1-5).

**Figure 1-5.** *Starting Azure Cloud Shell*

You will be prompted to create either a Linux or Power Shell window. Go with Linux, as the commands will be similar to what you use in your local terminal window (Figure 1-6).

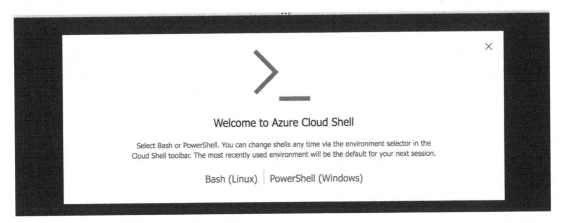

**Figure 1-6.** *Choosing between Bash (Linux) or PowerShell (Windows); go with the familiar, Linux*

It will also prompt you to create storage, which you will need in order to host the application (Figure 1-7). If this is your first time, you may see the option "**Free Trial**" in the drop-down. Either way, go with it and create storage.

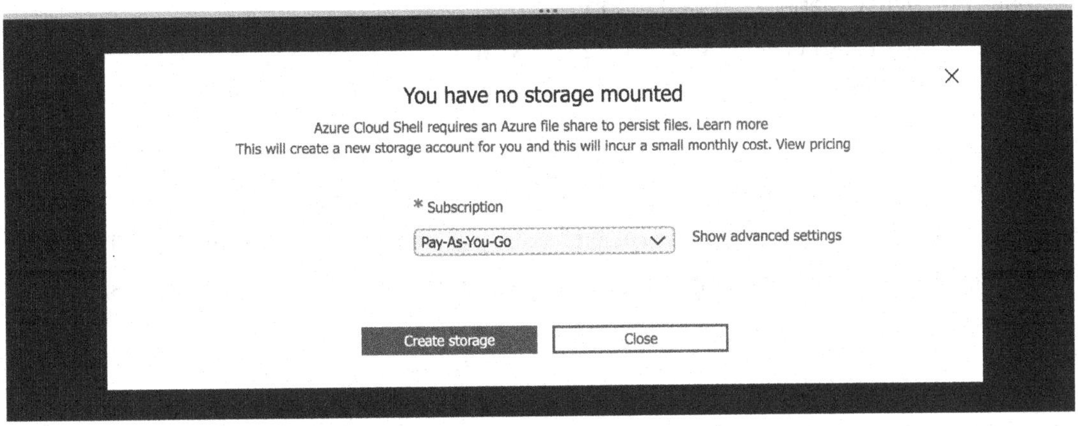

**Figure 1-7.** *You need to create a storage repository*

# Step 5: Create a Deployment User

You should now be in the Azure Cloud Shell. This user will have appropriate rights for FTP and local Git use. Here I set my user name to "**flaskuser11**" and password to "**flask123**"; come up with your own name and remember it, as you will need it later on (Listing 1-16).

***Listing 1-16.*** Creating a User

```
$ az webapp deployment user set --user-name <<REPLACE-WITH-YOUR-USER-NAME>>
--password flask123
```

The response from the command and most subsequent commands should look like the following screen shot. Look closely for any errors or issues and fix accordingly (Listing 1-17).

***Listing 1-17.*** Response Format from Azure's "**webapp**" Commands

```
manuel@Azure:~$ az webapp deployment user set --user-name flaskuser11
--password flask123
{
  "id": null,
  "kind": null,
  "name": "web",
  "publishingPassword": null,
  "publishingPasswordHash": null,
  "publishingPasswordHashSalt": null,
  "publishingUserName": "flaskuser11",
  "type": "Microsoft.Web/publishingUsers/web",
  "userName": null
}
```

Your output JSON should be full of nulls; if you see "**conflict,**" your "**user-name**" isn't unique and if you see "**bad request**," your password isn't compliant (it should be at least eight characters long and made up of a mix of characters, numbers, or symbols).

# Step 6: Create a Resource Group

Here we create a resource group for a location close to you–in my case "**West US**" (for locations, see https://azure.microsoft.com/en-us/regions/ or use the command "**az appservice list-locations --sku FREE**"–see Listing 1-18).

***Listing 1-18.*** Creating a Resource Group and Response

**Input:**

```
$ az group create --name myResourceGroup --location "West US"
```

**Output:**

```
manuel@Azure:~$ az group create --name myResourceGroup --location "West US"
{
  "id": "/subscriptions/1e9ea6de-d6b9-44a5-b319-68b0ab52c2bc/resource
  Groups/myResourceGroup",
  "location": "westus",
  "managedBy": null,
  "name": "myResourceGroup",
  "properties": {
    "provisioningState": "Succeeded"
  },
  "tags": null
}
```

# Step 7: Create an Azure Service Plan

Create your Azure service plan. Set the name to "**myAppServicePlan**" (it can be whatever you want; Listing 1-19).

***Listing 1-19.*** Creating a Service Plan and Successful Response

**Input:**

```
$ az appservice plan create --name myAppServicePlan --resource-group
myResourceGroup --sku FREE
```

**Truncated Output:**

```
manuel@Azure:~$ az appservice plan create --name myAppServicePlan
--resource-group myResourceGroup --sku FREE
{
  "adminSiteName": null,
  "appServicePlanName": "myAppServicePlan",
  "geoRegion": "West US",
  "hostingEnvironmentProfile": null,
  "id": "/subscriptions/1e9ea6de-d6b9-44a5-b319-68b0ab52c2bc/resource
  Groups/myResourceGroup/providers/Microsoft.Web/serverfarms/
  myAppServicePlan",
  "isSpot": false,
  "kind": "app",
  "location": "West US",
  "maximumNumberOfWorkers": 1,
  "name": "myAppServicePlan",
  "numberOfSites": 0,
  "perSiteScaling": false,
  "provisioningState": "Succeeded"
...
```

# Step 8: Create a Web App

Next, create a web app and set the name parameter to the name of your application (it has to be unique). I am setting mine to "**AmunateguiIntroWebApp**" and telling the web app that the code will be deployed via local Git (Listing 1-20).

*Listing 1-20.*  Creating a Web App (replace this with your app name)

```
$ az webapp create --resource-group myResourceGroup --plan myAppServicePlan
--name <<REPLACE-WITH-YOUR-APP-NAME>> --runtime "python|3.4" --deployment-
local-git
```

Check the large response string from the "**az web app create**" command and copy the link after "**Local git is configured with url of...**" or from the "**deploymentLocalGitUrl**" value–both are the same, so pick whichever is easiest. You will need this when you push your Flask files out to Azure (Listing 1-21).

*Listing 1-21.*  Copy Your Git URL; You Will Need It Later

Local git is configured with the URL **'https://flaskuser11@ amunateguiintrowebapp.scm.azurewebsites.net/AmunateguiIntroWebApp.git'**
```
{
    "availabilityState": "Normal",
    "clientAffinityEnabled": true,
    "clientCertEnabled": false,
    "cloningInfo": null,
    "containerSize": 0,
    "dailyMemoryTimeQuota": 0,
    "defaultHostName": "amunateguiintrowebapp.azurewebsites.net",
    "deploymentLocalGitUrl": "https://flaskuser11@amunateguiintrowebapp.scm.
    azurewebsites.net/AmunateguiIntroWebApp.git",
    ...
```

Extract the local Git configuration URL for your Azure project is (Listings 1-22 and 1-23).

*Listing 1-22.*  The Extracted Git URL in My Case

```
https://flaskuser11@amunateguiintrowebapp.scm.azurewebsites.net/
AmunateguiIntroWebApp.git
```

*Listing 1-23.*  Yours Will Look Like the Following

```
https://<<REPLACE-WITH-YOUR-USER-NAME>>@<<REPLACE-WITH-YOUR-APP-NAME>>.scm.
azurewebsites.net/<<REPLACE-WITH-YOUR-APP-NAME>>.git
```

# Check Your Website Placeholder

If everything worked, you should be able to visit the placeholder website. Replace "**<<REPLACE-WITH-YOUR-APP-NAME>>**" with the application name you created in the "**az webapp create**" step and drop it into your browser (Listing 1-24 and Figure 1-8).

*Listing 1-24.*  Checking Your Web Placeholder

```
http://<<REPLACE-WITH-YOUR-APP-NAME>>.azurewebsites.net
```

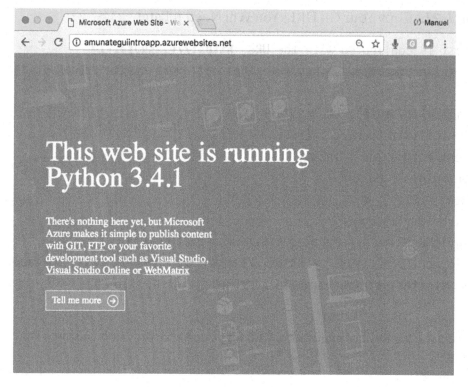

*Figure 1-8.* *Confirming that your site's placeholder is created and running*

If this didn't work, you need to check each step again and make sure you didn't miss one or if any returned an error that needs to be addressed.

## Step 9: Pushing Out the Web Application

Now go back to your local terminal/command window on your local computer pointing to the correct directory and with the initialized Git session we created earlier. Append the URL we saved previously with the location of your GIT repository to the **"add azure"** command (Listing 1-25).

*Listing 1-25.* Final Code Push to Azure

```
$ git remote add azure https://flaskuser11@amunateguiintrowebapp.scm.
azurewebsites.net/AmunateguiIntroWebApp.git
```

It may prompt for your password; make sure you use the one you created in the **"az webapp deployment user"** step (**"flask123"** in my case; Listing 1-26).

***Listing 1-26.*** Final Code Push to Azure

```
$ git push azure master
```

## Step 10: View in Browser

That's it! You can get back to your placeholder browser page and hit refresh (or open a new browser page and enter `http://<<REPLACE-WITH-YOUR-APP-NAME>>.azurewebsites.net` and you should see "**Hello World!**"; Figure 1-9)

# Hello World!

***Figure 1-9.*** *Flask application successfully running on Amazon Azure*

In case you are not seeing the "**Hello World!**" site, you can access the tail of the log directly in your command window– just swap the name for the web site name (in my case "**amunateguiintroapp**") and the group (in my case "**myResourceGroup**"); see Listing 1-27.

***Listing 1-27.*** Final Code Push to Azure

```
$ az webapp log tail --resource-group myResourceGroup --name
amunateguiintroapp
```

# Step 11: Don't Forget to Delete Your Web Application!

If you aren't using your web application anymore, don't forget to delete it. If you don't, the meter will keep running and eating credits or accruing cost. The easiest way to delete everything is to log into the Azure Dashboard and enter "**All resources**" in the search bar and delete everything you created (Figure 1-10).

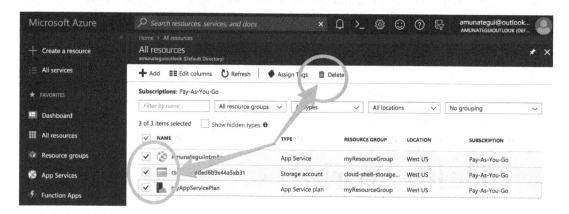

***Figure 1-10.***  *Deleting your application to not incur additional costs*

# Conclusion and Additional Information

Microsoft Azure is a powerful cloud provider with a lot of offerings. It runs simple Flask applications and deploys quickly. In order to load more complicated libraries, you will need the support of Python wheels (https://pythonwheels.com/).

For additional information, see the excellent post titled "Create a Python web app in Azure" on Microsoft Azure Docs, upon which this section was based: https://docs.microsoft.com/en-us/azure/app-service/app-service-web-get-started-python.

# Introducing Serverless Hosting on Google Cloud

Google Cloud is a powerful platform to build, manage, and deploy web applications. It integrates seamlessly with TensorFlow and its distributed graph mechanism (Figure 1-11).

***Figure 1-11.*** *Google Cloud*

Let's see how we can run our basic Flask application on Google Cloud's serverless App Engine. We'll keep the steps as simple as possible, as we'll go deeper into this provider in subsequent chapters.

---

**Note**   Download the files for Chapter 1 by going to `www.apress.com/9781484238721` and clicking the source code button and open the **"serverless-hosting-on-google-cloud"** folder.

---

## Step 1: Get an Account on Google Cloud

At the time of writing, Google is offering a convenient 12-month, $300 credit free trial to get you started. For more information, see `https://console.cloud.google.com/start`.

There are two types of App Engines you can opt for: the **"Standard Environment,"** which is simple but less customizable, and the **"Flexible Environment,"** which can handle more-or-less anything you throw at it. We'll stick with the simple one in this section, the **"Standard Environment."**

## Step 2: Download Source Files

Download the files for this chapter onto your local machine and navigate to the folder named **"serverless-hosting-on-google-cloud"** (Listing 1-28).

***Listing 1-28.*** All Files Needed for Our Web Application on Google Cloud

```
serverless-hosting-on-google-cloud/
├── app.yaml
├── appengine_config.py
├── main.py
└── requirements.txt
```

"**app.yaml**" declares where the controlling Flask Python application (in this case "**main**") resides along with the static and templates folder locations (Listing 1-29).

***Listing 1-29.*** A Look at app.yaml

```
runtime: python27
api_version: 1
threadsafe: true

handlers:
- url: /static
  static_dir: static
- url: /.*
  script: main.app
- url: /favicon.ico
  static_files: static/images/favicon.ico
  upload: static/images/favicon.ico

libraries:
- name: ssl
  version: latest
```

The "**appengine_config.py**" points to the lib folder to hold additional Python libraries during deployment (Listing 1-30).

***Listing 1-30.*** A Look at "**appengine_config.py**"

```
from google.appengine.ext import vendor
# Add any libraries installed in the "lib" folder
vendor.add('lib')
```

The "**requirements.txt**" holds Python library names and the version to be installed and added to the lib folder (Listing 1-31).

***Listing 1-31.*** Content of "**requirements.txt**"

```
Flask>=0.12
```

"**main.py**" is the brains of the Flask operations and holds all of the Python code and directives for each HTML page (Listing 1-32).

***Listing 1-32.***  Content of "**main.py**"

```python
from flask import Flask
app = Flask(__name__)

@app.route("/")
def hello():
    return "Hello World!"
```

# Step 3: Open Google Cloud Shell

Log into your instance of Google Cloud and select the project you want your App
Engine to live in (if you don't have one, see Creating and Managing Projects: `https://
cloud.google.com/resource-manager/docs/creating-managing-projects`). Start the
cloud shell command line tool by clicking the upper-right caret button. This will open
a familiar-looking command line window in the bottom half of the GCP dashboard
(Figure 1-12).

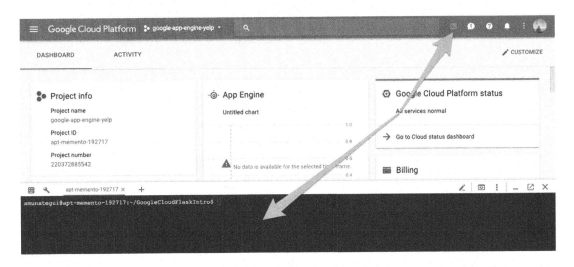

***Figure 1-12.***  *Starting the Google Cloud shell*

In the terminal section of the GCP dashboard, create a new folder called "**GoogleCloudFlaskIntro**" (Listing 1-33).

*Listing 1-33.*  Creating Directory in GCP

```
$ mkdir GoogleCloudFlaskIntro
$ cd GoogleCloudFlaskIntro
```

# Step 4: Upload Flask Files to Google Cloud

There are many ways to proceed, you can upload the files one-by-one, clone a GitHub repository, or you can zip them into one archive file and upload all of it in one go. We'll go with the latter. So zip the following four files only: "**app.yaml**," "**appengine_config.py**," "**main.py**," and "**requirements.txt**" into one archive file (Figure 1-13).

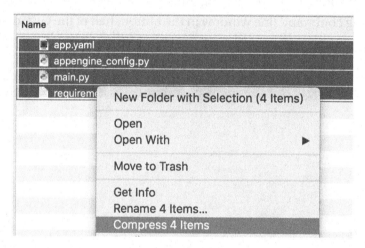

*Figure 1-13.*  *Zipping web application files for upload to Google Cloud*

Upload it using the "**Upload file**" option (this is found on the top right side of the shell window under the three vertical dots Figure 1-14).

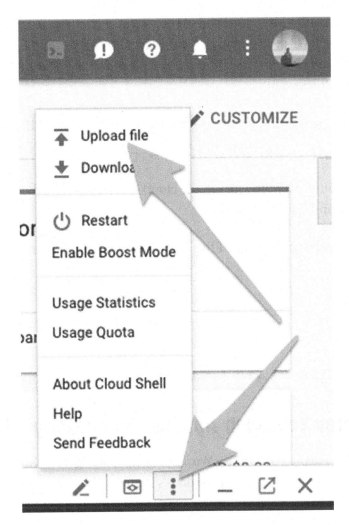

**Figure 1-14.**  *Uploading files via Google Cloud shell*

It will upload the file (in my case called "**Archive.zip**") into the root directory, so you will need to move it into the "**GoogleCloudFlaskIntro**" folder and unzip it (Listing 1-34).

**Listing 1-34.**  Moving and Unzipping Web Application Files

```
$ mv ../Archive.zip Archive.zip
$ unzip Archive.zip
```

Run the "**pip install -t lib**" command to install all the libraries in the "**requirements. txt**" file (Listing 1-35). This will create the necessary "**lib**" folder holding any needed Python libraries (you may get some complaints if you don't use "**sudo**," which gives you root rights, but don't abuse it!)

***Listing 1-35.*** Filling the lib Directory with Necessary Python Libraries

```
$ sudo pip3 install -t lib -r requirements.txt
```

At this point, your folder directory in the cloud should look like the following if you run the "**ls**" command (Listing 1-36).

***Listing 1-36.*** Checking Content of the "**GoogleCloudFlaskIntro**" Folder on GCP

**Input:**

```
$ ls
```

**Output:**

```
$ appengine_config.py app.yaml Archive.zip lib main.py requirements.txt
```

# Step 5: Deploy Your Web Application on Google Cloud

We are now ready to deploy the "**Hello World**" site. Call the "**gcloud app deploy command**" from within the dashboard's shell window under the "**GoogleCloudFlaskIntro**" folder. It will ask you to confirm that you do indeed want to deploy to the –and "**Y**"es we do ( Listing 1-37).

***Listing 1-37.*** Deploying the Web Application

**Input:**

```
$ gcloud app deploy app.yaml
```

**Truncated Output:**

```
...
File upload done.
Updating service [default]...done.
Setting traffic split for service [default]...done.
```

```
Deployed service [default] to [https://apt-memento-192717.appspot.com]
```

```
To view your application in the web browser, run:
$ gcloud app browse
```

If all goes well, you can call the convenience "**gcloud app browse**" command to get the full URL (Listing 1-38).

***Listing 1-38.*** Getting Our Web Application's URL Address

**Input:**

```
$ gcloud app browse
```

**Output:**

```
Did not detect your browser. Go to this link to view your app:
https://apt-memento-192717.appspot.com
```

Either click the link in the Google Cloud Shell or paste it in a browser (Figure 1-15).

# Hello World!

***Figure 1-15.*** *Flask application successfully running on Google Cloud*

## Step 6: Don't Forget to Delete Your Web Application!

If you aren't using your web application anymore, don't forget to delete it. If you don't, the meter will keep running and accrue cost. You can't just flat out delete an App Engine application if it is your only application (you are required to have a default application), instead you redirect traffic to a blank application.

In the Google Cloud shell, enter the Linux text editor "**vi**" (Listing 1-39).

***Listing 1-39.*** Editing the "**app.yaml**" File

```
$ vi app.yaml
```

This will open a small command line editor; then hit the "**i**" key to insert or edit the file (Listings 1-40 and 1-41).

***Listing 1-40.*** Scroll Down and Replace "**main.app**"

```
script: main.app
```

***Listing 1-41.*** With "**blank.app**"

```
script: blank.app
```

Your "**app.yaml**" file should now have its "**url**" handler pointing to "**blank.app**." This stops GCP from serving anything, as "**blank.app**" doesn't exist and will stop accruing charges (Listing 1-42).

***Listing 1-42.*** Pointing Our Yaml File to a Black Script

```
runtime: python27
api_version: 1
threadsafe: true

handlers:
- url: /static
  static_dir: static
- url: /.*
  script: blank.app
- url: /favicon.ico
  static_files: static/images/favicon.ico
  upload: static/images/favicon.ico

libraries:
- name: ssl
  version: latest
```

Click the escape key to get out of insert mode, and type "**wq**" for write and quit. Then redeploy your web application (Listing 1-43).

**_Listing 1-43._** Redeploying Blank Web Application

```
$ gcloud app deploy app.yaml
```

After the App Engine has had time to propagate, the URL should show an error instead of "**Hello World**" (Figure 1-16).

### Error: Server Error

The server encountered an error and could not complete your request.

Please try again in 30 seconds.

**_Figure 1-16._**  _The error confirms your site is down_

# Conclusion and Additional Information

GCP has a lot of features to offer and is tightly integrated with other Google offerings (like their great Cloud APIs at `https://cloud.google.com/apis/`) and TensorFlow. If you need to use more powerful Python libraries, you can switch from standard App Engine to Flexible.

For additional information, see the handy post titled "**Quickstart for Python App Engine Standard Environment**" on the Google Cloud Docs at: `https://cloud.google.com/appengine/docs/standard/python/quickstart`

By the way, if you have any issues with your Google Cloud App Engine, you can access the logs with the following command (Listing 1-44).

**_Listing 1-44._** Viewing Deployment Logs

```
$ gcloud app logs tail -s default
```

# Introducing Serverless Hosting on Amazon AWS

AWS Elastic Beanstalk is a simple yet powerful platform for deploying web applications. It comes with all the amenities like scaling, load balancing, monitoring, etc. It only charges for resources used (Figure 1-17).

***Figure 1-17.*** *Amazon Web Services*

Let's see how we can run our basic Flask application on Amazon's AWS Elastic Beanstalk. We'll keep the steps as simple as possible, as we'll look deeper into this provider in subsequent chapters.

---

**Note**    Download the files for Chapter 1 by going to `www.apress.com/9781484238721` and clicking the source code button, and open the **"serverless-hosting-on-amazon-aws"** folder.

---

## Step 1: Get an Account on Amazon AWS

Amazon AWS offers an "**AWS Free Tier**" account that allows you to try some of its services for free. For more information on creating an account, go to: `https://aws.amazon.com/free/`.

# Step 2: Download Source Files

Download the files for this chapter onto your local machine and navigate to the folder named "**serverless-hosting-on-amazon-aws**." The folder structure should look like Listing 1-45.

***Listing 1-45.*** All Files Needed for Our Web Application on AWS Elastic Beanstalk

```
serverless-hosting-on-amazon-aws/
├── application.py
└── requirements.txt
```

# Step 3: Create an Access Account for Elastic Beanstalk

Log into the AWS web console and go to the Identity and Access Management (IAM) console. A quick way there is to simply type "**IAM**" in the AWS services search box on the landing page. Select "**Users**" in the navigation section and click the "**Add user**" button (Figure 1-18).

***Figure 1-18.*** *Adding a user through the Access Management console*

Select a user name–here we enter "**ebuser**" and check "**Access type: Programmatic access**" (Figure 1-19).

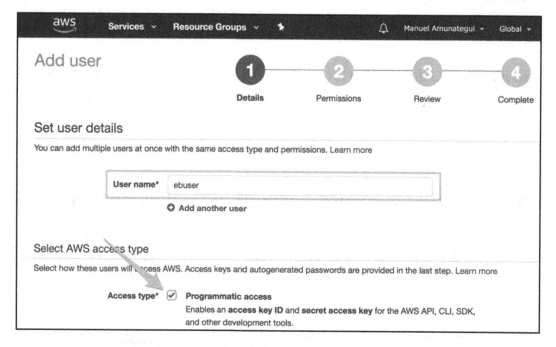

*Figure 1-19.*  *Adding correct access rights to "ebuser"*

Click the blue "**Next: Permissions**" button. This will take you to the "**Set permissions**" page; click the "**Add user to group**" large menu button then click "**Create group**." Create a group name, "**ebadmins**" in this case, and assign it the policy name "**WSElasticBeanstalkFullAccess**." Then click the "**Create group**" button to finalize the group (Figure 1-20).

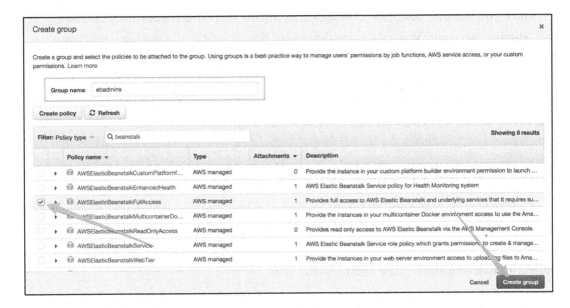

**Figure 1-20.**  *Create group with "WSElasticBeanstalkFullAccess" access*

Click the "**Next: review**" blue button and, on the following page, click the blue "**Create user**" button (Figure 1-21).

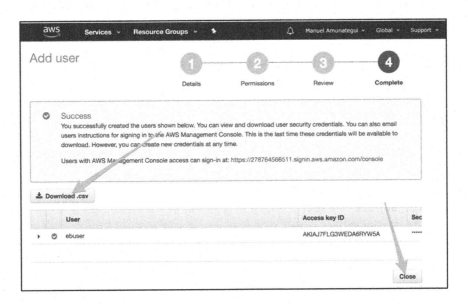

**Figure 1-21.**  *Download access key after successfully creating a user*

Once you see the "**Success**" message, this means you have successfully created the "**ebuser**" account. Make sure you download the "**.csv**" file to your local machine by clicking the "**Download .csv**" button. This file is important as it holds your key and secret code. Store it in a known location on your local machine as you will need that information to upload the web application code and to Secure Shell (SSH) into your EB (we won't need SSH in this section but will in subsequent ones).

## Step 4: Install Elastic Beanstalk (EB)

Start by creating a virtual environment to segregate installations. This isn't an obligation but it will help you keeps things clean and neat by separating this environment from the rest of your machine (if you haven't installed it yet, see the earlier section named "**Virtual Environments**"); see Listing 1-46.

*Listing 1-46.* Start a Virtual Environment

```
$ python3 -m venv amazon_aws_intro
$ source amazon_aws_intro/bin/activate
```

Install the "**awsebcli**" library to interact and manage our EB service on AWS (Listing 1-47 and 1-48).

*Listing 1-47.* For Mac and Linux Users

```
$ pip3 install awscli
$ pip3 install awsebcli
```

*Listing 1-48.* For Windows (if it complains about the "**user**" parameter, try without it)

```
$ pip3 install awscli --user
$ pip3 install awsebcli --user
```

# Step 5: EB Command Line Interface

It's time to initialize the Elastic Bean interface (Listing 1-49).

***Listing 1-49.*** Start the Elastic Beanstalk Command Line Interface

```
$ eb init -i
```

This will ask you a series of questions and you can go with most of the defaults. Under "**Enter Application name**," enter "**AWSBeanstalkIntroduction**" (Figure 1-22).

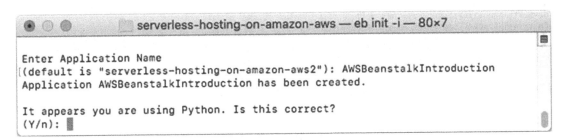

```
Enter Application Name
[(default is "serverless-hosting-on-amazon-aws2"): AWSBeanstalkIntroduction
Application AWSBeanstalkIntroduction has been created.

It appears you are using Python. Is this correct?
(Y/n):
```

***Figure 1-22.*** *Creating a new EB application*

If this is your first time running AWS on your computer, it will ask for your credentials. Open the "**credentials.csv**" that was downloaded on your machine when you created a user and enter the two fields required (Figure 1-23).

```
        serverless-hosting-on-amazon-aws — eb create serverless-hosting-on-amazo...
You have not yet set up your credentials or your credentials are incorrect
You must provide your credentials.
[(aws-access-id): AKIAIMAYE3RBMZ4ALVAQ                                    ]
[(aws-secret-key): We/Ft/WBAEqYAzVFhb6Z4nOtAhztN1w+wrueSffY              ]
```

***Figure 1-23.*** *Entering your credentials*

Go with the Python defaults (it needs to be a 3.x version); ignore warnings. Say yes to SSH (Figure 1-24).

```
Do you want to set up SSH for your instances?
[(Y/n): y
```

**Figure 1-24.**  *Turning on SSH settings*

Create a new key pair or select an existing one and keep going with the defaults. If you create a new key pair, it will save it in a folder and tell you the location (Figure 1-25).

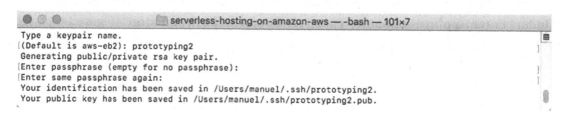

**Figure 1-25.**  *Create or reuse an RSA key pair*

Next, you need to create your EB (this has to be a unique name). This command will automatically zip up the data in the folder and upload it to the AWS cloud (Listings 1-50 and 1-51). This can take a few minutes, so be patient.

**Listing 1-50.**  Create Your EB and Upload it to AWS

```
$ eb create <<ENTER-YOUR-EB-NAME>>>
```

**Listing 1-51.**  My EB for This Example

```
$ eb create AWSBeanstalkIntroduction
```

# Step 6: Take if for a Spin

It takes a few minutes, and you should get a success message if all goes well. Then you can simply use the "**eb open**" command to view the web application live (Listing 1-52).

**Listing 1-52.**  Fire Up Your Web Application

```
$ eb open AWSBeanstalkIntroduction
```

It may take a little bit of time to run the application the first time around and may even timeout. Run the "**eb open**" one more time if it times out (Figure 1-26).

Hello World!

*Figure 1-26.* *Flask application successfully running on Amazon AWS*

If things don't go as planned, check out the logs for any error messages (Listing 1-53).

*Listing 1-53.* Access the Logs in Case of Problems

```
$ eb logs
```

# Step 7: Don't Forget to Turn It Off!

Finally, we need to terminate the Beanstalk instance to not incur additional charges. This is an important reminder that most of these cloud services are not free. It will ask you to enter the name of the environment; in my case it is "**AWSBeanstalkIntroduction**" (Listing 1-54).

*Listing 1-54.* Don't Forget to Terminate Your EB and Deactivate Your Virtual Environment

```
$ eb terminate AWSBeanstalkIntroduction
$ deactivate
```

It does take a few minutes but will take the site down. It is a good idea to double-check on your AWS dashboard that all services are indeed turned off. This is easy to do: simply log into your AWS account at https://aws.amazon.com/ and make sure that your EC2 and Elastic Beanstalk accounts don't have any active services you didn't plan on having (Figure 1-27). In case you see an instance that seems to keep coming back to life after each time you "**terminate**" it, check under EC2 "**Load Balancers**" and terminate those first, then terminate the instances again.

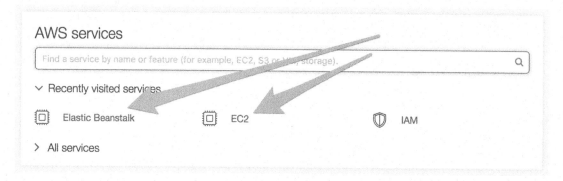

***Figure 1-27.***  *Checking the Amazon AWS dashboard that no unwanted services are still running*

## Conclusion and Additional Information

No doubt, AWS is the leader in the cloud space. It may not be the simplest or cheapest, but it would more than likely do anything you ask it to.

For additional information, see the handy post titled "**Deploying a Flask Application to AWS Elastic Beanstalk**" on the Amazon's AWS Docs at `https://docs.aws.amazon.com/elasticbeanstalk/latest/dg/create-deploy-python-flask.html`.

## Introducing Hosting on PythonAnywhere

PythonAnywhere is a great way to rapidly prototype your Python interactive ideas and models on the Internet. It is integrated and designed for anything Python! It isn't serverless in the classic sense, but it is a dedicated Python framework, it doesn't require a credit card to sign up, and it can craft a proof-of-concept in no time (Figure 1-28).

***Figure 1-28.***  *PythonAnywhere*

Proof is in the pudding; no code is needed for this project as PythonAnywhere already defaults to a "**Hello World**" example when you spin up an instance.

# Step 1: Get an Account on PythonAnywhere

Sign up for a free account on PythonAnywhere.com and log into it (you will have to confirm your email address).

# Step 2: Set Up Flask Web Framework

Let's create a web server on PythonAnywhere with the Flask web-serving platform. It is super easy to do. Under the "**Web**" tab, click the "**Add a new web app**" blue button. And accept the defaults until you get to the 'Select a Python Web framework' and click "**Flask**" and then the latest Python framework (Figure 1-29).

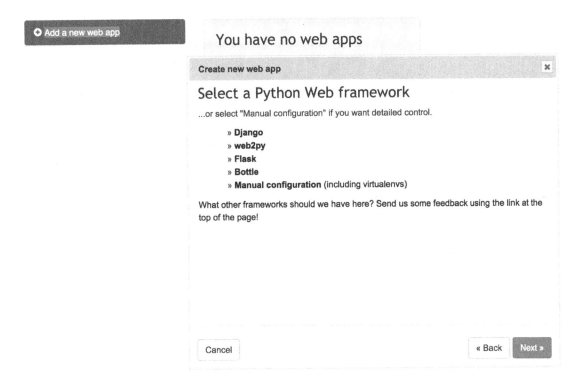

*Figure 1-29.* *Adding a new web app on PythonAnywhere*

You will get to the landing configuration page, hit the green "**Reload your account. pythonanywhere.com**" button and take your new URL for a spin (Figure 1-30).

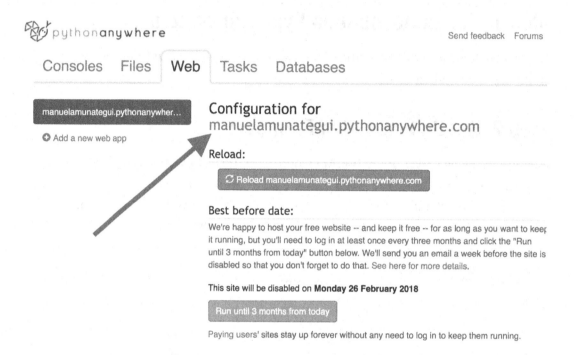

***Figure 1-30.*** *Accessing the website configuration under the Web tab*

You should see a simple but real web page with the "**Hello from Flask!**" message right out of the box (Figure 1-31).

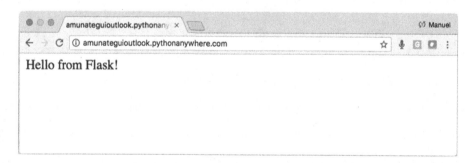

***Figure 1-31.*** *Flask application successfully running on PythonAnywhere*

## Conclusion and Additional Information

PythonAnywhere may not be a 100% serverless cloud provider but it is free for basic hosting. It is the easiest to work with and can be run directly from its online dashboard. This is a great option when traveling.

For a treasure-trove of help docs and step-by-step guides, see `https://help.pythonanywhere.com/pages/`.

# Summary

If you made it this far, great job! We've covered a very simple, stand-alone web application using Flask and deployed it on four cloud providers. Please expect some variance in the methodology of uploading web applications onto each cloud provider, as they do change things here and there from time to time. Keeping an eye on the documentation is critical.

It's time to roll up our sleeves and start tackling some more interesting and more involved web applications!

**CHAPTER 2**

# Client-Side Intelligence Using Regression Coefficients on Azure

Let's build an interactive web application to understand bike rental demand using regression coefficients on Microsoft Azure.

For our first project, we're going to model the Bike Sharing Dataset from the Capital Bikeshare System using regression modeling and learn how variables such as temperature, wind, and time affect bicycle rentals in the mid-Atlantic region of the United States (Figure 2-1).

© Manuel Amunategui, Mehdi Roopaei 2018
M. Amunategui and M. Roopaei, *Monetizing Machine Learning*, https://doi.org/10.1007/978-1-4842-3873-8_2

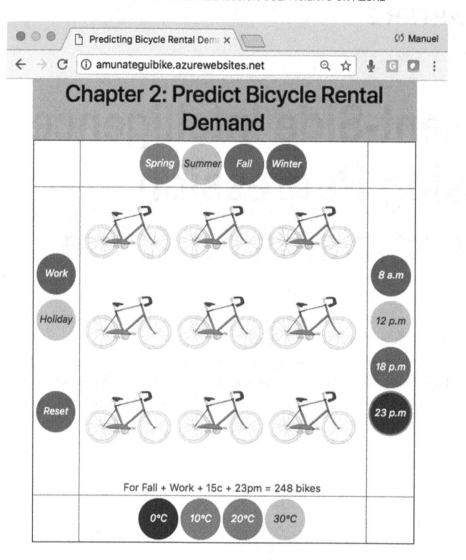

***Figure 2-1.***  *Our final web application for this chapter*

The data is graciously made available through the UCI Machine Learning Repository of the University of California, Irvine (https://archive.ics.uci.edu/ml/datasets/bike+sharing+dataset).

---

**Note**    Download the files for Chapter 2 by going to www.apress.com/9781484238721 and clicking the source code button. Open Jupyter notebook "**chapter2.ipynb**" to follow along with this chapter's content.

---

# Understanding Bike Rental Demand with Regression Coefficients

We're going to build a simple and visually intuitive way of interacting with different environmental factors and see how they affect bike rentals. This is a great way for users to confirm their intuitive assumptions of what would make people want to bicycle vs. not, and in some cases, surprise them too (like seeing more riders in the winter than in the summer–but I'll let you discover that on your own).

The "**brain**" behind this web application is a linear regression model. It has the ability of finding linear relationships between an outcome variable and historical data. We are going to leverage this skill by having it learn bike rental demand over time and under different environmental factors, and see if it can help us predict future demand.

Whenever you extend a Python model to the Wweb, it is critical to iron out all issues and bugs locally before adding the extra layers necessary to build it into a web application. Get all the easy issues resolved before moving anything to the cloud! Following this piece of advice will save you from many headaches.

# Exploring the Bike Sharing Dataset

Bike sharing is very popular albeit still new and experimental. Using a mobile phone, a rider can sign up online, download a phone application, locate bicycles, and rent one. This model creates an entire ecosystem where nobody needs to talk or meet in person to start enjoying this service. According to Hadi Fanaee-T of the Laboratory of Artificial Intelligence and Decision Support (from the liner notes on the UCI Machine Learning Repository's Dataset Information):

> *Opposed to other transport services such as bus or subway, the duration of travel, departure and arrival position is explicitly recorded in these systems. This feature turns [a] bike sharing system into a virtual sensor network that can be used for sensing mobility in the city. Hence, it is expected that most of [the] important events in the city could be detected via monitoring these data.*[1]

The download contains two datasets: "**hour.csv**" and "**day.csv**." See Table 2.1 for feature details.[2]

---

[1]Hadi Fanaee-T and Joao Gama, "Event Labeling Combining Ensemble Detectors and Background Knowledge," *Progress in Artificial Intelligence* 2, no. 2–3 (2013): pp. 113-127.

[2]https://archive.ics.uci.edu/ml/datasets/bike+sharing+dataset

*Table 2-1.* *Bike Sharing Data Legend*

| Feature Name | Description |
|---|---|
| instant | record index |
| dteday | date |
| season | season (1: spring, 2: summer, 3: fall, 4: winter) |
| yr | year (0: 2011, 1:2012) |
| mnth | month (1 to 12) |
| hr | hour (0 to 23) |
| holiday | whether day is holiday or not |
| weekday | day of the week |
| workingday | If day is neither weekend nor holiday it is 1, otherwise it is 0. |
| weathersit | 1.  Clear, Few clouds, Partly cloudy |
| | 2.  Mist + Cloudy, Mist + Broken clouds, Mist + Few clouds, Mist |
| | 3.  Light Snow, Light Rain + Thunderstorm + Scattered clouds, Light Rain + Scattered clouds |
| | 4.  Heavy Rain + Ice Pallets + Thunderstorm + Mist, Snow + Fog |
| temp | Normalized temperature in Celsius. The values are derived via $(t-t\_min)/(t\_max-t\_min)$, $t\_min = -8$, $t\_max = +39$ (only in hourly scale). |
| atemp | Normalized feeling temperature in Celsius. The values are derived via $(t-t\_min)/(t\_max-t\_min)$, $t\_min = -16$, $t\_max = +50$ (only in hourly scale). |
| hum | Normalized humidity. The values are divided by 100 (max). |
| windspeed | Normalized wind speed. The values are divided by 67 (max). |
| casual | count of casual users |
| registered | count of registered users |
| cnt | count of total rental bikes including both casual and registered |

# Downloading the Data from the UCI Machine Learning Repository

The dataset can be downloaded directly from UCI's repository using Python or manually at: `https://archive.ics.uci.edu/ml/datasets/bike+sharing+dataset`). The download contains the following three files:

- day.csv

- hour.csv

- Readme.txt

The daily set of bike rentals contains 731 rows and the hourly set, 17,379 records.

# Working with Jupyter Notebooks

Each chapter has a corresponding Jupyter notebook. Let's go over some of the basics to get started with the notebook for this chapter. Download the source files, open a terminal window, and navigate to that folder. In it, you should find two files and a folder (Figure 2-2).

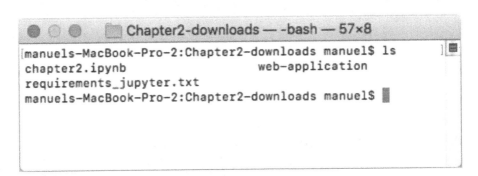

*Figure 2-2.* *Terminal window*

The "**requirements_jupyter.txt**" file contains the Python libraries necessary to run this chapter's Jupyter notebook. You can quickly install them by running the "**pip3**" command (Listing 2-1).

***Listing 2-1.***   Installing the Required Files to Run the Notebook

```
$ pip3 install -r requirements_jupyter.txt
```

The file named "**chapter2.ipynb**" is the actual Jupyter notebook for this chapter. There are different ways of starting a notebook, but one popular way is using the "**jupyter notebook**" command (Listing 2-2). If this doesn't work for you, please refer to the official Jupyter documentation.

***Listing 2-2.***   Starting the Jupyter notebook for This Chapter

```
$ jupyter notebook
```

This will open a browser window with a file-explorer dashboard pointing to the same folder where it was launched from. Go ahead and click the "**chapter2.ipynb**" link (Figure 2-3).

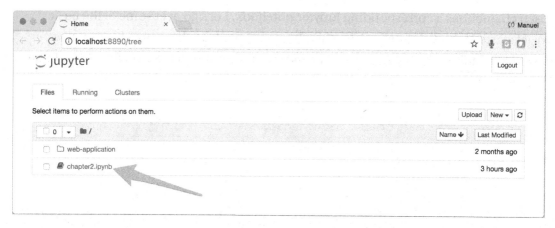

***Figure 2-3.***   *Jupyter's file explorer with this chapter's notebook link*

This will open a new tab and the corresponding notebook containing all the exploratory code needed to follow along with the chapter's content. All code in this book assumes Python 3.x; if you use another version you may have to tweak some parts of the code. Once you have opened the notebook, you are ready to go. Highlight the first box and hit the play button to run that portion of the code (Figure 2-4). If you see errors, please address them before continuing, as each code snippet builds upon the previous one (errors can be related to Python version compatibility issues or missing libraries that need to be installed).

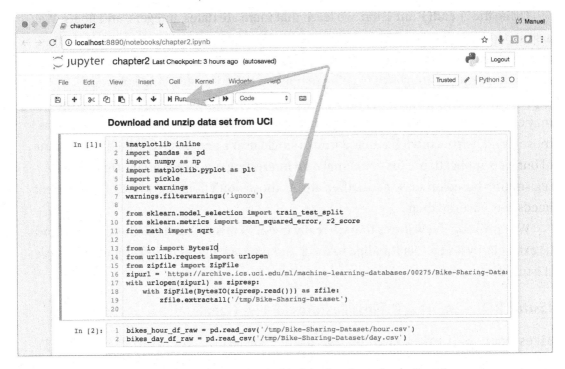

**Figure 2-4.**  *Jupyter notebook with code highlighted and ready to be run*

The corresponding Jupyter notebook for the chapter shows a way to download and unzip the data directly using Python (if you are having firewall issues, please download it manually).

# Exploring the Data

The Python Pandas "**head()**" function offers a glimpse into the first few rows of the data shown in Listing 2-3 and Figure 2-5.

***Listing 2-3.***  First Rows of the Dataset

```
bikes_hour_df_raw.head()
```

| | instant | dteday | season | yr | mnth | hr | holiday | weekday | workingday | weathersit | temp | atemp | hum | windspeed | casual | registered | cnt |
|---|---|---|---|---|---|---|---|---|---|---|---|---|---|---|---|---|---|
| 0 | 1 | 2011-01-01 | 1 | 0 | 1 | 0 | 0 | 6 | 0 | 1 | 0.24 | 0.2879 | 0.81 | 0.0 | 3 | 13 | 16 |
| 1 | 2 | 2011-01-01 | 1 | 0 | 1 | 1 | 0 | 6 | 0 | 1 | 0.22 | 0.2727 | 0.80 | 0.0 | 8 | 32 | 40 |
| 2 | 3 | 2011-01-01 | 1 | 0 | 1 | 2 | 0 | 6 | 0 | 1 | 0.22 | 0.2727 | 0.80 | 0.0 | 5 | 27 | 32 |
| 3 | 4 | 2011-01-01 | 1 | 0 | 1 | 3 | 0 | 6 | 0 | 1 | 0.24 | 0.2879 | 0.75 | 0.0 | 3 | 10 | 13 |
| 4 | 5 | 2011-01-01 | 1 | 0 | 1 | 4 | 0 | 6 | 0 | 1 | 0.24 | 0.2879 | 0.75 | 0.0 | 0 | 1 | 1 |

**Figure 2-5.**  *bike_df.head() output*

Using the "**head()**" function, we learn that there are dates, integers, and floats. We also see some redundant features like date (dteday) have already been categorized through "**season**," "**yr**," "**mnth**," "**hr**," etc. Therefore the "**dteday**" feature is an easy candidate to drop (though we're going to hold on to it for a little longer for our exploration needs). Some other features seem redundant, like "**temp**" and "**atemp**" and may call for closer inspection. We also drop the "**casual**" and "**registered**" features, as those won't help us model demand from a single user's perspective, which is the point of our web application. This could make an interesting outcome variable to forecast registration based on season, weather, etc. As those don't fit in the scope of our current needs, we will drop them.

We only keep the features that we really need, as this will remove clutter and afford us extra clarity and understanding to reach our data science and web application goals (Listing 2-4).

***Listing 2-4.*** Removing Useless Features for Our Ggoals

```
bikes_hour_df = bikes_hour_df_raw.drop(['casual', 'registered'], axis=1)
```

The Pandas "**info()**" function is also a great way of seeing the data types, quantities, and null counts contained in the data (Listing 2-5).

***Listing 2-5.*** Getting information about features

**Input:**

```
bikes_hour_df.info()
```

**Output:**

```
RangeIndex: 17379 entries, 0 to 17378
Data columns (total 15 columns):
instant      17379 non-null int64
dteday       17379 non-null object
season       17379 non-null int64
yr           17379 non-null int64
mnth         17379 non-null int64
hr           17379 non-null int64
holiday      17379 non-null int64
weekday      17379 non-null int64
```

```
workingday      17379 non-null int64
weathersit      17379 non-null int64
temp            17379 non-null float64
atemp           17379 non-null float64
hum             17379 non-null float64
windspeed       17379 non-null float64
cnt             17379 non-null int64
```

Using the "**info()**" function, we see that all the data currently held in memory is either a float or an integer, and that none of them are nulls. If we did happen to find nulls, date data types, or text data types, we would need to address them before moving on to modeling. The majority of models in existence require numerical data and that is what we have, so we're looking good so far.

# A Closer Look at Our Outcome Variable

Let's look at the outcome variable that we're going to use to train our model, "**cnt**," count of total rental bikes. Pandas "**describe()**" function is another go-to tool to understand quantitative data. Let's apply it to our outcome variable (also known as the model's label), as shown in Listing 2-6.

*Listing 2-6.* Number Summary of the Bike Rental Count "**cnt**" Feature

**Input:**

```
bikes_hour_df['cnt'].describe()
```

**Output:**

```
count    17379.000000
mean       189.463088
std        181.387599
min          1.000000
25%         40.000000
50%        142.000000
75%        281.000000
max        977.000000
Name: cnt, dtype: float64
```

We see that feature "**cnt**" ranges between a minimum of 1 and maximum of 977 counts. This means that each recorded hour has seen a minimum of 1 bike rental to a maximum of 977 bike rentals. We also see that the average rental count is 189.5.

We confirm that we are dealing with a continuous numerical variable where a linear regression (or a linear regression-like model) is the right choice to train and predict bicycle rental counts. Let's plot this feature to better understand the data (Listing 2-7 and Figure 2-6).

***Listing 2-7.*** Number Summary of the Bike Rental Count "**cnt**" Feature

```
fig,ax = plt.subplots(1)
ax.plot(sorted(bikes_hour_df['cnt']), color='blue')
ax.set_xlabel("Row Index", fontsize=12)
ax.set_ylabel("Sorted Rental Counts", fontsize=12)
ax.set_ylabel("Sorted Rental Counts", fontsize=12)
fig.suptitle('Outcome Variable - cnt - Rental Counts')
plt.show()
```

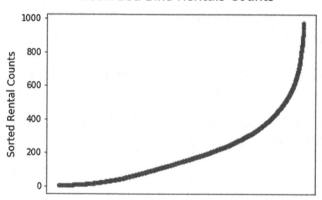

***Figure 2-6.*** *Sorted counts of bike rentals reveal that the majority of the rentals happen in the 0 to 400 range; values higher than that are either rare or outliers*

## Quantitative Features vs. Rental Counts

Let's create scatter plots of all our float data types. We'll plot them against rental counts to visualize potential relationships (Figures 2-7 and 2-8).

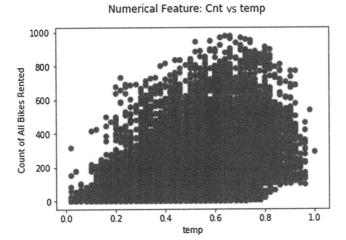

***Figure 2-7.*** *Count of all bikes rented vs. "**temp**" feature*

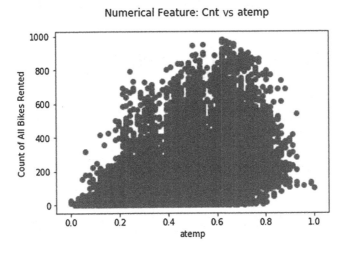

***Figure 2-8.*** *Count of all bikes rented vs. "**atemp**" feature*

We can see that there is a somewhat linear relationship between the number of bikes rented and temperature; the warmer it is, the more bikes get rented. We also see that both features–"**temp**" and "**atemp**" –have similar distributions and may present redundancy and even multicollinearity. To keep things clean, we will drop feature "**atemp**" (Figures 2-9 and 2-10).

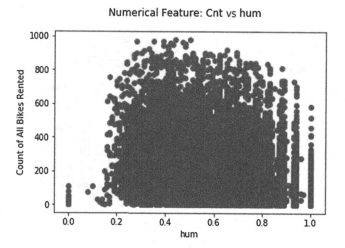

***Figure 2-9.***   *Count of all bikes rented vs. "**hum**" feature*

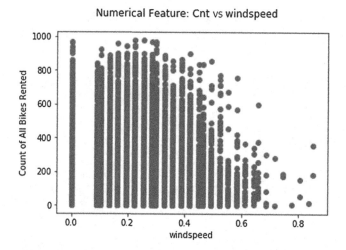

***Figure 2-10.***   *Count of all bikes rented vs. "**windspeed**" feature*

Feature "**hum**" or humidity looks like a big blob though the edges do show some sparseness. Feature "**windspeed**" does show an inverse linear relationship with rentals; too much wind and bike rentals don't seem to mix.

# Let's Look at Categorical Features

In this dataset, with the exception of bicycle rental counts "**cnt**," integer data are categorical features. Categorical data yields a lot of interesting tell-tales when viewed through histograms (distribution charts; Figure 2-11).

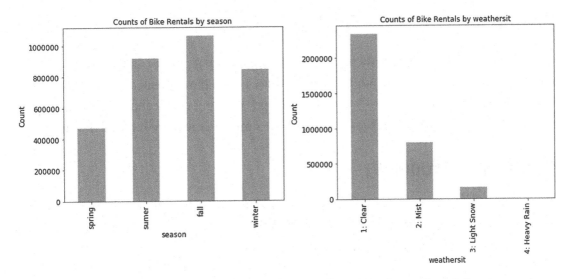

**Figure 2-11.**  *Counts of bike rentals by "**season**" and by "**weathersit**"*

Feature "**weathersit**" shows that people rent more bikes in nice weather and "**season**" shows that fall is the top season to rent bikes.

And finally, feature "**hr**," or rental hour, clearly shows peak office commute hours and afternoon rides as very popular bicycling times, and 4 AM is the least popular bicycling time (Figure 2-12).

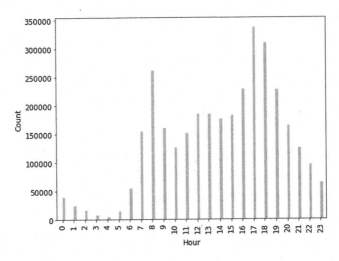

**Figure 2-12.**  *Total bike rentals by hour*

Even though we can learn a lot by eyeballing charts, more thorough and systematic testing is required to reach conclusive decisions about features to keep and features to drop.

# Preparing the Data for Modeling

In most data science projects, there is a data wrangling phase where data is assessed and cleaned in order to be "**model ready**." In this case we have already dropped some useless features, we have no nulls to deal with, and we won't worry about correlation or multicollinearity as we are only going to use four simple features in our final model.

# Regression Modeling

In statistical analysis, a regression model attempts to predict the relationships among variables. It will analyze how independent variables relate to dependent ones. A fitted model can be used to predict new dependent variables.

# Simple Linear Regression

A linear regression is probably the simplest modeling algorithm. It attempts to explain the relationship between one dependent variable and one or more independent variables. See the basic regression equation in Figure 2-13.

$$y = \beta_0 + \beta_1 x$$

***Figure 2-13.***  *The basic linear regression equation*

In the equation y = estimated dependent variable score, $\beta 0$ = constant, $\beta\_1$ = regression coefficient, and x = score on the independent variable.

# A Simple Model

Let's start with a simple nultilinear regression model where we input all variables and get a base root mean squared error (RMSE). RMSE expresses the error in units scaled to the outcome variable (also known as the y label), so it is easy to see how well the model does at learning/predicting bike rentals and the error becomes a form of confidence interval. You want the lowest possible RMSE score, so the goal is to keep tweaking the data and model until it stops going down. We'll base all our modeling efforts in this chapter on the Python scitkit-learn/sklearn library.[3] This is a phenomenal library that should satisfy most of Python users' modeling needs.

---

[3]http://scikit-learn.org/stable/

Even though we're only going to run a simple linear regression, we're going to leverage three functions from the sklearn library: "**train_test_split**" to create two random datasets from our original data and separate features from outcomes, "**linear_ model**" to run our model, and "**mean_squared_error**" to evaluate how well the model learned (Listing 2-8).

***Listing 2-8.*** Snippet of Code to Split the Dataset into Training and Testing Portions

```
# set outcome variable
outcome = 'cnt'

# create feature list
features = [feat for feat in list(bike_df_model_ready) if feat not in
[outcome, 'instant']]

# split data into train and test portions
from sklearn.model_selection import train_test_split
X_train, X_test, y_train, y_test = train_test_split(bike_df_model_
ready[features],
                                    bike_df_model_ready[['cnt']],
                                    test_size=0.3, random_state=42)
```

The "**train_test_split()**" function will split the data into two random datasets using a seed. Setting the "**random_state**" seed parameter is a good idea whenever you are testing different approaches and want to ensure that you are always using the same splits for fair comparison. The "**test_size**" parameter sets the size of the test split. Here we set it to .3, or 30%, thus it will randomize the data and assign 70% of the data to the training set and 30% to the testing set (Listing 2-9).

***Listing 2-9.*** Linear Regression Code

```
from sklearn import linear_model
model_lr = linear_model.LinearRegression()

# train the model on training portion
model_lr.fit(X_train, y_train)
```

We declare a "**LinearRegression()**" model then call function "**fit()**" to train the model using the training data and training labels. Model "**model_lr**" is now trained and ready to predict (Listing 2-10).

*Listing 2-10.* Predict and Get the RMSE Score

**Input:**

```
predictions = model_lr.predict(X_test)
from sklearn.metrics import mean_squared_error
print("Root Mean squared error: %.2f" % sqrt(mean_squared_error(y_test,
predictions)))
```

**Output:**

```
Root Mean squared error: 143.08
```

Finally, we call function "**predict()**" using the 30% of the data earmarked for testing and feed the predicted labels into function "**mean_squared_error()**" to get the root mean squared error score. We get an RMSE of **143.08** and we will use that as our base benchmark score. This is what we get with the current seeded split (the seed we applied on the train_test_split function to make sure we always get the same split each time) and all the features we've selected so far. One way to interpret the score, as it is in the same scale as our outcome variable, is that our model predictions are off by 143 bikes. Considering that the mean bike rental demand per hour is approximately 190, our model does a better job than simply picking the overall mean value of bike rentals. But let's see if we can improve on this.

# Experimenting with Feature Engineering

Let's see if we can get a better score by experimenting with a few different techniques, including polynomials, nonlinear modeling, and leveraging time-series.

# Modeling with Polynomials

Applying polynomial transformations to a series of values can allow for better separation during linear regression. This is very easy to do in the Python's "**sklearn**" library (Listing 2-11).

***Listing 2-11.*** Create Polynomial Features

```
from sklearn.preprocessing import PolynomialFeatures
poly = PolynomialFeatures(2)
X_train = poly.fit_transform(X_train)
X_test = poly.fit_transform(X_test)
```

Let's transform all features to the 2nd degree (Listing 2-12).

***Listing 2-12.*** 2nd-Degree Polynomials

**Input:**

```
print("Root Mean squared error with PolynomialFeatures set to 2 degrees:
%.2f" % sqrt(mean_squared_error(y_test, predictions)))
```

**Output:**

```
Root Mean squared error with PolynomialFeatures set to 2 degrees: 122.96
```

Now transform all features to the 3rd degree (Listing 2-13).

***Listing 2-13.*** 3rd-Degree Polynomials

**Input:**

```
print("Root Mean squared error with PolynomialFeatures set to 3 degrees:
%.2f" % sqrt(mean_squared_error(y_test, predictions)))
```

**Output:**

```
Root Mean squared error with PolynomialFeatures set to 3 degrees: 111.65
```

And now transform all features to the 4th degree (Listing 2-14).

***Listing 2-14.*** 4th-Degree Polynomials

**Input:**

```
print("Root Mean squared error with PolynomialFeatures set to 4 degrees:
%.2f" % sqrt(mean_squared_error(y_test, predictions)))
```

**Output:**

```
Root Mean squared error with PolynomialFeatures set to 4 degrees: 114.84
```

As you can see, applying polynomials to a dataset is extremely easy with sklearn's **"PolynomialFeatures()"** function. The score does improve using the 2nd and 3rd degree but then degrades beyond that point.

# Creating Dummy Features from Categorical Data

Another approach worth trying is to dummify categorical data. This means creating separate columns for each category. Take feature **"weathersit"**: this isn't a continuous variable, instead it is an arbitrary category. If you feed it into a model as such, it will consider it as linear numerical data, and this doesn't really make sense in this case; adding 1 to **"mist & cloud"** doesn't equal **"snow."** The model will do a better job on **"weathersit"** by creating four new columns: **"clear,"** **"mist,"** **"snow,"** and **"rain"** and assign each a binary true/false value.

This is easy to do with the Pandas function **"get_dummies()."** We abstract the code into a function that will make our web application easier to create (Listing 2-15).

***Listing 2-15.*** Abstracting the Code to Create Dummy Data

```
def prepare_data_for_model(raw_dataframe,
                           target_columns,
                           drop_first = False,
                           make_na_col = True):

    # dummy all categorical fields
    dataframe_dummy = pd.get_dummies(raw_dataframe, columns=target_columns,
    drop_first=drop_first,
dummy_na=make_na_col)
    return (dataframe_dummy)
```

This will break each category out into its own column. In the code snippet below, we ask to **"dummify"** the following three columns: **"season,"** **"weekday,"** and **"weathersit"** (Listing 2-16).

***Listing 2-16.*** Dummify Categorical Columns

```
bike_df_model_ready = prepare_data_for_model(bike_df_model_ready,
        target_columns = ['season', 'weekday', 'weathersit'], drop_first
        = True)
```

After applying the function to the dataset, each weather category is now in a separate column (minus the first column, which is redundant—if it isn't "**weathershit_2**," "**weathershit_3**," or "**weathershit_4**," then we infer it is "**weathershit_1**"; Listing 2-17 and Figure 2-14).

**Listing 2-17.** A Look at the Dummified Weather Field

```
bike_df_model_ready[['weathersit_2.0', 'weathersit_3.0',
'weathersit_4.0']].head()
```

| | weathersit_2.0 | weathersit_3.0 | weathersit_4.0 |
|---|---|---|---|
| **0** | 0 | 0 | 0 |
| **1** | 0 | 0 | 0 |
| **2** | 0 | 0 | 0 |
| **3** | 0 | 0 | 0 |
| **4** | 0 | 0 | 0 |

**Figure 2-14.** *A look at the dummied columns of feature "weathersit"*

So, does creating dummies out of the categorical data help the model or not? (Listing 2-18)

**Listing 2-18.** RMSE after Dummying the Categorical Data

**Input:**

```
print("Root Mean squared error: %.2f" % sqrt(mean_squared_error(y_test,
predictions)))
```

**Output:**

```
Root Mean squared error: 139.40
```

This isn't very impressive, and certainly not enough to justify all that extra work. Llet's move on and try other techniques.

## Trying a Nonlinear Model

As a final modeling experiment, let's run our dummied data into a **"Gradient Boosting Regressor"** model from sklearn. Switching from one model to another in the sklearn package is trivial, and we only need to load the appropriate model in memory and change two lines (Listing 2-19).

*Listing 2-19.*  Using a GBM Model

**Input:**

```
from sklearn.ensemble import GradientBoostingRegressor
model_gbr = GradientBoostingRegressor()
model_gbr.fit(X_train, np.ravel(y_train))
predictions = model_gbr.predict(X_test)
print("Root Mean squared error: %.2f" % sqrt(mean_squared_error(y_test,
predictions)))
```

**Output:**

```
Root Mean squared error: 68.13
```

Wow, that is the lowest RMSE score we've seen yet; we've cut our error rate in two!

## Even More Complex Feature Engineering—Leveraging Time-Series

Here is one last feature engineering experiment; this idea comes from data scientists over at Microsoft.[4] The data is a ledger of bike rentals over time, so it is a time-series dataset. Whenever your dataset records events over time, you want to take that into account as an added feature. For example, an event that happened an hour ago is probably more important than one that happened a year ago. Time can also capture trends, changing needs and perceptions, etc. We want to create features that capture all those time-evolving elements!

---

[4]http://blog.revolutionanalytics.com/2016/05/bike-rental-demand.html

For each row of data, we'll add two new features: the sum of bicycle rentals for the previous hour, and the sum of bicycle rentals from two hours ago. The intuition here is that if we want to understand the current bicycling mood, we can start by looking at what happened an hour ago. If the rentals were great one hour ago, they're probably going to be good now. This time element can be seen as a proxy to prosperous or calamitous times, good or bad weather, etc.

To create a sum of bicycles per date and hour, we use Pandas extremely powerful "**groupby()**" function. We extract three fields, "**dteday**," "**hr**," and "**cnt**" and group the count by date and hour (Listing 2-20 and Figure 2-15).

**Listing 2-20.** Looking at Rental Counts in the Previous Period

```
bikes_hour_df_shift = bikes_hour_df[['dteday','hr','cnt']].
groupby(['dteday','hr']).sum()
bikes_hour_df_shift.head()
```

| | | cnt |
|---|---|---|
| **dteday** | **hr** | |
| **2011-01-01** | **0** | 16 |
| | **1** | 40 |
| | **2** | 32 |
| | **3** | 13 |
| | **4** | 1 |

**Figure 2-15.** *Shifting the date to create new look-back features*

This function tallies the counts by hour and date. Next, we create two new features, one shifted forward 1 row and the other 2 rows, thus giving the current row the total bike rentals for the past hour and the hour past that. Finally we add it all back to our main data frame using Pandas "**merge()**" command (Listing 2-21).

***Listing 2-21.*** Playing with Time Shifts

```
# prior hours
bikes_hour_df_shift = bikes_hour_df[['dteday','hr','cnt']].
groupby(['dteday','hr']).sum().reset_index()
bikes_hour_df_shift.sort_values(['dteday','hr'])

# shift the count of the last two hours forward so the new count can take
in consideration how the last two hours went
bikes_hour_df_shift['sum_hr_shift_1'] = bikes_hour_df_shift.cnt.shift(+1)
bikes_hour_df_shift['sum_hr_shift_2'] = bikes_hour_df_shift.cnt.shift(+2)

# merge the date and hour counts back to bike_df_model_ready
bike_df_model_ready =  pd.merge(bikes_hour_df, bikes_hour_df_
shift[['dteday', 'hr', 'sum_hr_shift_1', 'sum_hr_shift_2']], how='inner',
on = ['dteday', 'hr'])
```

After we split this new data and run it into a Gradient Boosted Model (GBM) for regression (sklearn's GradientBoostingRegressor), we calculate the RMSE score over the test dataset (Listing 2-22).

***Listing 2-22.*** RMSE from Time Shifts

**Input:**

```
from sklearn.ensemble import GradientBoostingRegressor
model_gbr = GradientBoostingRegressor()
model_gbr.fit(X_train, np.ravel(y_train))
predictions = model_gbr.predict(X_test)

print("Root Mean squared error: %.2f" % sqrt(mean_squared_error(y_test,
predictions)))
```

**Output:**

```
Root Mean squared error: 44.43
```

Wow, crazy, an RMSE of 44.43; even better!!!

# A Parsimonious Model

Unfortunately, it isn't always about the best score. Here, we need a simple model in order to predict using a regression equation. This isn't something that can easily be done with complicated models or overly engineered features. GBM isn't a linear model and doesn't give us a handy and lightweight regression equation. Also, the time shifts we created previously require that we have total counts for the two previous hours of prediction, something that our web visitors won't benefit from because we don't have access to live data.

This is an important lesson when your goal is to create web applications: if the most accurate prediction comes from a complicated modeling technique, it just won't translate well into a production pipeline.

# Extracting Regression Coefficients from a Simple Model—an Easy Way to Predict Demand without Server-Side Computing

A linear regression model is not the most powerful model out there, nor does it advertise itself as such, but it can distill fairly complex data down to an extremely simple and clear linear representation. And it is this simple representation that will fuel our application.

A powerful product of regression modeling is the learned model's coefficients. This is even more powerful in the context of a web application where we can rely solely on the coefficients and a simple regression equation to estimate future bike rental demand. This can potentially enable applications to make complicated decisions entirely on the client's front end–lightweight, fast, and useful!

In order to end up with a small set of coefficients and a simple regression equation, we need to train and test a regression model first. Only when we are happy with the score, the features used, and quality of the predictions do we extract the coefficients. We are also going to pare down the features fed into the model to the essential and illustrative ones. Here is one of the lessons of building web applications: we have to balance the best modeling scores with production realities. If you build a phenomenal model but nobody can operate it or it cannot be run in a timely fashion in production, it is a failure.

We are going to keep things simple for the sake of our web application and only use four features: "**season**," "**hr**," "**holiday**," and "**temp**." These are features that users can understand, are easy to acquire, and by only having four of them our model will be fast. Let's first model these features individually and check their R-squared score.

# R-Squared

R-squared is a statistical measure of how close the data is to the fitted regression line. It is also known as the coefficient of determination, or the coefficient of multiple determination for multiple regression. The definition of R-squared is fairly straightforward; it is the percentage of the response variable variation that is expected by a linear model (Listing 2-23).

*Listing 2-23.* R-squared Formula

```
R-squared = Expected variation / Total variation
```

The coefficient of determination can be thought of as a percent. It gives you an idea of how many data points fall within the results of the line formed by the regression equation. The higher the coefficient, the more points fall within the line. If the coefficient is 0.80, then 80% of the points should fall within the regression line.

We want to see that the R-squared is as close to 1 (or 100%) with no negative numbers. The calculation is easy to do with sklearn's "**r2_score**" function (Listing 2-24).

*Listing 2-24.* R-squared Score Over Our Features

**Input:**

```
from sklearn.metrics import r2_score
for feat in features:
    model_lr = linear_model.LinearRegression()
    model_lr.fit(X_train[[feat]], y_train)
    predictions = model_lr.predict(X_test[[feat]])
    print('R^2 for %s is %f' % (feat, r2_score(y_test, predictions)))
```

**Output:**

```
R^2 for hr is 0.160161
R^2 for season is 0.034888
R^2 for holiday is -0.001098
R^2 for temp is 0.154656
```

Every R-squared is positive, and we see that "**hr**" and "**temp**" explain more variance than "**season**" and "**holiday**." Keep in mind that we are looking at each feature separately; a good next step not covered here would be to calculate the R-squared score of all of them together (or an adjusted R-square to handle multiple features).

As shown earlier, we will also "**dummify**" the "**season**" variable. If we rerun the R-squared calculating loop on all features including the "**dummified**" ones, we get the following scores (Listing 2-25).

***Listing 2-25.*** R-squared Score over Dummified Features

```
R^2 for hr is 0.156594
R^2 for holiday is 0.001258
R^2 for temp is 0.154471
R^2 for season_1 is 0.053717
R^2 for season_2 is 0.003657
R^2 for season_3 is 0.016976
R^2 for season_4 is 0.001111
```

Everything is positive, so we're looking good to use that set of features in our final web application.

# Predicting on New Data Using Extracted Coefficients

Now that we have the model's coefficients, we can predict new rental counts using the regression equation. The regression equation is the equation of the line-of-best-fit from our regression model. The formula is common and can be seen in most books covering statistics (Figure 2-16).

$$y = \beta 0 + \beta 1 \, x$$

***Figure 2-16.*** *The regression equation*

"**y**" is the dependent variable, or what we're trying to predict, in our case, the number of bike rentals, a is the intercept, "**β**" is the slope of the line and "**x**" is the independent variable. In the case of a multiple linear regression, you simply add more independent variables.

One important note is that this formula and its coefficients represent the smallest error possible from the data used in the model. So, we cannot really change all the independent variables at once when we inject new data into it. This is an important point to remember, though we will allow the user to play around with all sorts of environmental settings to affect the number of bike rentals, we will also have a "**reset**" button to reset all variables back to their original mean.

After we run our final model, we need to extract intercept and coefficients. This is trivial to do with sklearn's "**linear_model**" function and only requires calling the "**intercept_**" and "**coef_**" parameters to get them (Listings 2-26 and 2-27, and Figure 2-17).

***Listing 2-26.*** Getting the Intercepts

**Input:**

```
from sklearn import linear_model
model_lr = linear_model.LinearRegression()
model_lr.fit(X_train, y_train)
print('Intercept: %f' % model_lr.intercept_)
```

**Output:**

```
Intercept: -121.029547
```

***Listing 2-27.*** Getting the Coefficients

```
feature_coefficients = pd.DataFrame({'coefficients':model_lr.coef_[0],
                                     'features':X_train.columns.values})

feature_coefficients.sort_values('coefficients')
```

| | coefficients | features |
|---|---|---|
| 5 | -41.245562 | season_3 |
| 1 | -23.426176 | holiday |
| 4 | -1.624812 | season_2 |
| 3 | 3.861149 | season_1 |
| 0 | 8.631624 | hr |
| 6 | 39.009224 | season_4 |
| 2 | 426.900259 | temp |

***Figure 2-17.*** *A look at the coefficients from the linear-regression model*

We can then assign those constants to our web application so that it can in turn make predictions on bike rental demand (Listing 2-28).

***Listing 2-28.*** Creating Constants out of Extracted Coefficients

```
INTERCEPT = -121.029547
COEF_HOLIDAY = -23.426176    # if day is holiday or not
COEF_HOUR = 8.631624         # hour (0 to 23)
COEF_SEASON_1 = 3.861149     # 1: spring
COEF_SEASON_2 = -1.624812    # 2: summer
COEF_SEASON_3 = -41.245562   # 3: fall
COEF_SEASON_4 = 39.009224    # 4: winter
COEF_TEMP = 426.900259       # normalized temp in Celsius -8 to +39
```

We also need to get the mean historical values in order to build our regression equation. If the values are categorical, then we pick the highest mean and set that to 1 and the other to 0 (as we do with holiday and season) (Listing 2-29).

***Listing 2-29.*** Setting our Feature Means

```
MEAN_HOLIDAY = 0    # if day is holiday or not
MEAN_HOUR = 11.6    # hour (0 to 23)
MEAN_SEASON_1 = 0   # 1: spring
MEAN_SEASON_2 = 0   # 2: summer
MEAN_SEASON_3 = 1   # 3: fall
MEAN_SEASON_4 = 0   # 4: winter
MEAN_TEMP = 0.4967  # normalized temp in Celsius -8 to +39
```

We now have all we need to predict new rental counts. Let's see how many rentals we get at 9 AM while all other values are held constant around their mean (Listing 2-30).

***Listing 2-30.*** Let's Make a Prediction Using Our Extracted Coefficients

**Input:**

```
rental_counts = INTERCEPT + (MEAN_HOLIDAY * COEF_HOLIDAY) \
        + (9 * COEF_HOUR) \
        + (MEAN_SEASON_1 * COEF_SEASON_1)   + (MEAN_SEASON_2 * COEF_
          SEASON_2) \
        + (MEAN_SEASON_3 * COEF_SEASON_3)   + (MEAN_SEASON_4 * COEF_
          SEASON_4) \
        + (MEAN_TEMP * COEF_TEMP)

print('Estimated bike rental count for selected parameters: %i' %
int(rental_counts))
```

**Output:**

```
Estimated bike rental count for selected parameters: 171
```

And the result is 171 bikes rented at 9 AM (your results may vary slightly). We will allow users to change multiple features at a time, but keep in mind that too many changes from the original equation value may degrade the model.

# Designing a Fun and Interactive Web Application to Illustrate Bike Rental Demand

Now for the fun part, let's design our web application. We always need to have the end goal in mind–what is it that we want to share with others and what will others want to see?

We're going to design an interactive web application that is going to allow users to customize environmental variables (time, holiday, temperature, and season) and get visual feedback on the numbers of bicycles rented.

This application needs to be visually compelling in order to attract users and keep them interested as they play with it. This means that as much thought needs to be invested into the message, UI, visuals, and interactive controls as was put into gathering data and modeling.

# Abstracting Code for Readability and Extendibility

As with most of my web applications, I try to abstract the code into logical modules. One module would be the process of collecting the user data and another, the brains, which would build the regression equation, run it, and return a bike rental prediction. Keeping the code in logical units will drastically simplify your life as you build and debug a web application. This will allow you to unit test each module to make sure everything works accordingly or as a process of elimination when things don't.

In the case of this chapter's web application, most of the "**brains**" will reside directly in the main HTML page. Flask is a web-serving framework and mostly used to retrieve, analyze, and serve back customized content. In this chapter, all we need is the regression equation to predict bicycle rental demand, so there really isn't much of a case for going back and forth between the user and the web server (and this is the only chapter where we do that; in all others the brain will clearly reside on the server, not the client's web page).

# Building a Local Flask Application

Before mounting the code into the cloud, it is important to run things locally; this will save you both headaches and time. First, let's perform a simple Flask exercise on our local machine. If you have never run a Flask application locally, you will need to install the following Python libraries using "**pip3 install**" or whatever tool your OS and Python versions support.

- Flask

Once you have installed Flask, open a text editor, type in the following code, and save it as "**hello.py**" (Listing 2-31).

***Listing 2-31.*** Simple Flask Script

```
from flask import Flask
app = Flask(__name__)
@app.route('/')
def hello_world():
    return 'Hello, World!'
```

Then open a terminal/command window and enter the following command on the Mac or Windows (Listings 2-32 and 2-33).

***Listing 2-32.*** On the Mac

```
$ export FLASK_APP=main.py
$ flask run
```

***Listing 2-33.*** On Windows

```
$ export FLASK_APP= main.py
$ python -m flask run
```

You should see something like Figure 2-18.

```
● ○ ●       test — IPython: chapter-1/python-anywhere-model — flask run — 80×24
[^Cmanuels-MacBook-Pro-2:test manuel$ export FLASK_APP=hello.py        ]
[manuels-MacBook-Pro-2:test manuel$ flask run                         ]
 * Serving Flask app "hello"
 * Running on http://127.0.0.1:5000/ (Press CTRL+C to quit)
```

***Figure 2-18.*** *Terminal/command window displaying local URL for Flask application*

Then copy the URL "http://127.0.0.1:5000/" (or whatever is stated in the terminal window) into your browser and you should see the web application appear. Plenty more examples and tips can be found at the source, the official Flask quick start guide: "http://flask.pocoo.org/docs/0.12/quickstart".

So, what just happened here? If you are new to web-serving frameworks, this may seem a little daunting (and keep in mind that Flask is one of the simplest frameworks out there). Let's break down the approach step-by-step.

In the preceding "**Hello world**" example, everything happened on the hypothetical web-server side (which is really just your local machine). Its job is to process commands and spit out consumable HTML back to the requesting client's web page.

First, we load the Flask library in memory (Listing 2-34).

***Listing 2-34.*** Import Flask

```
from flask import Flask
```

Then we instantiate the Flask session (Listing 2-35).

***Listing 2-35.*** Instantiate Flask

```
app = Flask(__name__)
```

Finally, we create a function to do something and we decorate it with a routing parameter, so it knows which commands it will process from the web client. In this case, the '/' simply means either the root page or the root "**index.html**" page session (Listing 2-36).

***Listing 2-36.*** Flask Function to Handle Traffic Coming from Root URL

```
@app.route('/')
def hello_world():
    return 'Hello, World!'
```

Obviously, very rarely will the function be this simple; it most likely will call a database or a Representational State Transfer (REST) API call to gather custom information and fire it back to the client's web page via an HTML template. This entire process allows intelligent, customized data to be created and then wrapped into a sophisticated looking web page. For all intents and purposes, this will look like a handcrafted page, though it was dynamically created by Flask. We will use Flask throughout this book and you will have a strong grasp on this tool if you work your way through each chapter.

The final bit of code is only used in local mode (i.e., run from your local machine) and run's the web server code and, in this case, turns on the debug flag (Listing 2-37).

*Listing 2-37.* Automatically Running the Flask Application Locally

```
if __name__=='__main__':
        app.run(debug=True)
```

# Downloading and Running the Bike Sharing GitHub Code Locally

Download the files for this chapter if you haven't already done so and navigate to the "**web-application**" folder. Your folder should look like Listing 2-38.

*Listing 2-38.* Web Application Files

```
web-application
├── appengine_config.py
├── main.py
├── requirements.txt
├── app.yaml
├── static
└── images
                ├── bike_zero.png
                ├── bike_one.png
                ├── bike_four.png
                ├── bike_nine.png
                └── bike_sixteen.png
└── templates
        └── index.html
```

Once you have downloaded and unzipped everything, open a command line window, and change the drive into the "**web-application**" folder and install all the required Python libraries by running the "**pip install -r**" command (Listing 2-39).

*Listing 2-39.* Installing Requirements

```
$ pip3 install -r requirements.txt
```

Then run the same commands you ran for the "**Hello World**" experiment (running "**python3 main.py**" will do the trick also; Figure 2-19).

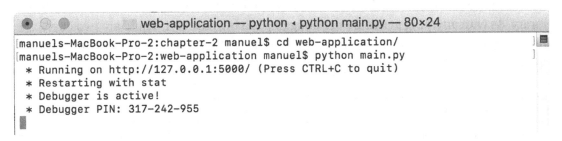

```
[manuels-MacBook-Pro-2:chapter-2 manuel$ cd web-application/
[manuels-MacBook-Pro-2:web-application manuel$ python main.py
 * Running on http://127.0.0.1:5000/ (Press CTRL+C to quit)
 * Restarting with stat
 * Debugger is active!
 * Debugger PIN: 317-242-955
```

***Figure 2-19.*** *Starting the local web server on this chapter's web application*

It should look like the following screen shot in Figure 2-20.

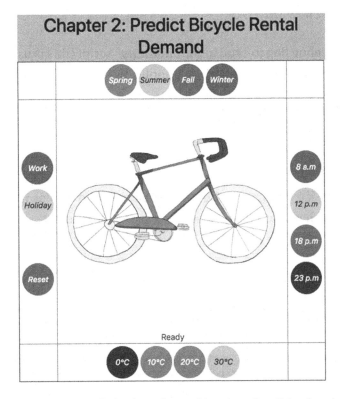

***Figure 2-20.*** *Local rending of Flask web application for this chapter*

# Debugging Tips

If you do not see the screen shot, then your system has an issue or it's missing a file or library. As with anything in this field, debugging is a big part of it. There are two easy things you can do to help out. If this is a Flask issue and your browser looks like Figure 2-21, do the following steps.

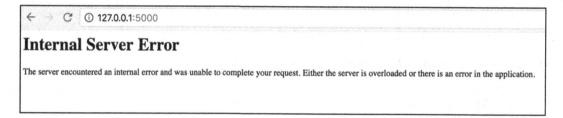

*Figure 2-21.*  *Local web site error*

Turn the Flask debug flag to True in the "**main.py**" script (this is usually at the end of the file). This only works when running your application locally (Listing 2-40).

*Listing 2-40.*  Web Application Files

```
if __name__=='__main__':
        app.run(debug=True)
```

If the issue is Flask related, the debugger will catch it and display it in the browser, and will return a much more helpful message as seen in Figure 2-22.

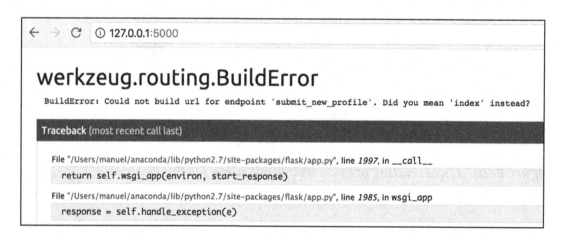

*Figure 2-22.*  *Flask error log in web browser*

And you will also see the Flask error message in the terminal/command window regardless of the debug flag, as shown in Figure 2-23.

```
●  ●  ●                    🔲 google-cloud-model — IPython: Users/manuel — -bash — 115×43
[ * Running on http://127.0.0.1:5000/ (Press CTRL+C to quit)
[127.0.0.1 - - [24/Feb/2018 09:05:56] "GET / HTTP/1.1" 200 -
127.0.0.1 - - [24/Feb/2018 09:07:32] "GET / HTTP/1.1" 200 -
127.0.0.1 - - [24/Feb/2018 09:08:17] "GET / HTTP/1.1" 200 -
127.0.0.1 - - [24/Feb/2018 09:08:18] "GET / HTTP/1.1" 200 -
127.0.0.1 - - [24/Feb/2018 09:08:39] "GET / HTTP/1.1" 200 -
^Cmanuels-MacBook-Pro-2:google-cloud-model manuel$
manuels-MacBook-Pro-2:google-cloud-model manuel$
manuels-MacBook-Pro-2:google-cloud-model manuel$ python main.py
 * Running on http://127.0.0.1:5000/ (Press CTRL+C to quit)
[2018-02-24 09:08:52,204] ERROR in app: Exception on / [GET]
Traceback (most recent call last):
  File "/Users/manuel/anaconda/lib/python2.7/site-packages/flask/app.py", line 1982, in wsgi_app
    response = self.full_dispatch_request()
```

***Figure 2-23.*** *Flask error log in terminal/command window*

After you fix all Flask issues, you may still have some front-end bugs to address. Most browsers will offer some debugging tools. Figure 2-24 shows an example of how to get the JavaScript debugger up and running in Chrome (you should be easily able to find the equivalent in whatever browser brand you use).

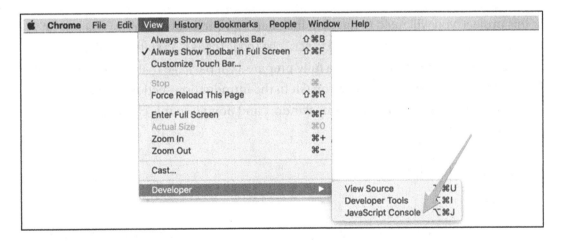

***Figure 2-24.*** *Accessing the JavaScript Console in Google Chrome*

This will open a nifty little debugging center to the right of the web page listing any errors or warnings. It is always a good idea to check it just in case there are warning messages. The same goes with testing your web application in different browser brands and formats such as computers, phones, and tablets (Figure 2-25).

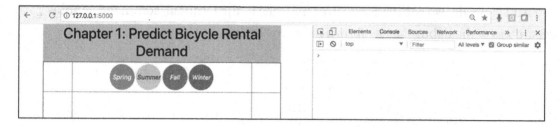

*Figure 2-25.*  *JavaScript Console in Google Chrome in action*

# Microsoft Azure—Mounting a Web Application for the First Time

We're ready to export our model to Azure. You will need an account on Microsoft Azure, and at the time of this writing Microsoft offers a convenient $200 credit and 30-day trial on all services and 12 months access. For more information, see `https://azure.microsoft.com/en-us/free/`.

# Git—Getting All Projects in Git

For this project, you will need to have Git installed on your local machine (you can find the install binaries at `https://www.git-scm.com/downloads`). As stated earlier, Git is a source-code versioning tool and it is a fully prepared Git package that we will push out to Microsoft Azure (see the brief primer on Git in the introduction section).

Open your terminal/command-line window and point it to this chapter's "**web-application**" folder (Listing 2-41).

*Listing 2-41.*  Code Input

```
$ git init
```

It is a great idea to run "**git status**" a couple of times throughout to make sure you are tracking the correct files (Listing 2-42).

*Listing 2-42.*  Running "**git status**"

**Input:**

```
$ git status
```

**Output:**

```
Untracked files:
  (use "git add <file>..." to include in what will be committed)

        main.py
        ptvs_virtualenv_proxy.py
        requirements.txt
        static/
        templates/
        web.3.4.config
```

Add all the web-application files from the "**web-application**" file and check "**git status**" again (Listing 2-43).

***Listing 2-43.***  Adding Web Application Files to Git

**Input:**

```
$ git add .
$ git status
```

**Output:**

```
 Changes to be committed:
  (use "git rm --cached <file>..." to unstage)

        new file:    main.py
        new file:    ptvs_virtualenv_proxy.py
        new file:    requirements.txt
        new file:    static/images/bike_four.png
        new file:    static/images/bike_nine.png
        new file:    static/images/bike_one.png
        new file:    static/images/bike_sixteen.png
        new file:    static/images/bike_zero.png
        new file:    templates/index.html
        new file:    web.3.4.config
```

Do a local Git commit and add a comment that makes sense, in case you need to revisit past actions in the future (Listing 2-44 and Figure 2-26).

***Listing 2-44.*** Git Commit

```
$ git commit -am "bike rental web application commit"
```

```
[master (root-commit) 027b037] bike rental web application commit
 11 files changed, 467 insertions(+)
 create mode 100644 __pycache__/hello.cpython-36.pyc
 create mode 100644 main.py
 create mode 100644 ptvs_virtualenv_proxy.py
 create mode 100644 requirements.txt
 create mode 100644 static/images/bike_four.png
 create mode 100644 static/images/bike_nine.png
 create mode 100644 static/images/bike_one.png
 create mode 100644 static/images/bike_sixteen.png
 create mode 100644 static/images/bike_zero.png
 create mode 100644 templates/index.html
 create mode 100644 web.3.4.config
```

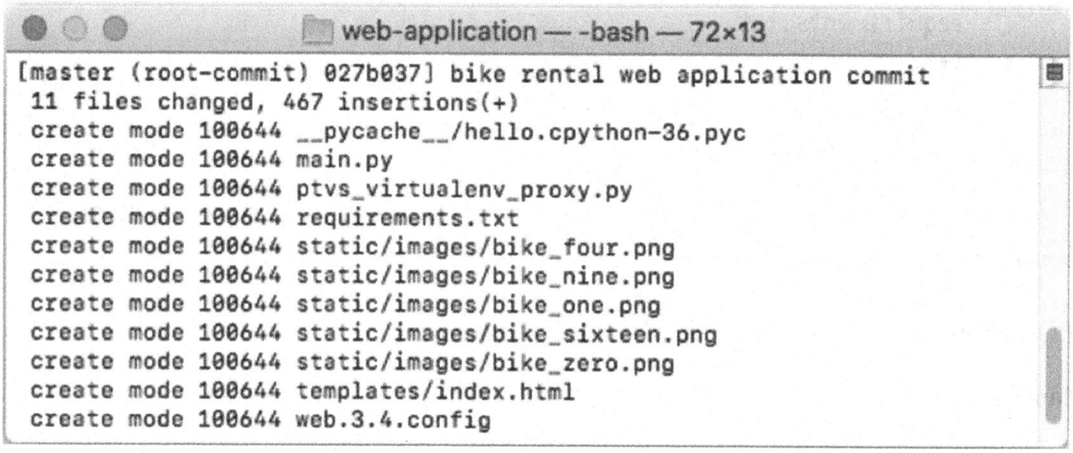

***Figure 2-26.*** *Committed data ready for Azure upload*

For more information on the Git Deployment to Azure App Service, see https://docs.microsoft.com/en-us/azure/app-service/app-service-deploy-local-git.

# The azure-cli Command Line Interface Tool

We will rely on the "**azure-cli**" tool to get us up and running, as it is a convenient way to start and control web instances (for more information on setting this up, see the official docs at https://docs.microsoft.com/en-us/cli/azure/get-started-with-azure-cli).

For Mac:

```
$ brew update && brew install azure-cli
```

For all other Operating Systems, refer to the official documentation: https://docs.microsoft.com/en-us/cli/azure/install-azure-cli.

# Step 1: Logging In

After installing the "**azure-cli**" command-line tool (or using the Azure Cloud Shell directly if the local command-line tool is giving you trouble), create an "**az**" session (Listing 2-45).

***Listing 2-45.*** Logging into Azure from azure-cli

**Input:**

```
az login
```

**Output:**

```
To sign in, use a web browser to open the page https://microsoft.com/
devicelogin and enter the code BTJMDCR34 to authenticate.
```

Follow the instructions, point a browser to the givenURL address, and enter the code accordingly (Figure 2-27).

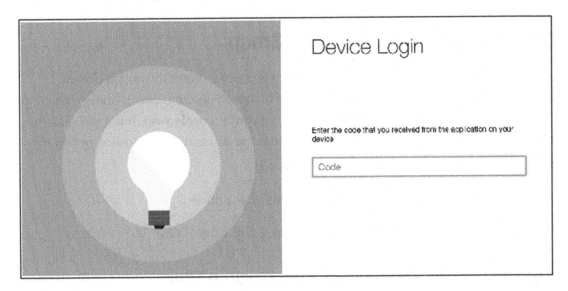

***Figure 2-27.*** *Authenticating session*

If all goes well (i.e., you have an Azure account in good standing), it will connect the azure-cli terminal to the cloud server automatically. Also, once you are authorized, you can safely close the browser window. Make sure your command-line tool is pointing to this chapter's "**web-application**" folder.

# Step 2: Create Credentials for Your Deployment User

This user will have appropriate rights for FTP and local Git use. Here I set the user-name to "**flaskuser11**" and password to "**flask123**". You should only have to do this once, then you can reuse the same account. In case it gives you trouble, simply create a different user name (or add a number at the end of the user name and keep incrementing it like I do (Listing 2-46).

***Listing 2-46.***  Creating a User

```
$ az webapp deployment user set --user-name <<REPLACE-WITH-YOUR-USER-NAME>>
--password flask123
```

As you proceed through each "**azure-cli**" step, you will get back JSON replies confirming your settings. In the case of the "**az webapp deployment**," most should have a null value and no error messages. If you have an error message, then you have a permission issue that needs to be addressed ("**conflict**" means that name is already taken so try another, and "**bad requests**" means the password is too weak).

# Step 3: Create your Resource Group

This is going to be your logical container. Here you need to enter the region closest to your location (see https://azure.microsoft.com/en-us/regions/). Going with "**West US**" for this example isn't a big deal even if you're worlds away, but it will make a difference in a production setting where you want the server to be as close as possible to your viewership for best performance (Listing 2-47).

***Listing 2-47.***  Creating a Resource Group and Response

```
$ az group create --name myResourceGroup --location "West US"
```

# Step 4: Create Your Azure App Service Plan

Here I set the name to "**myAppServicePlan**" and select a free instance (sku; Listing 2-48).

***Listing 2-48.***  Creating a Service Plan and Successful Response

```
$ az appservice plan create --name myAppServicePlan --resource-group
myResourceGroup --sku FREE
```

# Step 5: Create Your Web App

Your "**webapp**" name needs to be unique, and make sure that you "**resource-group**" and "**plan**" names are the same as what you set in the earlier steps. In this case I am going with "**amunateguibike**" (Listing 2-49).

***Listing 2-49.***   Creating a Web App

```
$ az webapp create --resource-group myResourceGroup --plan myAppServicePlan
--name amunateguibike --runtime "python|3.4" --deployment-local-git
```

For a full list of supported runtimes, see Listing 2-50.

***Listing 2-50.***   List of Supported runtimes

```
$ az webapp list-runtimes
```

The output of "**az webapp create**" will contain an important piece of information that you will need for subsequent steps. Look for the line "**deploymentLocalGitUrl**" (Figure 2-28).

```
● ○ ●                          🗔 web-application — -bash — 102×11
Local git is configured with url of 'https://flaskuserX@amunateguibike.scm.azurewebsites.net/amunategu
ibike.git'
[{
  "availabilityState": "Normal",
  "clientAffinityEnabled": true,
  "clientCertEnabled": false,
  "cloningInfo": null,
  "containerSize": 0,
  "dailyMemoryTimeQuota": 0,
  "defaultHostName": "amunateguibike.azurewebsites.net"
  "deploymentLocalGitUrl": "https://flaskuserX@amunateguibike.scm.azurewebsites.net/amunateguibike.git
```

***Figure 2-28.***   *Truncated output of Git URL from "**deploymentLocalGitUrl**"*

For extracting the local Git configuration URL for your Azure project instance, see Listings 2-51 and 2-52.

***Listing 2-51.*** The Extracted Git URL in My Case

```
https://flaskuser11@amunateguibike.scm.azurewebsites.net/amunateguibike.git
```

***Listing 2-52.*** Yours Will Look Like the Following

```
https://<<REPLACE-WITH-YOUR-USER-NAME>>@<<REPLACE-WITH-YOUR-APP-NAME>>.scm.
azurewebsites.net/<<REPLACE-WITH-YOUR-APP-NAME>>.git
```

# Step 6: Push git Code to Azure

Append the URL we saved previously with the location of your GIT repository to the "**add azure**" command (Listing 2-53).

***Listing 2-53.*** Final Code Push to Azure

```
# if git remote already exits, run 'git remote remove azure'
$ git remote add azure https://flaskuser11@amunateguibike.scm.
azurewebsites.net/amunateguibike.git
```

It may prompt for your password; make sure you use the one you created in the "**az webapp deployment user**" step ("**flask123**" in my case; Listing 2-54).

***Listing 2-54.*** Final Code Push to Azure

```
$ git push azure master
```

That's it! You can get back to your placeholder browser page and hit refresh or open a new browser page and enter http://amunateguibike.azurewebsites.net

(or in your case http://<<REPLACE-WITH-YOUR-APP-NAME>>.azurewebsites.net) and you should see "**Predict Bicycle Rental Demand**" (Figure 2-29).

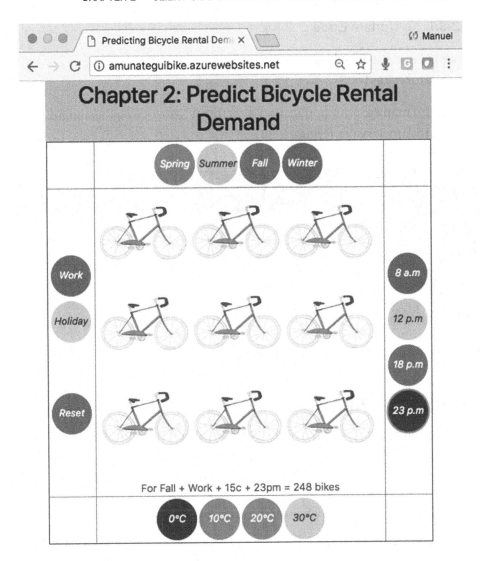

**Figure 2-29.**  *Enjoy the fruits of your hard work–The "**Predict Bicycle Rental Demand**" web application!*

On the other hand, if the azure-cli returns error messages, you will have to address them (see the troubleshooting section). Anytime you update your code and want to redeploy it, send a "**push**" command (Listing 2-55).

***Listing 2-55.*** To Update Code

```
$ git commit -am "updated output"
$ git push azure master
```

You can also manage your application directly on Azure's web dashboard. Log into Azure and go to App Services (Figure 2-30).

***Figure 2-30.*** *Microsoft Azure dashboard*

# Important Cleanup!

This is a critical step; you should never leave an application running in the cloud that you don't need, as it does incur charges (or use up your free credits if you are on the trial program). If you don't need it anymore, take it down (Listing 2-56 and Figure 2-31).

***Listing 2-56.*** Don't Forget to Delete Your Azure Instance When Done!

```
$ az group delete --name myResourceGroup
```

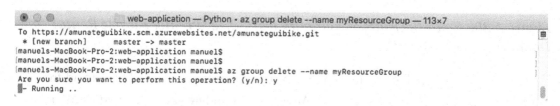

***Figure 2-31.*** *Deleting the web application from the Azure cloud*

Or delete it using Azure's web dashboard under "**App Services.**"

# Troubleshooting

It can get convoluted to debug web application errors. One thing to do is to turn on logging through Azure's dashboard (Figure 2-32).

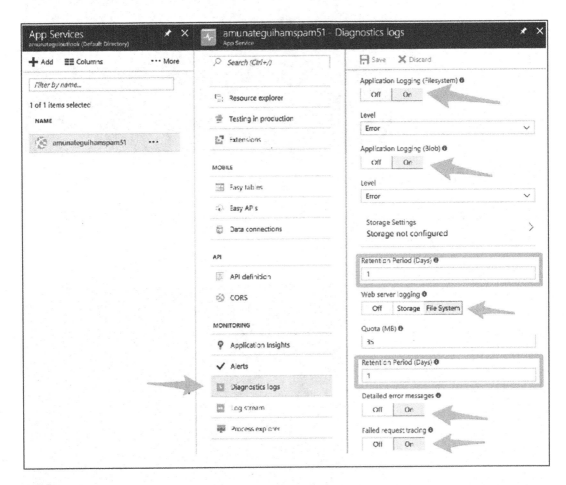

***Figure 2-32.*** *Turning on Azure's diagnostics logs*

Then you turn the logging stream on to start capturing activity (Figure 2-33).

***Figure 2-33.***  *Capturing log information*

You can also check your file structure using the handy Console tool built into the Azure dashboard (Figure 2-34).

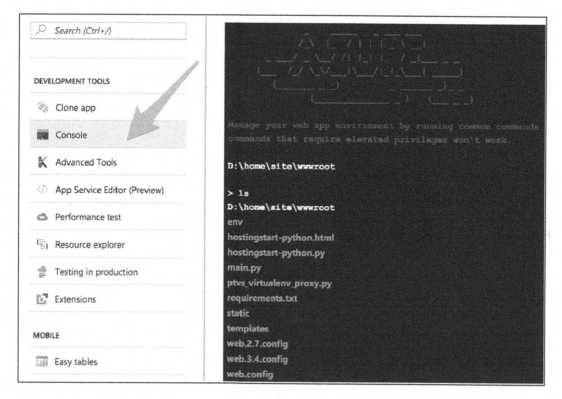

***Figure 2-34.***  *Azure's built-in command line tool*

You can even check if your "**requirement.txt**" file works by using the install command (Listing 2-57).

***Listing 2-57.*** Running Commands in the Azure Console

```
> env\scripts\pip install -r requirements.txt
```

# Steps Recap

1. Point your terminal/command window to the right directory with the chapter's web application files (and confirm that it runs locally).

   ```
   $ cd chapter-2/web-application
   ```

2. Git commit all files.

   ```
   $ git init
   ```

   ```
   $ git add .
   ```

   ```
   $ git commit -am "bike rental commit"
   ```

3. Log into the Azure command line interface and authenticate the session.

   ```
   $ az login
   ```

4. Prepare the Azure web application.

   ```
   $ az webapp deployment user set --user-name flaskuser11
   --password flask123
   ```

   ```
   $ az group create --name myResourceGroup --location
   "West US"
   ```

   ```
   $ az appservice plan create --name myAppServicePlan
   --resource-group myResourceGroup --sku FREE
   ```

   ```
   $ az webapp create --resource-group myResourceGroup
   --plan myAppServicePlan --name amunateguibike --runtime
   "python|3.4" --deployment-local-git
   ```

5.  Push the web application to the Azure cloud.

```
$ git remote add azure https://flaskuser11@amunateguibike.scm.
azurewebsites.net/amunateguibike.git
```

```
$ git push azure master
```

6.  Open the URL in a browser window and enjoy!

```
http://<<WEBAPP NAME>>.azurewebsites.net
```

7.  Terminate instance!

```
$ az group delete --name myResourceGroup
```

# What's Going on Here? A Look at the Scripts and Technology Used in Our Web Application

Let's do a brief flyover of our web application's code. There are two important files: "**main.py**," which is the web-serving controlling script and the template file "**index. html**," which is the face of our web application. As most of the processing happens directly in "**index.html**," we'll spend most of our time there looking at the HTML and JavaScript running this web application.

## main.py

Under normal circumstances, this would be the brains behind the operation. It can do about anything a standalone Python script can, with the addition of being able to generate content for web pages. In this chapter, there really isn't much going on here except for passing average feature values and the model's intercept and coefficients to the template. In this book, we will use both "**main.py**" and "**application.py**." There isn't a right or wrong way of naming the controlling Python web-serving file—the only exception being some reserved words. If you do opt for a custom name, you will need to update the YAML file and/or the Web Server Gateway Interface configuration file. Also, some cloud providers default to different application names; Google Cloud defaults to "**main**" and Azure defaults to "**application**." But in either case, it is possible to customize and change.

The one interesting thing that "**main.py**" is doing here is transmitting starting values for the intercept, coefficients, and mean values.

The decorator "**@app.route**" will route any traffic calling the root URL or with the file name "**index.html**." It then simply passes all the default values to the "**index.html**" template (Listing 2-58).

*Listing 2-58.* Routing to the "**index.html**"

```
@app.route("/", methods=['POST', 'GET'])
def index():
        # on load set form with defaults
    return render_template('index.html',
                        mean_holiday = MEAN_HOLIDAY,
                        mean_hour = MEAN_HOUR,
                        mean_sesaon1 = MEAN_SEASON_1,
                        mean_sesaon2 = MEAN_SEASON_2,
                        mean_sesaon3 = MEAN_SEASON_3,
                        mean_sesaon4 = MEAN_SEASON_4,
                        mean_temp = MEAN_TEMP,
                        model_intercept = INTERCEPT,
                        model_holiday = COEF_HOLIDAY,
                        model_hour = COEF_HOUR,
                        model_season1 = COEF_SEASON_1,
                        model_season2 = COEF_SEASON_2,
                        model_season3 = COEF_SEASON_3,
                        model_season4 = COEF_SEASON_4,
                        model_temp = COEF_TEMP)
```

Flask uses a technology called "**Jinja2**" to inject those variables directly into the HTML template form. If you look at the return statement, it calls Flask's "**render_template**" function and passes the intended variables to "**index.html**."

All a template needs to do to receive those variables is use the double curly bracket command. To see all of this, refer to the web-application "**index.htm**" full script (Listing 2-59).

*Listing 2-59.* Using Jinja2 to Set Python Variables into JavaScript

```
<SCRIPT>
        var HOLIDAY = {{mean_holiday}}    // day is holiday or not
        var HOUR = {{mean_hour}}          // hour (0 to 23)
        var HOUR = {{mean_hour}}          // hour (0 to 23)
        var SEASON_1 = {{mean_sesaon1}}   // 1:spring
        var SEASON_2 ={{mean_sesaon2}}    // 2:summer
        var SEASON_3 = {{mean_sesaon3}}   // 3:fall
        var SEASON_4 = {{mean_sesaon4}}   // 4:winter
        var TEMP = {{mean_temp}}          // norm temp in Celsius -8 to +39
        var INTERCEPT = {{model_intercept}}
        var COEF_HOLIDAY = {{model_holiday}}  // day is holiday or not
        var COEF_HOUR = {{model_hour}}        // hour (0 to 23)
        var COEF_SEASON_1 = {{model_season1}} // 1:spring
        var COEF_SEASON_2 = {{model_season2}} // 2:summer
        var COEF_SEASON_3 = {{model_season3}} // 3:fall
        var COEF_SEASON_4 = {{model_season4}} // 4:winter
        var COEF_TEMP = {{model_temp}}        // norm temp in Celsius -8 to +39
...
```

# /static/ folder

The static folder, as its name implies, holds static, nonchanging files. This is where you store images, files, and other shareable data for our web application.

# /templates/index.html folder and script

The templates folder holds all the templates required for our web application. In the subsequent chapters there are usually two html files, an "**index.html**" and a response html file. It is better to break these files apart instead of trying to cram everything into a single html file with complex "**if then**" forks.

Most of the action in this chapter happens inside "**index.html**," so let's take a deeper look. Open the full "**index.html**" file in your editor to follow along. As mentioned earlier, the "**brains**" of this chapter aren't Flask but the "**index.html**" front-end page–and mostly all inside of the JavaScript snippet at the end of the page. JavaScript brings a great level of interactivity to a web page.

In the case of this web application, it listens for button click events and recalculates bike rental demand by running the regression equation with the selected features. It doesn't end there; once it gets a new demand estimate, it will weigh that number and decide which bicycle picture collage to show–if it's a small estimate it returns a single bicycle, if it's a huge one, all sixteen.

First the user clicks a feature button to get a new bike rental estimate using that particular feature (Listing 2-60).

***Listing 2-60.*** Calling for Predictions Using the HTML "**\<button\>**" event.

```
<button type="button" onclick="calculateBikeDemand(this)" id="season_
spring" class="btn btn-info btn-circle btn-xl"><i class="fa fa-
check">Spring</i></button>
```

The "**onclick()**" function inside the "**\<button\>**" tag will send the Id "**season_spring**" to the main JavaScript function "**calculateBikeDemand()**." This is telling the function that the user wants to recalculate the regression equation with the season's variable set as seen in Listing 2-61.

***Listing 2-61.*** The "**calculateBikeDemand()**" Modeling JavaScript Function

```
function calculateBikeDemand(elem) {

        // apply new value to stored variables
        ...

        // recalculate the regression equation
        rental_counts = INTERCEPT + (HOLIDAY * COEF_HOLIDAY)
                + (HOUR * COEF_HOUR)
                + (SEASON_1 * COEF_SEASON_1)  + (SEASON_2 * COEF_SEASON_2)
                + (SEASON_3 * COEF_SEASON_3)  + (SEASON_4 * COEF_SEASON_4)
                + (TEMP * COEF_TEMP)if (rental_counts < 0)

        // figure out which image to show
        if (rental_counts < 0) {
                bike_out = 'static/images/bike_sixteen.png'
                if (rental_counts < 0) {
                        bike_out = 'static/images/bike_zero.png'
                } else if (rental_counts < 100) {
```

```
                               bike_out = 'static/images/bike_one.png'
                   } else if (rental_counts < 200) {
                               bike_out = 'static/images/bike_four.png'
                   } else if (rental_counts < 300) {
                               bike_out = 'static/images/bike_nine.png'}

        // build a new string that is readable with variables select by user
        // and new bike rental estimate
        output = 'For ' + season + ' + ' + holiday + ' + ' + temp + ' +
        ' + hour + ', demand = ' + Math.round(rental_counts) + ' bikes';

        // inject new value and image source directly into the HTML tag
        document.getElementById("query").innerHTML = output;
        document.getElementById('bike_out').src = bike_out
}
```

# Conclusion

That's it for our first project! Though this was a simple one with little back and forth between the web-client and web-server, it fulfills the definition of a real web application. In this chapter we introduced the concept of extending standalone scripts into interactive web applications by using Flask and web controls. We also saw how easy Python and Python libraries can communicate with the Flask web framework, making the leap into web computing almost seamless.

The process started by planning what our web application should be and what would be of interest to the viewer. This step can't be emphasized enough: if it isn't of interest to anybody, then there is no need to bother building it. Too often we start by modeling and then attempt to retrofit it to make it work around a web application story.

We then explored the Bike Sharing Dataset from Capital Bikeshare System,[5] experimented with different modeling approaches to predict rental demand under environmental factors and chose the final features and model coefficients to use in our web application.

---

[5]https://www.capitalbikeshare.com/system-data

We ran a local version of the Flask application and finally deployed it to the Microsoft Azure cloud. If you follow these steps in this order, you should be fine. Always start by designing the web application story, build and run as much of it as you can locally, and only then deploy to the cloud.

# Additional Resources

If you want to learn more about Flask, Google it; there's so much material on this topic.

For more information on:

- **Flask**: Go to the source, the official portal: `http://flask.pocoo.org/`

- **CSS**: See the tutorial on the great w3schools site (CSS and everything else web related): `www.w3schools.com/css/`

- **Bootstrap**: See the portal: `https://getbootstrap.com/`

- **JQuery**: Check out their portal: `https://jquery.com/`

- **YAML**: Check out their streamlined portal: `http://yaml.org/`

- **Jinja2**: See the official documentation: `http://jinja.pocoo.org/docs/`

# CHAPTER 3

# Real-Time Intelligence with Logistic Regression on GCP

Let's understand who survived the Titanic shipwreck by building an interactive passenger profile builder on Google Cloud.

In this chapter, we revisit the classic and dramatic Titanic dataset, favored by modeling text books and educational blogs all over the world. We will analyze the passenger manifest and attempt to understand why some did, and others didn't, survive this tragic accident. We will explore the dataset, prepare it for modeling, and extend it into an interactive web application that will allow users to create a fictional passenger, tweak parameters, and visualize how well he or she fared on the voyage (Figure 3-1).

© Manuel Amunategui, Mehdi Roopaei 2018
M. Amunategui and M. Roopaei, *Monetizing Machine Learning*, https://doi.org/10.1007/978-1-4842-3873-8_3

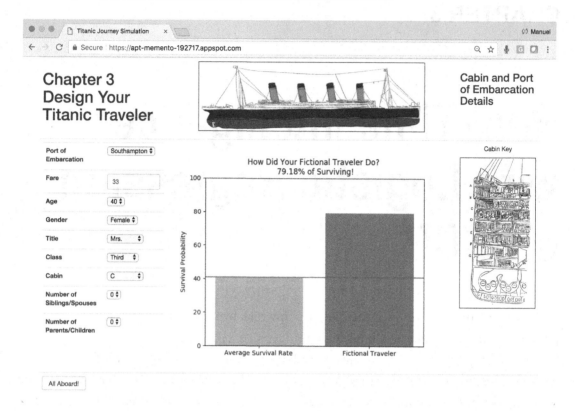

**Figure 3-1.** *The final web application for this chapter*

The Vanderbilt University Department of Biostatistics[1] is graciously hosting the data (along with many other interesting datasets) and can be conveniently downloaded using a direct call from the Pandas Python library.

---

**Note**    Download the files for Chapter 3 by going to `www.apress.com/9781484238721` and clicking the source code button. Open Jupyter notebook "**chapter3.ipynb**" to follow along with this chapter's content.

---

[1]`http://biostat.mc.vanderbilt.edu/wiki/Main/DataSets`

# Planning our Web Application

Our first step is to get our Titanic concept fully thought out and working as a web application on our local machine.

The brain behind our application is a simple logistic regression model from the **"sklearn Python library."**[2] After we prepare the data by removing nonpredictive features, creating dummy columns for categorical data, and applying basic natural-language processing on text fields, we train the model to identify the common patterns around those who survived the trip and those who didn't.

We verify the accuracy of our model by following the standard modeling procedure of splitting the data into chunks, one to train the model and the other to validate it. Once we are satisfied with the model's abilities, we have it predict the probability of survival on a fictional passenger. This probability, a number between 0 and 1, represents the chance for the fictional passenger to make it out alive from the shipwreck–the closer to 1, the better the chances of surviving.

Finally, we will abstract and generalize this entire process to run in the constructor of our web application. This means that the entire process of ingesting, preparing, and modeling data will only happen once during deployment of the model (and whenever the web server is rebooted). This ensures that when a user wants to interact with the web application, it quickly yields a prediction from the trained model because it is already loaded in memory. But we're jumping the gun here; let's first finish the local version of the project.

# Data Wrangling

As this is a classic data science exercise and a famously recorded event, we already know (or have an intuition of) which features are of interest in understanding who survived the voyage (see the Vanderbilt Biostat for additional information[3]). Go ahead and download the files for this chapter into a folder called **"chapter-3."** Open up the Jupyter notebook to follow along. Let's look at the data legend from the host (Table 3-1).

---

[2]http://scikit-learn.org/stable/modules/generated/sklearn.linear_model.
LogisticRegression.html

[3]http://biostat.mc.vanderbilt.edu/wiki/pub/Main/DataSets/Ctitanic3.html

**Table 3-1.** *Dataset Legend*

| Feature | Description |
| --- | --- |
| pclass | Ticket class comprised of 3 levels |
| sex | Gender |
| age | Age |
| sibsp | Number of siblings/spouses aboard |
| parch | Number of parents/children aboard |
| fare | Passenger fare |
| cabi | Cabin number |
| embarked | Point of embarkation |
| name | Passenger name |

A first pass at exploring data can be done programmatically using the Pandas "**head()**" function that will return the top five rows (Listing 3-1 and Figure 3-2). You can also use the "**tail()**" function to see the bottom five.

**Listing 3-1.** Quick Look at the Top Five Rows

```
titanic_df.head()
```

| | pclass | survived | name | sex | age | sibsp | parch | ticket | fare | cabin | embarked | boat | body | home.dest |
| --- | --- | --- | --- | --- | --- | --- | --- | --- | --- | --- | --- | --- | --- | --- |
| 0 | 1 | 1 | Allen, Miss. Elisabeth Walton | female | 29.00 | 0 | 0 | 24160 | 211.3375 | B5 | S | 2 | NaN | St Louis, MO |
| 1 | 1 | 1 | Allison, Master. Hudson Trevor | male | 0.92 | 1 | 2 | 113781 | 151.5500 | C22 C26 | S | 11 | NaN | Montreal, PQ / Chesterville, ON |
| 2 | 1 | 0 | Allison, Miss. Helen Loraine | female | 2.00 | 1 | 2 | 113781 | 151.5500 | C22 C26 | S | NaN | NaN | Montreal, PQ / Chesterville, ON |
| 3 | 1 | 0 | Allison, Mr. Hudson Joshua Creighton | male | 30.00 | 1 | 2 | 113781 | 151.5500 | C22 C26 | S | NaN | 135.0 | Montreal, PQ / Chesterville, ON |
| 4 | 1 | 0 | Allison, Mrs. Hudson J C (Bessie Waldo Daniels) | female | 25.00 | 1 | 2 | 113781 | 151.5500 | C22 C26 | S | NaN | NaN | Montreal, PQ / Chesterville, ON |

**Figure 3-2.** *The first five rows of the raw data*

The data does conform to Vanderbilt Biostat's data legend. Things that should jump out at you are the missing values in the "**body**" column, a lot of text data in the "**name**" and "**home.dest**" columns, and some mix content in the "**cabin**" column.

Functions "**info()**," "**describe()**," and "**isnull()**" are also key for quick data exploration. It is highly recommended to run these whenever facing a new dataset or after any data transformation work.

The Pandas "**info()**" function tells you the data types and non-null counts contained in the dataset (Listing 3-2).

*Listing 3-2.*  Quick Look Feature Data Types

**Input:**

```
titanic_df.info()
```

**Output:**

```
<class 'pandas.core.frame.DataFrame'>
RangeIndex: 1309 entries, 0 to 1308
Data columns (total 1 columns):
pclass        1309 non-null int64
survived      1309 non-null int64
name          1309 non-null object
sex           1309 non-null object
age           1046 non-null float64
sibsp         1309 non-null int64
parch         1309 non-null int64
ticket        1309 non-null object
fare          1308 non-null float64
cabin          295 non-null object
embarked      1307 non-null object
boat           486 non-null object
body           121 non-null float64
home.dest      745 non-null object
dtypes: float64(3), int64(4), object(7)
memory usage: 143.2+ KB
```

Data labeled as "**non-null object**" from the "**info()**" function output can be considered as text based, and we need to figure out what type of text it is. This is a bit of a subjective art, as there are various ways to approach this (more on this shortly).

The Pandas "**describe()**" function gives you an aggregate summary of all quantitative fields. Right off the bat, we can see that the "**survived**" feature has a mean of 0.38. This means that only 38% of the passengers survived and, as we will use that feature as our outcome label to train the model, that the dataset is skewed toward nonsurvivors (i.e., most passengers did not survive the voyage; Listing 3-3 and Figure 3-3).

***Listing 3-3.*** Summary of Quantitative Data

```
titanic_df.describe()
```

| | pclass | survived | age | sibsp | parch | fare | body |
|---|---|---|---|---|---|---|---|
| **count** | 1309.000000 | 1309.000000 | 1046.000000 | 1309.000000 | 1309.000000 | 1308.000000 | 121.000000 |
| **mean** | 2.294882 | 0.381971 | 29.881138 | 0.498854 | 0.385027 | 33.295479 | 160.809917 |
| **std** | 0.837836 | 0.486055 | 14.413493 | 1.041658 | 0.865560 | 51.758668 | 97.696922 |
| **min** | 1.000000 | 0.000000 | 0.170000 | 0.000000 | 0.000000 | 0.000000 | 1.000000 |
| **25%** | 2.000000 | 0.000000 | 21.000000 | 0.000000 | 0.000000 | 7.895800 | 72.000000 |
| **50%** | 3.000000 | 0.000000 | 28.000000 | 0.000000 | 0.000000 | 14.454200 | 155.000000 |
| **75%** | 3.000000 | 1.000000 | 39.000000 | 1.000000 | 0.000000 | 31.275000 | 256.000000 |
| **max** | 3.000000 | 1.000000 | 80.000000 | 8.000000 | 9.000000 | 512.329200 | 328.000000 |

***Figure 3-3.*** *Description output of the titanic data frame*

The "**isnull()**" function can be wrapped into a counter to find out how many missing values we are dealing with. Books have been written on the topic of imputation and how best to deal with missing data in modeling scenarios. Here we will drop a few features and impute others (Listing 3-4 and Figure 3-4).

***Listing 3-4.*** Analyzing Missing Data

```
titanic_missing_count = titanic_df.isnull().sum().sort_
values(ascending=False)
pd.DataFrame({'Percent Missing':titanic_missing_count/len(titanic_df)})
```

| | Percent Missing |
|---|---|
| body | 0.907563 |
| cabin | 0.774637 |
| boat | 0.628724 |
| home.dest | 0.430863 |
| age | 0.200917 |
| embarked | 0.001528 |
| fare | 0.000764 |
| ticket | 0.000000 |
| parch | 0.000000 |
| sibsp | 0.000000 |
| sex | 0.000000 |
| name | 0.000000 |
| survived | 0.000000 |
| pclass | 0.000000 |

*Figure 3-4.* *Percent missing data per feature in the titanic data frame*

Upon analyzing the function outputs, we gather that the data includes numerical, categorical, and text-based features. The dataset contains a total of 1,309 rows. We also see that 90% of the entries in feature "**body**" are missing; this makes for an easy feature to drop. There are other features that we will ignore, as they are either hard to work with or of little help to model survivorship (such as the passenger's last name).

# Dealing with Categorical Data

Working with categorical data is an important topic, and some aspects are subjective while others aren't. A good rule of thumb is to find out how many values for a particular text-based column are unique or repeated. Checking the frequency is a good place to start. Let's combine Pandas "**groupby()**" and "**count()**" functions call on feature "**cabin.**" This will tell if values are shared among passengers and thus should be considered categories or free-form text entries (Listing 3-5 and Figure 3-5).

*Listing 3-5.* Counting Repeats in the "**cabin**" Feature

```
titanic_feature_count = titanic_df.groupby('cabin')['cabin'].count().reset_
index(name = "Group_Count")
titanic_feature_count.sort_values('Group_Count', ascending=False).head(20)
```

| | cabin | Group_Count |
| --- | --- | --- |
| 80 | C23 C25 C27 | 6 |
| 184 | G6 | 5 |
| 47 | B57 B59 B63 B66 | 5 |
| 60 | B96 B98 | 4 |
| 183 | F4 | 4 |
| 181 | F33 | 4 |
| 180 | F2 | 4 |
| 79 | C22 C26 | 4 |
| 117 | D | 4 |
| 102 | C78 | 4 |

*Figure 3-5.* *Cabins are concatenated and can benefit from being decoupled*

The resulting output confirms that these values are categories, as they are often repeated (this follows the intuition that there can be more than one passenger per cabin). Let's run the same experiment on another text-based column, the "**name**" feature (and we already know how many repeats we'll see; Listing 3-6 and Figure 3-6).

***Listing 3-6.*** Counting Repeats in the "**name**" Feature

```
titanic_feature_count = titanic_df.groupby('name')['name'].count().reset_
index(name = "Group_Count")
titanic_feature_count.sort_values('Group_Count', ascending=False).head(10)
```

|  | name | Group_Count |
| --- | --- | --- |
| **261** | Connolly, Miss. Kate | 2 |
| **638** | Kelly, Mr. James | 2 |
| **0** | Abbing, Mr. Anthony | 1 |
| **879** | O'Brien, Mrs. Thomas (Johanna "Hannah" Godfrey) | 1 |
| **877** | O'Brien, Mr. Thomas | 1 |
| **876** | Nysveen, Mr. Johan Hansen | 1 |
| **875** | Nysten, Miss. Anna Sofia | 1 |
| **874** | Nye, Mrs. (Elizabeth Ramell) | 1 |
| **873** | Novel, Mr. Mansouer | 1 |
| **872** | Nourney, Mr. Alfred ("Baron von Drachstedt") | 1 |

***Figure 3-6.*** *How many times is a name repeated?*

As intuition would have it, the majority of names are unique. This just isn't a good column to be considered categorical nor is it a good column for modeling in general (you need repeating data to find patterns, and unique text entries don't serve that purpose in their raw state).

Categories can be found in both Integer and Text data types; therefore, it is important to consider each feature individually and determine how best to model. For example, the feature "**sex**" is categorical and binary, so fractions of that data won't help us. We will change it to a single column named "**isFemale**." The feature "**cabin**" does repeat as shown before, but we can get it to repeat much more with a little help (Listing 3-7).

***Listing 3-7.*** **"Head()"** of Cabin Features

**Input:**

```
titanic_df['cabin'].head()
```

**Output:**

```
0          B5
1    C22 C26
2    C22 C26
3    C22 C26
4    C22 C26
Name: cabin, dtype: object
```

By extracting a sample of that feature, we notice a commonality with the data: each number is preceded by a letter representing the ship deck level (Figure 3-7).

***Figure 3-7.*** *Cutout of the Titanic with labels representing the cabin levels (Illustration by Lucas Amunategui)*

So, one way to leverage that data is to take the first letter and drop the number. This gives us an interesting feature to work with: what floor was the passenger on and was there a relationship between the distance from the floor and the lifeboats on the top deck regarding survivorship? Maybe there is, and you can find out by using the application we are about to build (Listing 3-8).

***Listing 3-8.*** Using Only the First Character from Each Cabin Name

**Input:**

```
titanic_df['cabin'] = titanic_df['cabin'].replace(np.NaN, 'U')
titanic_df['cabin'] = [ln[0] for ln in titanic_df['cabin'].values]
titanic_df['cabin'] = titanic_df['cabin'].replace('U', 'Unknown')
titanic_df['cabin'].head()
```

**Output:**

```
0    B
1    C
2    C
3    C
4    C
Name: cabin, dtype: object
```

And let's do a "**groupby()**" and "**count()**" as we did previously to count frequency (Listing 3-9 and Figure 3-8).

***Listing 3-9.*** Counting Each Cabin First Letter Groups

```
titanic_feature_count = titanic_df.groupby('cabin')['cabin'].count().reset_
index(name = "Group_Count")
titanic_feature_count.sort_values('Group_Count', ascending=False).head(10)
```

| | cabin | Group_Count |
|---|---|---|
| 8 | Unknown | 1014 |
| 2 | C | 94 |
| 1 | B | 65 |
| 3 | D | 46 |
| 4 | E | 41 |
| 0 | A | 22 |
| 5 | F | 21 |
| 6 | G | 5 |
| 7 | T | 1 |

*Figure 3-8.* *Total counts of our new "cabin" feature*

Wow, there are a lot more repeats than previously observed. This should be more useful as a feature than in its previous format. We will do the same to the passenger's "**name**" feature and extract only the title: "**Mrs.**" or "**Mr.**", etc. and drop the other parts (see the Jupyter notebook for more details).

This isn't to say that raw, free-form text (the nonrepeating kind) can't be useful for modeling. On the contrary, most data in the world is unstructured and very rich in potential–think doctor notes, or store reviews! This can be modeld but requires more advanced approaches such as natural language processing, singular vector decomposition, word vectoring, etc. We will see some of these techniques in later chapters of the book.

# Creating Dummy Features from Categorical Data

Once we have identified our categorical features and transformed those that needed transformation, we still need to turn them into a numerical form so our models can use them. One great function from the Pandas library is "**get_dummies().**" This will break each category out into its own column.

Here is an example of using the "**get_dummies()**" function and its output (Listing 3-10 and Figure 3-9).

***Listing 3-10.*** Creating Dummies out of the "**cabin**" Feature

```
pd.get_dummies(titanic_df['cabin'], columns=['cabin'], drop_first=False).
head(10)
```

| | A | B | C | D | E | F | G | T | Unknown |
|---|---|---|---|---|---|---|---|---|---|
| 0 | 0 | 1 | 0 | 0 | 0 | 0 | 0 | 0 | 0 |
| 1 | 0 | 0 | 1 | 0 | 0 | 0 | 0 | 0 | 0 |
| 2 | 0 | 0 | 1 | 0 | 0 | 0 | 0 | 0 | 0 |
| 3 | 0 | 0 | 1 | 0 | 0 | 0 | 0 | 0 | 0 |
| 4 | 0 | 0 | 1 | 0 | 0 | 0 | 0 | 0 | 0 |
| 5 | 0 | 0 | 0 | 0 | 1 | 0 | 0 | 0 | 0 |
| 6 | 0 | 0 | 0 | 1 | 0 | 0 | 0 | 0 | 0 |
| 7 | 1 | 0 | 0 | 0 | 0 | 0 | 0 | 0 | 0 |
| 8 | 0 | 0 | 1 | 0 | 0 | 0 | 0 | 0 | 0 |
| 9 | 0 | 0 | 0 | 0 | 0 | 0 | 0 | 0 | 1 |

***Figure 3-9.*** *The cabin feature transformed into binary data (dummified)*

It takes a feature and breaks out each unique value into a separate column and drops the original. Keep in mind that not all noncontinuous numbers should be made into dummy fields. Think about zip codes in the United States; if you dummify them, you will be adding an additional 43,000 features of extremely sparse data to your dataset–not always a good idea. In the case of zip codes, a better approach may be larger categorical groupings like modeling at the town or state level.

# Modeling

Keeping things simple for our second project will allow us to focus on the big picture and spend equal time on each piece involved in building a web application. We will use the Logistic Regression model from the "**sklearn**" library. If you recall from the last chapter, we used a linear regression, which attempts to predict a continuous variable. A logistic regression, on the other hand, attempts to predict a binary outcome such as true or false, happy or sad, etc. These are both extremely common models, but you need to use the correct one depending on the type of outcome variable you are trying to model and predict.

# Train/Test Split

We leverage the "**train_test_split()**" function from sklearn that will split the data into two random datasets with seed. Setting the "**random_state**" parameter is a good idea whenever you are testing different approaches and want to ensure that you are always using the same splits for fair comparison. The "**test_size**" parameter sets the size of the test split. Here we set it to .5, or 50%, thus it will randomize the data and split it in half between training and testing. We'll use the training portion to model the data, and the testing portion to evaluate how well our model performed. It is easy to use (Listing 3-11 and Figure 3-10).

*Listing 3-11.* Splitting the Data into Train and Test portions

```
from sklearn.model_selection import train_test_split
features = [feat forfeat in list(titanic_ready_df) if feat != 'survived']
X_train, X_test, y_train, y_test = train_test_split(titanic_ready_
df[features],

titanic_ready_df[['survived']], test_size=0.5, random_state=42)
print(X_train.head(3))
```

|      | age  | sibsp | parch | fare   | isfemale | pclass_Second | pclass_Third | \ |
|------|------|-------|-------|--------|----------|---------------|--------------|---|
| 455  | 63.0 | 1     | 0     | 26.000 | 0        | 1             | 0            |   |
| 83   | 64.0 | 1     | 1     | 26.550 | 1        | 0             | 0            |   |
| 1228 | 31.0 | 0     | 0     | 7.925  | 0        | 0             | 1            |   |

|      | pclass_nan | cabin_B | cabin_C | ... | embarked_S | embarked_Unknown | \ |
|------|------------|---------|---------|-----|------------|------------------|---|
| 455  | 0          | 0       | 0       | ... | 1          | 0                |   |
| 83   | 0          | 1       | 0       | ... | 1          | 0                |   |
| 1228 | 0          | 0       | 0       | ... | 1          | 0                |   |

|      | embarked_nan | title_Master. | title_Miss. | title_Mr. | title_Mrs. | \ |
|------|--------------|---------------|-------------|-----------|------------|---|
| 455  | 0            | 0             | 0           | 1         | 0          |   |
| 83   | 0            | 0             | 0           | 0         | 1          |   |
| 1228 | 0            | 0             | 0           | 1         | 0          |   |

|      | title_Rev. | title_Unknown | title_nan |
|------|------------|---------------|-----------|
| 455  | 0          | 0             | 0         |
| 83   | 0          | 0             | 0         |
| 1228 | 0          | 0             | 0         |

[3 rows x 28 columns]

***Figure 3-10.*** *Training split of the titanic data frame ready for modeling*

It splits out the outcome variables into "**y_train**" and "**y_test**"; the model will only have access to "**y_train**" (Listing 3-12).

***Listing 3-12.*** Top Outcome Values from Training Dataset

**Input:**

```
print(y_train.head(3))
```

**Output:**

|      | survived |
|------|----------|
| 455  | 0        |
| 83   | 1        |
| 1228 | 1        |

# Logistic Regression

It is time to decide on what model to use and set it up. As we are predicting a binary outcome, whether a passenger survived the voyage or not, a logistic regression is a good and lightweight choice–perfect for a web application. It is always good to keep the endgame in mind (Listing 3-13).

***Listing 3-13.*** Sklearn's Logistic Regression Model

```
from sklearn.linear_model import LogisticRegression
lr_model = LogisticRegression()
# ravel() simply creates a flattened array
lr_model.fit(X_train, y_train.values.ravel())
```

The data is fairly straightforward, and the patterns of survivorship are well known and can stand out with most modeling algorithms. The Sklearn Logistic Regression model makes it very easy to peek into the model's resulting coefficients to help us interpret which features are deemed important for surviving the Titanic trip (Listing 3-14).

***Listing 3-14.*** Extracting Our Model's Coefficients

```
coefs = pd.DataFrame({'Feature':features, 'Coef':lr_model.coef_[0]})
coefs.sort_values('Coef', ascending=False)
```

Figure 3-11 shows the top-positive and bottom-negative influencers in predicting survivorship. Clearly, on this particular trip, you were better off being female and rich than male and poor (see the notebook for the full list of coefficients).

| Positive Features | | | Negative Features | | |
|---|---|---|---|---|---|
| | Coef | Feature | | Coef | Feature |
| 4 | 1.863107 | isfemale | 7 | -0.414656 | pclass_Third |
| 22 | 1.617701 | title_Master. | 27 | -0.487802 | title_Unknown |
| 8 | 0.955656 | cabin_A | 18 | -0.559317 | embarked_Q |
| 25 | 0.947049 | title_Mrs. | 1 | -0.562287 | sibsp |
| 11 | 0.770398 | cabin_D | 10 | -0.660546 | cabin_C |
| 17 | 0.732904 | embarked_C | 24 | -0.943036 | title_Mr. |
| 5 | 0.535326 | pclass_First | 16 | -1.074717 | cabin_Unknown |

***Figure 3-11.*** *Top-positive and bottom-negative feature influencers*

# Predicting Survivorship

Once we have a trained model, we can validate its accuracy by using the testing portion of the data we earmarked earlier to validate the model (using the "**train_test_split()**" function mentioned). This data should never be used in the training phase! It therefore guaranties a fresh look at the model and an objective way of getting a performance score (Listing 3-15).

***Listing 3-15.*** Predict Using the Testing Portion of the Dataset

**Input:**

```
y_pred = lr_model.predict(X_test)
print('Accuracy of logistic regression classifier on test set: {:.2f}%'
      .format(lr_modl.score(X_test, y_test)*100))
```

**Output:**

```
Accuracy of logistic regression classifier on test set: 79.35%
```

The model with the validation dataset scored almost an 80% accuracy rate in predicting who did survive the trip. This isn't a bad score, considering we are using a simple model and a small dataset.

We're almost done with the Python script; we just need to make sure we can predict using fictional data. This is an important step, as we want our users to be able to come up with their own data and run it through the trained model. OK, so let's try a 50-year old male in third class score (Listing 3-16).

***Listing 3-16.*** Setting Up a Custom Prediction by Creating a Fictional Passenger

```
x_predict_pclass = 'Third'
x_predict_is_female=0
x_predict_age=50
x_predict_sibsp=3
x_predict_parch = 0
x_predict_fare = 200
x_predict_cabin = 'A'
x_predict_embarked = 'Q'
x_predict_title = 'Mr.'
```

If you are familiar with this dataset and historic event, you know that our fictional passenger won't fare well. After we run it through the model, we present the results using a simple comparative chart that shows the average survival rate next to our fictional passenger. Being male and in third class is a bad combination (Figure 3-12).

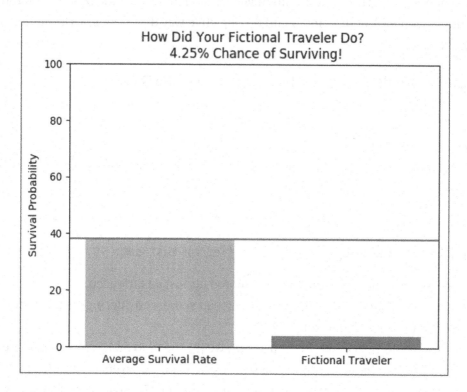

***Figure 3-12.*** *Average survival rate vs. our fictional traveler*

# Abstracting Everything in Preparation for the Cloud

Now that we have confirmed that our model works, that we can create fictional passengers and predict their probability of survivorship, we can package the code into two functions for our web application. Keeping things clean and neat will reduce the complexities and debugging headaches in the process of moving from a stand-alone scripting project and into a cloud-based environment.

# Function startup()

We create a "**startup()**" function that will take care of loading the data in memory, perform all the feature engineering, create dummy columns, and train the model. This function only gets called once when the web server is brought online and whenever it is rebooted. Preloading as much of the work as possible into memory will offer a faster and more responsive experience with the users interacting with the web application.

# Function submit_new_profile()

The second function is called "**submit_new_profile()**." This function handles the new fictional passenger profile, formats the data into the same shape as the real training data, creates the needed dummy columns, and asks the model to predict and yield a probability of survivorship.

That's it; most of the brain processing that we need will be handled by those two functions. All the rest of the code is used for communicating between the web server and the HTML page, displaying results, and making the whole thing look professional. But we're jumping ahead of ourselves; let's now get more acquainted with Flask.

A great reason for using Flask is that it allows us to link stand-alone Python scripts functions to server-side web controls without leaving the Python language. This makes passing data between a model and the web a whole lot easier!

# Interactivity with HTML Forms

Besides Flask, a critical front-end web technology is the "**HTML Form**."[4] Though this is basic stuff, it is the critical link between a user and our application. The HTML Form allows the user to interact with information on the web page, then hit the submit button to send that customized data back to the Flask web server (Listing 3-17).

*Listing 3-17.* Interacting with Users

```
<FORM id='submit_params' method="POST" action="{{ url_for('submit_new_
profile') }}">
...
```

---

[4]https://www.w3schools.com/html/html_forms.asp

```
<SELECT class="selectpicker" name="selected_embarked">
            <option value="{{selected_embarked}}" selected>{{selected_
            embarked}}</option>
            <option>Cherbourg</option>
            <option>Queenstown</option>
</SELECT>
...
<BUTTON type="submit">Submit</BUTTON>
</FORM>
```

# Creating Dynamic Images

Here we use an important technique to translate images created on the fly in Python into strings, so they can be dynamically fed and understood by an HTML interpreter. This is offered through the "**base64**" Python module:

> *This module provides data encoding and decoding as specified in RFC 3548. This standard defines the Base16, Base32, and Base64 algorithms for encoding and decoding arbitrary binary strings into text strings that can be safely sent by email, used as parts of URLs, or included as part of an HTTP POST request. The encoding algorithm is not the same as the uuencode program.*[5]

In the following simplified code snippet, we create an image in Python using the "**matplotlib.pyplot**" library (Listing 3-18).

*Listing 3-18.* Creating Dynamic Images

```
import matplotlib.pyplot as plt
fig = plt.figure()
plt.bar(y_pos, performance, align='center', color = colors, alpha=0.5)
img = io.BytesIO()
plt.savefig(img, format='png')
img.seek(0)
plot_url = base64.b64encode(img.getvalue()).decode()
```

---

[5]https://docs.python.org/2/library/base64.html

Then the variable "**plot_url**" can be injected into the HTML code using Flask Jinja2 template notation as such (Listing 3-19).

**Listing 3-19.** Plotting the Dynamic Image from Flask to Jinja2

```
model_plot = Markup('<img src="data:image/png;base64,{}">'.format(plot_url))
...
<div>{{model_plot}}</div>
```

And if you look at the HTML source output, you will see that the HTML image tag is made up of long string of characters (drastically truncated in the image shown). The interpreter will know how to translate that into an image (Figure 3-13).

```
<img src="data:image/
png;base64,6BCOHDnyk+pt7/tycHD4Scd5UC+/
/DLeeust1NfX46uvvkJSUhJsbGyQnJzcKe2hJ09
wcDDWrl2rWtfyeba1tf3J9T+MOh4mIyMjnDt3Tl
...
nOzc3F22+/jW+++UZZ17V v2oHfv3hg/
fjwGDBiAxMRE3Lhxgz2CEluzZg3Cw8Mxbdo09O/
ARERERJJhACQiIiKSDAMgERERkWT+B5qsMW4gCB
j4AAAAAElFTkSuQmCC">
```

**Figure 3-13.** *Image transformed into string of characters*

# Downloading the Titanic Code

Let's download the files for Chapter 3 and save them on your local machine if you haven't already done so. Once you have downloaded, open a command line window, and change drive to the "**web-application**" folder. The folder structure should look like Listing 3-20.

***Listing 3-20.*** Web Application Files

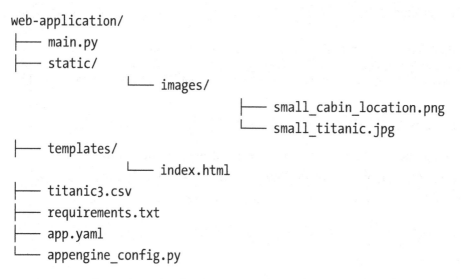

```
web-application/
├── main.py
├── static/
                └── images/
                                ├── small_cabin_location.png
                                └── small_titanic.jpg
├── templates/
                └── index.html
├── titanic3.csv
├── requirements.txt
├── app.yaml
└── appengine_config.py
```

Once you have downloaded and unzipped everything, open a command line window, change drive into the "**web-application**" folder, and install all the required Python libraries by running the "**pip install -r**" command (Listing 3-21).

***Listing 3-21.*** Installing Requirements

```
$ pip3 install -r requirements.txt
```

As with the previous local Flask applications, run the "**Python3 main.py**" command. It should look like the following screen shot in Figure 3-14.

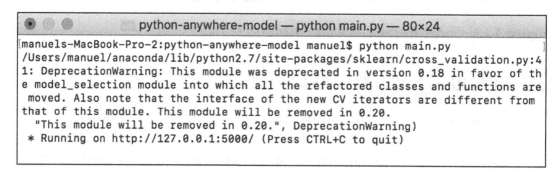

***Figure 3-14.*** *Command/terminal window output stating URL address of local Flask web page*

Then copy the URL "**http://127.0.0.1:5000/**" (or whatever is stated in the terminal window) into your browser and you should see the Titanic web application appear. If it doesn't, open "**main.py**" in your favorite code editor (sublime on the Mac or notepad++ on Windows are my favorites) and switch the Boolean flag on the last line to "**True.**" Rerun it and address whatever issues the logger complains about.

See the local version of the web application in Figure 3-15.

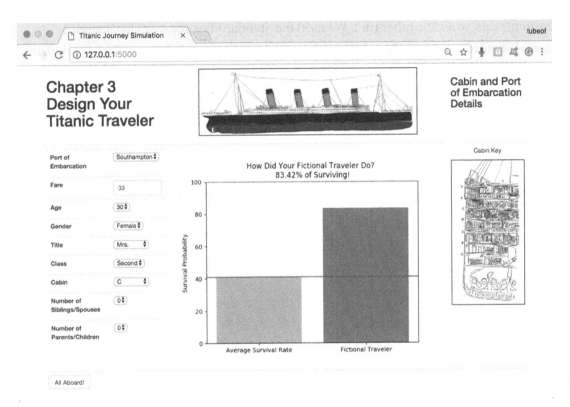

***Figure 3-15.***  *Local version of our web application*

# Google Cloud Flexible App Engine

Now, this is the other fun part, getting our application into the serverless cloud for the world to see! In the introduction chapter we looked at the Standard App Engine, this time around we'll have to use the Flexible App Engine in order to run more sophisticated Python libraries.

# Google App Engine

The Google App Engine is serverless, so you don't have to think about any of the hardware behind your web application. You don't have to know what OS your application is running under, it will scale accordingly, Google will take care of security patches, and you only pay for what you consume.

There are two types of App Engines you can opt for: one is very simple but less customizable, while the other isn't. We used the Standard Environment in Chapter 1; here we'll need to use the Flexible Environment due to the need of certain Python libraries (Figure 3-16).

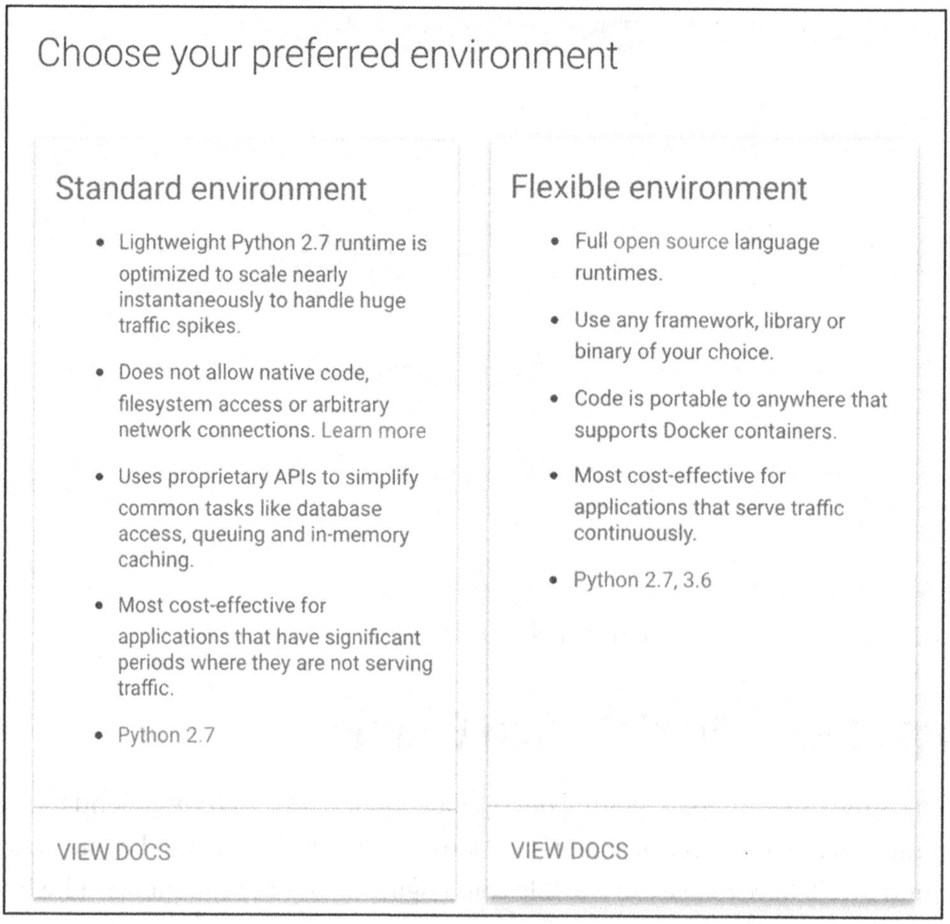

**Figure 3-16.** *Differences between App Engine's Standard and Flexible environments (may have changed by the time you read this)*

# Deploying on Google App Engine

There are multiple ways of deploying a web application on the Google App Engine. In this chapter we'll use the built-in shell terminal on the dashboard itself. This is something you use for quick jobs; for longer ones, you will need to open a terminal session from your local computer and initiate a connection to your Google Cloud account with the appropriate authentication. The other ways include linking a GitHub (or BitBucket) directly into your Google Cloud account, using a terminal session directly off your local machine, and there is also an experimental code editor directly in the dashboard (see `https://cloudplatform.googleblog.com/2016/10/introducing-Google-Cloud-Shels-new-code-editor.html`).

If you don't already have an account on Google Cloud, you can go to Google Cloud Getting Started (`https://console.cloud.google.com/getting-started`) and set one up. At the time of this writing, Google is offering a 12 month and $300 credit to get you started (Figure 3-17).

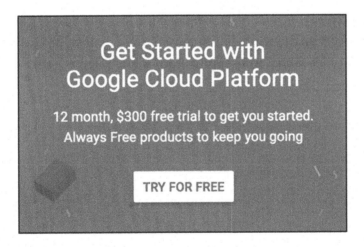

***Figure 3-17.***  *Google Cloud Platform special offerings*

# Step 1: Fire Up Google Cloud Shell

Log into your instance of Google Cloud and create or select the project you want your App Engine to reside in. Start the cloud shell command line tool by clicking on the upper-right caret button. This will open a familiar-looking command line window in the bottom half of the GCP dashboard (Figure 3-18).

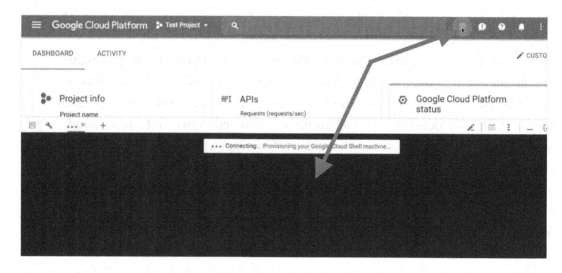

***Figure 3-18.*** *Accessing the Google Cloud shell*

# Step 2: Zip and Upload All Files to the Cloud

There are many ways to proceed: you can upload the files one by one, clone a GitHub repository, or you can zip them into one archive file and upload the zip. We'll go with the latter. So, zip the 11 files in the "**web-application**" folder (Figure 3-19).

| Name | ^ | Date Modified | Size | Kind |
|------|---|---------------|------|------|
| app.yaml | | Today at 9:17 PM | 487 bytes | YAML |
| appengine_config.py | | Feb 24, 2018 at 11:07 AM | 108 bytes | Python Script |
| main.py | | Today at 10:32 PM | 7 KB | Python Script |
| requirements.txt | | Today at 10:13 PM | 227 bytes | Plain Text |
| static | New Folder with Selection (11 Items) | at 5:24 PM | -- | Folder |
| ima | | at 8:37 PM | -- | Folder |
| | Open | at 6:40 PM | 713 KB | PNG image |
| | Open With ▶ | at 6:42 PM | 192 KB | JPEG image |
| templat | Move to Trash | at 5:03 PM | -- | Folder |
| inde | | 3:47 PM | 6 KB | HTML |
| titanic3 | Get Info | PM | 68 KB | Comm...et (.csv) |
| | Rename 11 Items... | | | |
| | Compress 11 Items | | | |
| | Duplicate | | | |

***Figure 3-19.*** *Zipping web application files for upload to Google Cloud*

Upload it using the "**Upload file**" option (this is found on the top right side of the shell window under the three vertical dots; Figure 3-20).

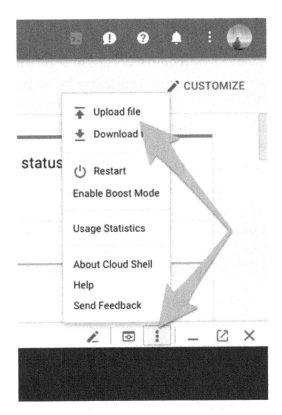

*Figure 3-20.* *Uploading files via Google Cloud shell*

# Step 3: Create Working Directory on Google Cloud and Unzip Files

Once the file is successfully uploaded, create a new directory called "**chapter-3**" for example, then move the compressed files into it and unzip them (Listing 3-22).

*Listing 3-22.* Getting the GCP Directory Ready for Deployment

```
$ mkdir chapter-3
$ cd chapter-3/
$ mv ../Archive.zip Archive.zip
$ unzip Archive.zip
```

## Step 4: Creating Lib Folder

We're almost there. If you look in the requirements.txt file, you will see one or more Python libraries that are required to run the application. When you build your own application, this is where you list all the libraries needed; you then run the script to actually install them the lib folder. A word of caution, the Standard Environment version of the Google App Engine only supports a minimal set of libraries; for anything more complicated, you will need to use the Flexible Environment (this is because it needs to be closer to the Python interpreter). So, run the following command to install all the needed additional libraries to the lib folder. When you deploy your web app, the lib folder will travel along with the needed libraries (Listing 3-23).

*Listing 3-23.* Loading All Required Python Libraries into the "**lib**" Folder

```
$ pip install -t lib -r requirements.txt
```

## Step 5: Deploying the Web Application

Finally, deploy it to the world using the deploy command. It will prompt a confirmation screen in order to proceed (Listing 3-24).

*Listing 3-24.* Deploying the Web Application to the Cloud

```
$ gcloud app deploy app.yaml
```

That is it! Sit back and let the serverless tool deploy our site. This is the Flexible App Engine, so it can take up to 20 minutes to be fully deployed. Once it is done setting everything, it will offer a clickable link to jump directly to the deployed web application or you can get there with the "**browse**" command (Listing 3-25).

*Listing 3-25.* Getting the Location URL of Our Web Application

```
$ gcloud app browse
```

Enjoy the fruits of your labor, and make sure to experiment with the web application by designing different passengers (Figure 3-21).

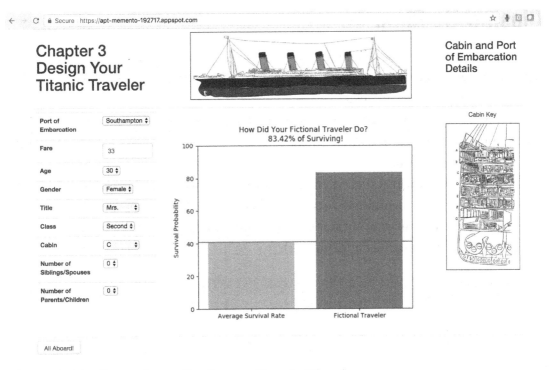

***Figure 3-21.*** *Our web application on Google Cloud*

# Troubleshooting

There will be cases where you will have issues and the Google Cloud logs will be your best friends. You can easily reach them either directly in the Google Cloud dashboard or with the logs URL (Listing 3-26).

***Listing 3-26.*** GCP Log Page

```
https://console.cloud.google.com/logs
```

Or you can stream the log's tail by calling the "**app logs tail**" command in the cloud shell (Listing 3-27).

***Listing 3-27.*** Following Deployment Logs

```
$ gcloud app logs tail -s default
```

# Closing-Up Shop

One last thing: before we are done with our web application: don't forget to stop or delete your App Engine Cloud instance. Even if you are using free credits, the meter is still running and there is no need to waste money or credits.

Things are a little different with the Flexible App Engine over the Standard one, as the Flexible costs more money. So, it is important to stop it if you aren't using it. Also, this can all be conveniently done via the Google Cloud dashboard.

Navigate to App Engine, then Versions. Click on your active version and stop it (Figure 3-22). If you have multiple versions, you can delete the old ones; you won't be able to delete the default one, but stopping it should be enough (if you really don't want any trace of it, just delete the entire project).

*Figure 3-22.* *Stopping and/or deleting your App Engine version*

# What's Going on Here?

Let's take a brief look at some noteworthy pieces in the code.

## main.py

The "**main.py**" file is a bit different than the Jupyter notebook we tackled for the chapter. It is always better to not rely too much on processing when running a web application. That is why we did away with any Pandas code, opting instead to use NumPy arrays (Listing 3-28).

***Listing 3-28.*** Creating a Matrix Array from a CSV File

```
from numpy import genfromtxt
titanic_array = genfromtxt('titanic3.csv', delimiter=',')
```

For example, when a user designs a new passenger profile, those values are simply added to the list (in the correct order) and fed into the logistic regression model directly. The fields aren't being dummified, instead they are created in a dummy state from the start (basically, we are dummying them manually; Listing 3-29).

***Listing 3-29.*** Dummying Categories Manually

```
if (selected_cabin=='B'):
    cabin_B = 1
if (selected_cabin=='C'):
    cabin_C = 1
if (selected_cabin=='D'):
    cabin_D = 1
if (selected_cabin=='E'):
    cabin_E = 1
if (selected_cabin=='F'):
    cabin_F = 1
if (selected_cabin=='G'):
    cabin_G = 1
if (selected_cabin=='T'):
    cabin_T = 1
if (selected_cabin=='Unknown'):
    cabin_Unknown = 1

user_designe_passenger = [[age, sibsp, parch, fare, isfemale, pclass_
Second, pclass_Third, pclass_nan, cabin_B, cabin_C, cabin_D, cabin_E,
cabin_F, cabin_G, cabin_T, cabin_Unknown, cabin_nan, embarked_Q,
embarked_S, embarked_Unknown, embarked_nan, title_Master, title_Miss,
title_Mr, title_Mrs, title_Rev, title_Unknown, title_nan]]
```

In this snippet of code, the "**cabin**" feature is stored directly as an integer. This saves us a few processing steps, such as avoiding the call to the Pandas "**get_dummies()**" function.

# app.yaml

YAML is a serialization language relied upon by many frameworks to configure and store program settings. The "**app.yaml**" file holds configuration settings such as setting the App Engine environment to "**flex**," the name of the Python starter script to "**main**," and information on hardware needed (Listing 3-30).

***Listing 3-30.*** A Look Inside the App Engine Flexible "**app.yaml**" File

```
runtime: python
env: flex
entrypoint: gunicorn -b :$PORT main:app

runtime_config:
  python_version: 3

# This sample incurs costs to run on the App Engine flexible environment.
# The settings below are to reduce costs during testing and are not
  appropriate
# for production use. For more information, see:
# https://cloud.google.com/appengine/docs/flexible/python/configuring-your-
  app-with-app-yaml
manual_scaling:
  instances: 1
resources:
  cpu: 1
  memory_gb: 0.5
  disk_size_gb: 10
```

The App Engine Flexible requires a disk size with a minimum of 10GB of space.

For more information on the "**yaml**" file for App Engines, see https://cloud.google.com/appengine/docs/flexible/python/configuring-your-app-with-app-yaml.

# appengine_config.py & lib folder

The "**appengine_config.py**" file and the "**lib**" folder work together to handle all extra Python libraries needed to get the web application running (Listing 3-31).

***Listing 3-31.*** A Look Inside the "**appengine_config.py**" Script

```
from google.appengine.ext import vendor

# Add any libraries installed in the "lib" folder
vendor.add('lib')
```

The "**lib**" folder is populated with needed Python libraries by calling "**pip install**" (Listing 3-32).

***Listing 3-32.*** Populating the "**lib**" Folder

```
pip install -t lib -r requirements.txt
```

If you look inside the "**lib**" folder after running this command, you will see all sorts of Python libraries deployed and ready to serve. This folder will get deployed with all your web application files to get them to function properly. The "**appengine_config.py**" does a whole lot more than what was shown here for the Google App Engine; see the official docs for more details.

---

**Note**   For more information on the appengine_config.py, see the Google docs at: `https://cloud.google.com/appengine/docs/standard/python/tools/appengineconfig`.

---

# requirements.txt

Here is a look at all the Python libraries needed to get the Titanic web application up and running (your version numbers will vary; Listing 3-33).

**Listing 3-33.** Python Libraries Needed to Run Our Web Application

```
click==6.7
Flask==0.12.2
itsdangerous==0.24
Jinja2==2.10
MarkupSafe==1.0
numpy==1.14.2
scikit-learn
scipy
python-dateutil==2.7.2
pytz==2018.4
six==1.11.0
Werkzeug==0.14.1
Pillow>=1.0
matplotlib
gunicorn>=19.7.1
```

# Steps Recap

Let's power through the steps needed to get the Titanic web application deployed on Google Cloud.

1.  Check that the web application runs locally, zip up all the web application files, create destination folder on the Google Cloud, and unzip the files:

    ```
    $ mkdir chapter-3
    $ cd chapter-3/
    $ mv ../Archive.zip Archive.zip
    $ unzip Archive.zip
    ```

2.  Create "**lib**" folder:

    ```
    pip install -t lib -r requirements.txt
    ```

3.  Deploy the web application:

    ```
    gcloud app deploy app.yaml
    ```

4. Get the URL to your web application:

```
gcloud app browse
```

Close up shop, go to your GCP dashboard into App Engine, and terminate any running version.

# Conclusion

Regarding the Titanic dataset, we learned that being rich and female gave you the best odds of survival while poor and male, the worst.

Even though this project was fairly straightforward, this chapter introduced a lot of concepts and new technologies. The first takeaway is to always think a couple of steps ahead whenever you are developing local Python ideas and models, to foresee ways to extended to the cloud. This includes keeping things simple, working on intuitive concepts, and keeping the code clean and efficient.

# CHAPTER 4

# Pretrained Intelligence with Gradient Boosting Machine on AWS

What makes a top-rated wine? Find out with a hard-to-resist real-time web dashboard on Amazon Web Services.

In this chapter, we are going to learn about wine quality with the help of the powerful "**Gradient Boosting Classifier**"[1] algorithm from the "**sklearn**" library. It can classify data into multiple classes, and that is what we'll use to group our wines into "**quality**" buckets. We will highlight that power in our web dashboard with the help of real-time sliders (Figure 4-1).

---

[1]http://scikit-learn.org/stable/modules/generated/sklearn.ensemble.
GradientBoostingClassifier.html

© Manuel Amunategui, Mehdi Roopaei 2018
M. Amunategui and M. Roopaei, *Monetizing Machine Learning*, https://doi.org/10.1007/978-1-4842-3873-8_4

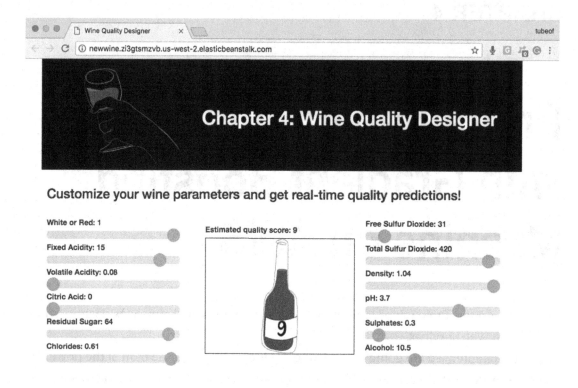

***Figure 4-1.***  *The final web application for this chapter*

This will invite visitors to interact with the model in a fun, responsive, and educational way. The data was collected for a paper called "**Modeling wine preferences by data mining from physicochemical properties.**"[2] It is graciously made available through the UCI Machine Learning Repository of the University of California, Irvine.[3]

---

**Note**    Download the files for Chapter 4 by going to `www.apress.com/ 9781484238721` and clicking the source code button. Open Jupyter notebook "**chapter4.ipynb**" to follow along with this chapter's content.

---

[2]P. Cortez, A. Cerdeira, F. Almeida, T. Matos, and J. Reis, "Modeling Wine Preferences by Data Mining from Physicochemical Properties," *Decision Support Systems* 47, no. 4 (2009): 547-553.
[3]`https://archive.ics.uci.edu/ml/index.php`

# Planning our Web Application: What Makes a Top-Rated Wine?

As is the case with any web application, it is critical to get the simple issues figured out first before extending it out onto the web. We will start by exploring the data, experimenting with the modeling, building a local Flask application, and only once everything is in working order will we then extend it to Amazon Web Services (AWS) Elastic Beanstalk.[4]

# Exploring the Wine-Quality Dataset

The Wine-Quality dataset can be downloaded directly from the UCI Machine Learning Repository using the Python "**Pandas**" library. It is made up of two datasets, 1,599 instances of red wine and 4,898 instances of white wine. The data represents chemical readings "related to red and white variants of the Portuguese "**Vinho Verde**" wine."[5] Go ahead and download the files for this chapter into a folder called "**chapter-4**." Fire up the Jupyter notebook to follow along.

According to UCI's data description, there are 11 attributes based on physicochemical tests and one output column based on sensory data:

- **Input**
  - fixed acidity
  - volatile acidity
  - citric acid
  - residual sugar
  - chlorides
  - free sulfur dioxide
  - total sulfur dioxide
  - density

---

[4]https://aws.amazon.com/elasticbeanstalk/
[5]https://archive.ics.uci.edu/ml/machine-learning-databases/wine-quality/
  winequality.names

- pH

- sulphates

- alcohol

- **Output**

  - quality (score between 0 and 10)

We will create a new feature called "**color**" to describe whether the wine is white or red, and concatenate both datasets into a single one. Being good web citizens, we also save a local copy of the finished and combined dataset for our web application, so we don't hit the servers every time a user interacts with it (Listing 4-1).

*Listing 4-1.* Create a New Wine Color Feature and Concatenate White and Red Together

```
white['color'] = 0
red['color'] = 1
wine_df = pd.concat([white, red], ignore_index=True)
```

Now that we have our dataset ready to go, let's dig into it and see what we have (Listing 4-2).

*Listing 4-2.* A look at the Feature Data Types

**Input:**

```
wine_df.info()
```

**Output:**

```
<class 'pandas.core.frame.DataFrame'>
RangeIndex: 6497 entries, 0 to 6496
Data columns (total 13 columns):
fixed acidity          6497 non-null float64
volatile acidity       6497 non-null float64
citric acid            6497 non-null float64
residual sugar         6497 non-null float64
chlorides              6497 non-null float64
free sulfur dioxide    6497 non-null float64
```

```
total sulfur dioxide     6497 non-null float64
density                  6497 non-null float64
pH                       6497 non-null float64
sulphates                6497 non-null float64
alcohol                  6497 non-null float64
quality                  6497 non-null int64
color                    6497 non-null int64
dtypes: float64(11), int64(2)
memory usage: 659.9 KB
```

The "**info()**" function tells us a lot about the data. We see that we have 13 columns, all floats except for two integers, "**quality**" and "**color**." The "**color**" feature is the one we added and keeps track of whether the wine is red or white. The "**quality**" feature is the outcome label and represents the quality level of a particular wine. This is an important feature, as it clusters the data by quality and is what our model will attempt to learn (Listing 4-3).

***Listing 4-3.*** Total Rating Counts of Wine Quality in Wine Data Frame

**Input:**

```
wine_df['quality'].value_counts()
```

**Output:**

```
6     2836
5     2138
7     1079
4      216
8      193
3       30
9        5
Name: quality, dtype: int64
```

The "**value_counts()**" function counts the frequency of values for a particular categorical feature. In the case of "**quality**," we see that there are 7 different quality types and that most are of quality "**6**." This feature could potentially be used as a continuous variable, meaning we are assuming some form of linearity between the lowest and highest quality. As this quality is based on a vote, we just can't assume it is a continuous numerical scale, along with the fact that some numbers are missing (the official data description states that quality is between 0 and 10, but we only see numbers between 3 and 9).

We'll play it safe and assume it is a categorical variable and use a multiclassification model instead of a regression model.

Another great way of visualizing a categorical variable is to use a histogram plot. This can easily be done within the Pandas library learn (Listing 4-4 and Figure 4-2).

***Listing 4-4.*** Histogram of Wine Quality Ratings by Groups and Votes

```
wine_df['quality'].hist()
plt.suptitle('Historgram of Wine Quality')
plt.xlabel('Quality Groups')
plt.ylabel('Number of Votes')
plt.show()
```

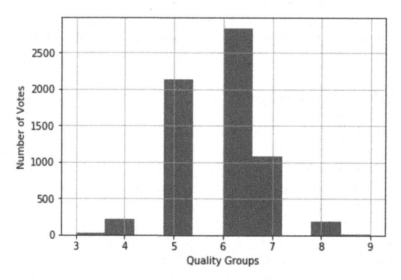

***Figure 4-2.*** *Histogram of wine quality ratings in wine data frame*

It's the same information as "**value_counts()**" but in an easier way to digest. We see a normal distribution in the middle ranges, which is intuitive as most wines are average and few are either very bad or very good.

According to the "**info()**" function, there are no null values nor text or text-based categorical data. The two integer features, "**quality**" and "**color**," are numerical categories and should be treated as such during modeling.

# Working with Imbalanced Classes

Referring to the wine quality histogram, we see that most of the quality resides within buckets 5, 6, and 7. This will make predicting edge quality buckets more challenging, as the model won't benefit from sufficient cases to learn from. For critical modeling projects, you would either balance the dataset by removing some of the middle classes or get more edge cases. In the Jupyter notebook for this chapter you see the rebalancing process.

Another approach is to remove weaker features. In some cases weak features can confuse the model, and by removing them you not only improve the score but make the model run faster. This can easily be done with tree-based models that return some form of variable importance. You get the list of features sorted in descending value of importance and try the model with only the best feature. You keep adding features until the score doesn't improve anymore and you end up with a good set of features to work with (this is known as forward-feature selection).

Let's see what would happen if we capped all classes to a maximum of 500 rows of data (Listing 4-5 and Figure 4-3).

*Listing 4-5.* Numerical Distribution of Capped Wine Quality

**Input:**

```
wine_balanced_df['quality'].value_counts()
```

**Output:**

```
7     500
6     500
5     500
4     216
8     193
3      30
9       5
Name: quality, dtype: int64
```

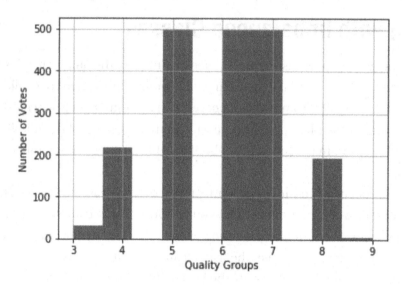

***Figure 4-3.*** *Histogram of capped wine quality*

It does flatten out around the center classes, but the edges are still extremely imbalanced. A much better way to proceed is to reclassify the quality class. We will group them down into only three quality groupings. Quality classes 3, 4, 5 will now belong to group 3, quality 6 will stay with 6, and quality classes 7, 8, 9 will belong to class 9 (Figure 4-4).

| Old Quality | New Quality |
|---|---|
| 3,4,5 | 3 |
| 6 | 6 |
| 7,8,9 | 9 |

***Figure 4-4.*** *Simplified wine quality super-groups*

In essence, we are creating three super groups: "**bad**," "**average**," and "**good**." After applying these new groupings, we get a much more balanced distribution (Listing 4-6 and Figure 4-5).

***Listing 4-6.*** Aggregating Wine Quality Down to Three Groups

**Input:**

```
wine_df['quality'].value_counts()
```

**Output:**

```
6    2836
3    2384
9    1277
Name: quality, dtype: int64
```

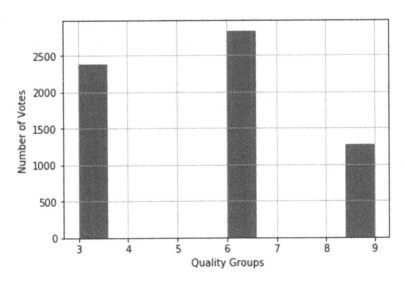

***Figure 4-5.*** *Histogram distribution of wine quality in smaller set of buckets*

When we plot the quality groupings in a histogram chart, the classes are much better balanced, ranging from 1,200 to 2,800.

# Modeling with Gradient Boosting Classifiers

Another way to improve working with an unbalanced dataset is to use models that can deal with them. A very popular one is sklearn's "**Gradient Boosting Classifier**," which is a powerful tree-based boosted model. It creates groups of trees and optimizes them according to their predictive strengths.

To use the sklearn GradientBoostingClassifier algorithm, you will need to install "**scikit-learn**" and "**scipy**" if you have never used them before (use the installation tool appropriate to your OS and software). To make things easier, you can simply run the "**requirements_jupyter.txt**" file containing all the necessary Python libraries for this chapter. You can quickly install them by running the "**pip3**" command (Listing 4-7).

*Listing 4-7.* Installing the Required Files to Run the Notebook

```
$ pip3 install -r requirements_jupyter.txt
```

We cast the "**quality**" feature as a categorical type the Pandas "**Categorical()**" function. This will allow us to use the "**cat.codes**" of that feature instead of the actual values. We do this because the real quality categories are 3-6-9, and by using the "**cat. codes**" we shift the range down between 0 and 2. We then use the sklearn "**train_test_ split**" functions (like we did in the previous chapter) to randomly split the data into a training chunk and a testing/validation chunk. If you look at the code, we set "**test_size**" to 0.2, meaning we are allocating 20% of the data for testing, and we set a seed using "**random_state**" to guarantee that our splits are always the same (Listing 4-8).

*Listing 4-8.* Preparing Our Training and Testing Datasets

```
from sklearn.model_selection import train_test_split
wine_df['quality'] = pd.Categorical(wine_df['quality'])wine_df['quality_
class'] = wine_df['quality'].cat.codes
outcome = 'quality_class'
outcome_buckets = len(set(wine_df['quality_class']))
X_train, X_test, y_train, y_test = train_test_split(wine_df[features],
wine_df[outcome], test_size=0.2, random_state=42)
```

And now we can feed that training data into the "**GradientBoostingClassifier**" for modeling (Listing 4-9).

*Listing 4-9.* Modeling with GBM

```
from sklearn.ensemble import GradientBoostingClassifier
gbm_model = GradientBoostingClassifier(random_state=10, learning_rate=0.1,
max_depth=10)
gbm_model.fit(X_train[features], y_train)
```

As you can see from the snippet, "**GradientBoostingClassifier**" takes various parameters, and playing around with them is important and referred to as hyper-parameter tuning.

Here are some of the critical parameters to tune the GBM model (for more detailed information, see the scikit-learn help[6])

- **learning_rate:** The learning rate determines the contribution of each tree.

- **n_estimators:** The number of boosting stages to perform

- **max_depth:** Maximum depth of the regression estimators

- **max_features:** Number of features to consider in each split

- **random_state:** The seed to use for reproducibility

The best way to train a "**GradientBoostingClassifier**" model is to run it multiple times with different parameter settings. See if adding the "**n_estimators**" parameter increases accuracy or not, and how a larger or smaller "**learning_rate**" or "**max_depth**" affects accuracy. For those who want to delve deeper into model tuning, there are many additional tools to help, such as cross-validation, hyper-parameter tuners, etc. One of my favorite aspects of the GradientBoostingClassifier is that it is very fast and can handle fairly large datasets, so you can easily write your own looping mechanism to try all sorts of variations and compare accuracy.

See the documentation for more granular details on Python API Reference.[7]

# Evaluating the Model

If you refer to the Jupyter notebook for this chapter, you will see that after we run the model, we evaluate it by asking it to predict wine quality on the out-of-sample data–our 20% chunk of test data (Listing 4-10). This is an important concept to remember: the model never gets to see the testing data, and this allows us to evaluate the model's performance with a fresh set of data.

---

[6]http://scikit-learn.org/stable/modules/generated/sklearn.ensemble. GradientBoostingClassifier.html
[7]http://scikit-learn.org/stable/modules/generated/sklearn.ensemble. GradientBoostingClassifier.html

***Listing 4-10.*** Predicting Using the Testing Data Split

```
preds = gbm_model.predict_proba(X_test)
```

The "**preds**" variable contains a list of three probabilities for every row, describing the probability of belonging to each of the wine quality classes (that's a mouthful). The probability of each row sums up to 1. For example, let's look at the predicted probabilities for the first row (Listing 4-11).

***Listing 4-11.*** Looking at One Wine's Prediction

**Input:**

```
preds[0]
```

**Output:**

```
array([0.50623207, 0.48718144, 0.00658649])
```

For the data at row 0 (i.e., the chemical readings for that particular wine), the model predicted the highest probability around index number 0 at 0.56, so quality bucket "**3**." This can be easily done using NumPy's "**argmax**" to get the index of the largest number in that list and then using that index position to get the bucket number (Listing 4-12).

***Listing 4-12.*** Using NumPy's "**argmax()**" Function to Get the Largest Number in the Array

**Input:**

```
print('Argmax: %i' % np.argmax(preds[0]))
print('Quality class: %i' % list(wine_df['quality'].cat.categories)
[np.argmax(preds[0])])
```

**Output:**

```
Argmax: 0
Quality class: 3
```

We can do the same for every wine in the test data and compare it with the actual "**ground truth**" classes using the sklearn.metrics "**precision_score()**" function. This will return a number between 0 and 1, where 0 is the worst score and 1, the best. Precision is better than accuracy in this case because we aren't talking about a simple binary prediction of a well-balanced dataset. Instead we want to know how well the model chooses between three quality classes for a particular wine on an imbalanced dataset–no easy feat! (Listing 4-13)

*The precision is the ratio tp / (tp + fp) where tp is the number of true positives and fp the number of false positives. The precision is intuitively the ability of the classifier not to label as positive a sample that is negative.*[8]

***Listing 4-13.*** Get the Highest Probability for Each Predicted Quality Class

**Input:**

```
from sklearn.metrics import precision_score
best_preds = np.asarray([np.argmax(line) for line in preds])
print ("Precision_score: %0.2f" % precision_score(y_test, best_preds,
average='macro'))
```

**Output:**

```
Precision_score: 0.74
```

Another useful way of looking at the big picture with multiclass models is by using a confusion matrix. This will plot a large square matrix that shows the model's best predictions against the ground truth (Listing 4-14 and Figure 4-6). Using a graphical confusion matrix instead of a printed one comes in handy when dealing with large amounts of data. The code to produce these graphical matrices comes directly from scikit-learn.org help files on confusion matrices.[9]

***Listing 4-14.*** Wine-Quality Predictions Shown on a Confusion Matrix

```
from sklearn.metrics import confusion_matrix
cnf_matrix = confusion_matrix(y_test, best_preds)
plt.figure()
plot_confusion_matrix(cnf_matrix, classes=set(wine_df['quality']),
                      title='Confusion matrix, without normalization')
plt.show()
```

---

[8]http://scikit-learn.org/stable/modules/generated/sklearn.metrics.precision_score.html
[9]http://scikit-learn.org/stable/auto_examples/model_selection/plot_confusion_matrix.html

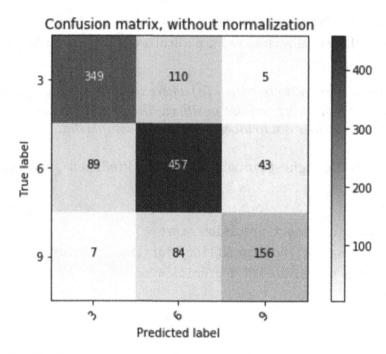

***Figure 4-6.*** *Predictions vs. actual confusion matrix*

It is worth spending a little time analyzing this chart. The y-axis represents the ground truth and the x-axis represents the best predictions from the GradientBoostingClassifier model.

In an ideal situation, all numbers would be 0 except for a single diagonal line going from the chart's top left all the way to the bottom right. That would mean all the predictions are correct (see the fabricated chart in Figure 4-7).

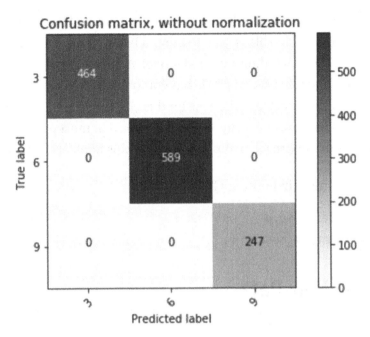

***Figure 4-7.*** *What a perfect confusion matrix looks like (if you see this level of perfection for a real model, be suspicious, as modeling is never perfect)*

Our model's confusion matrix tells us a few interesting things about our data. First, it is doing a pretty good job, as the diagonal line going from top left to bottom right does contain the biggest numbers. Where the data falls away from the diagonal line (i.e., incorrect predictions), it still stays close to its group. This is why the edges, where the model predicted 3 and it actually was 9 or the model predicted 9 and it was actually 3, are very small. This can be seen as a small consolation: whenever the prediction is wrong, it probably isn't that far off and the incorrect prediction is only as far as the adjacent bucket.

## Persisting the Model

In Chapter 3, the constructor of our web application trained the model (whenever the web server is restarted, it trains it). Here we're going to pretrain it, pickle it, and use that as our modeling engine. It isn't often that you have a dataset and model that are small enough that it's OK to train it directly in the cloud and on the web server. Most models are big and take a long time to train, or even take special hardware. By saving a copy of the fully trained model, we can then move it wherever we need it and in a ready-to-predict state. "**Pickling**," if you are not familiar with that Python term, is a tool to save an object in its current state to file.[10] Just like a comma delimited file (CSV) is used to move data around, a pickled file is used to move a Python object around. You can pickle anything you want, including a trained model. A note on Python versions, AWS Beanstalk defaults to Python 3.x, and pickle has compatibility issues between Python 2.x and 3.x versions so please stick to Python 3.x for this chapter.

This is called model persistence and it should be used in most production scenarios, and we will also use this approach in most chapters of this book[11] (Listing 4-15).

***Listing 4-15.*** Pickling Our Trained GBM Model

```
with open('gbm_model_dump.p', 'wb') as f:
        pickle.dump(gbm_model, f, 2)
```

## Predicting on New Data

Just like we did in the previous chapter, we need to make sure we can run our model and extract predictions on new data. This is an important step in building an interactive web application where the goal of the application is to offer new predictions based on user-inputted data.

To get us started, we calculate the mean values for each feature and use those values to predict the quality of the wine (Listing 4-16).

***Listing 4-16.*** Get Mean Values of Each Feature and Store in Data Frame

```
fixed_acidity = 7.215307
volatile_acidity = 0.339666
citric_acid = 0.318633
residual_sugar = 5.443235
chlorides = 0.056034
free_sulfur_dioxide = 30.525319
total_sulfur_dioxide = 115.744574
density = 0.994697
pH = 3.218501
sulphates = 0.531268
```

---

[10]https://wiki.python.org/moin/UsingPickle
[11]http://scikit-learn.org/stable/modules/model_persistence.html

```
alcohol = 10.491801
color = 0

# create data set of new data
x_test_tmp = pd.DataFrame([[fixed_acidity,
        volatile_acidity,
        citric_acid,
        residual_sugar,
        chlorides,
        free_sulfur_dioxide,
        total_sulfur_dioxide,
        density,
        pH,
        sulphates,
        alcohol,
        color]], columns = X_test.columns.values)
```

After creating a new data frame to store the customized wine chemical readings, we pass it to the model's "**predict**" function. We can add any outcome variable, as the model will ignore it when it makes a prediction (Listing 4-17).

***Listing 4-17.*** Predict the Quality of a Wine Based on Our Mean Values

```
preds = gbm_model.predict_proba(x_test_tmp)
```

Because we used mean values as our new wine, the predicted quality value should be close to the ground truth/actual wine quality–we basically created a boring wine, neither good nor bad (Listing 4-18).

***Listing 4-18.*** Get Wine-Quality Prediction

**Input:**

```
print(('Predicted wine quality: %i') % list(wine_df['quality'].cat.
categories)[np.argmax(preds)])
print(('Actual mean wine quaity: %0.2f') % np.mean(wine_df['quality'].
values))
```

**Output:**

```
Predicted wine quality: 6
Actual mean wine quality: 5.49
```

Not bad! The model predicted group 6 and the mean is 5.49. Keep in mind that this model will do better predicting average wines vs. edge ones. So, it shouldn't be a surprise that it nailed this. If we look at the prediction array, we see that the model struggled a tiny bit with putting this wine in quality 3 or quality 6 (35% 3, and 64% 6) but that quality group 6 won out (Listing 4-19).

*Listing 4-19.* Predicted Probabilities of Our Average Wine

```
array([[0.34124871, 0.63933304, 0.01941825]])
```

# Designing a Web Application to Interact and Evaluate Wine Quality

Building a fully functioning Flask version on our local model is a common theme and a proper next step throughout all chapters in this book. With enough practice you may skip this step but, in the meantime, it will save you plenty of time and headaches to iron out issues locally than on the cloud.

A good first step is to generalize the code into a big function. This will allow us to pass it new values and get a nice prediction in return with as little hassle as possible.

Once you have downloaded all the files for this chapter, open a command line window, and change the drive to the "**web-application**" folder. Your folder should look like Listing 4-20. Here we are showing the hidden folder "**.ebextensions**" needed for AWS EB. You can either use it as-is or create your own in the "**Fix the WSGIApplicationGroup**" section (don't worry about this when running the local version of the site, as it isn't affected by this fix).

*Listing 4-20.*  Web Application Files

```
web-application
├── application.py
├── requirements.txt
├── static \
                └── images
                                ├── quality_wine_logo.jpg
                                ├── wine_red_9.jpg
                                ├── wine_white_9.jpg
                                ├── wine_red_3.jpg
                                ├── wine_white_3.jpg
                                ├── wine_red_6.jpg
                                └── wine_white_6.jpg
├── pickles
                └── gbm_model_dump.p
└──templates
                └── index.html
└── .ebextensions <-- hidden folder
                └── wsgi_fix.config
```

Of note here, in some chapters we call our main Flask Python file "**main.py**" but here we use "**application.py**"–this is Amazon's Elastic Beanstalk default naming convention (there are always ways around this, but it requires editing the configuration file). From here on we'll work in a virtual environment.

# Introducing AJAX – Dynamic Server-Side Web Rendering

In this chapter, we're going to start using Ajax, a really cool technology that will allow us to update web content without rebuilding the entire page. This works great for highly interactive web applications that perform a lot of micro-updates. Ajax is also very easy to use and consists of two pieces, a front-end script function calling "**$.ajax**" and a back-end Flask function to catch and process the calls. We will dive a little deeper into Ajax in Chapter 10.

# Working in a Virtual Environment—a Sandbox for Experimentation, Safety and Clarity

Using a virtual environment offers many advantages:

- Creates an environment with no installed Python libraries

- Knows exactly which Python libraries are required for your application to run

- Keeps the rest of your computer system safe from any Python libraries you install in this environment

- Encourages experimentation

To start a virtual environment, you use the "**venv**" command. If it isn't installed on your computer, it is recommended you do so (it is available through installs with pip3, conda, brew, etc). For more information on installing virtual environments for your OS, see the "**venv - Creation of virtual environments**" user guide: `https://docs.python.org/3/library/venv.html`

In the command line window, navigate to the "**web-application**" folder if you aren't already there. Call the Python 3 "**venv**" function on the command line to create a sandbox area in Python 3 for our development work and a folder called "**wineenv**" (Listing 4-21).

***Listing 4-21.*** Starting a Virtual Environment

```
$ python3 -m venv wineenv
$ source wineenv/bin/activate
```

You are now ready to work in your virtual environment. Let's see if we can run the web application locally by calling "**python3**" (or use the commands that work for your OS; Listing 4-22).

***Listing 4-22.*** Run a local version of the web application

```
$ python3 application.py
```

It won't work, as we are in a clean virtual environment with no specialized Python libraries loaded. You will need to install all the libraries it is complaining about (substitute with the appropriate installation commands for your OS and software). The easiest is to "**pip3 install**" the included "**requirements.txt**" file (Listing 4-23).

***Listing 4-23.*** Installing Requirements

```
$ pip3 install -r requirements.txt
```

We have a comprehensive requirements.txt file containing all the Python libraries needed to run our web application. This really isn't of use for this local version of our web application but will be required for our cloud-based version.[12] Whenever you deploy your application, whether on Amazon, Google, or Microsoft's cloud, it uses the requirements file to install all the needed Python libraries wherever it runs your web application from.

Getting back to our local version experiment, run the same commands you ran for our previous Flask experiments (Listing 4-24).

***Listing 4-24.*** Run a Local Version of the Web Application

```
$ python3 application.py
```

Then copy the URL: "http://127.0.0.1:5000/" (or whatever is stated in the terminal window) into your browser and you should see the web Wine Quality Designer application appear (Figure 4-8). Hopefully it worked; if not, read the output errors and address them accordingly (remember that this chapter requires Python 3.x).

---

[12]https://docs.aws.amazon.com/elasticbeanstalk/latest/dg/python-configuration-requirements.html

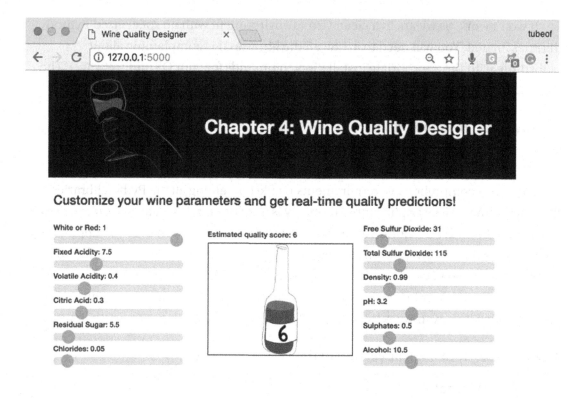

***Figure 4-8.***   *Web application running on local server*

# Amazon Web Services (AWS) Elastic Beanstalk

For our cloud-based portion of this chapter, we are going to host our application on Amazon's Elastic Beanstalk. It's a convenient hosting solution that packages your site, deploys, scales, and balances it automatically. It offers logging, traffic and health monitoring stats in a convenient web-based dashboard to keep you informed on what is going on. This should allow you to focus on your application and forget about site administration almost entirely.

You will need an Amazon Web Service account and setup security credentials. If you already have an Amazon.com account, you should be able to transfer it with little trouble to AWS. If you are new to this service, you can create a free-tier account that will give you access to basic Beanstalk features. Go to AWS Free Tier (`https://portal.aws.amazon.com/gp/aws/developer/registration/index.html`).

# Create an Access Account for Elastic Beanstalk

Once you have your AWS account, it is time to setup your security credentials in order to interact with AWS Elastic Beanstalk from your computer. A great guide to step you through that permission process is Amazon's Getting Started tutorial (https://aws.amazon.com/getting-started/tutorials/set-up-command-line-elastic-beanstalk/).

Log into the AWS web console and go to the Identity and Access Management (IAM) console. A quick way there is to simply type "**IAM**" in the AWS services search box on the landing page. Select "**Users**" in the navigation section and click the "**Add user**" button (Figure 4-9).

***Figure 4-9.*** *Creating a user in AWS*

Select a user name—here we enter "**ebuser**" and check "**Access type: Programmatic access**" (Figure 4-10).

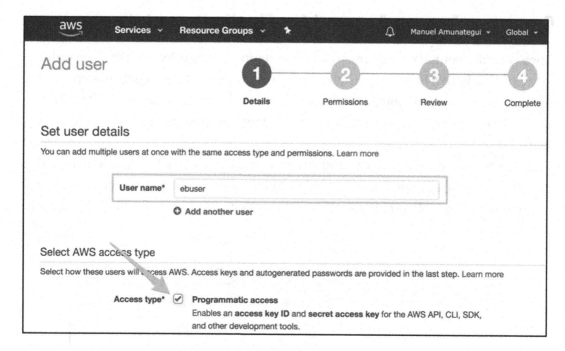

***Figure 4-10.*** *Assigning programmatic access to our new user*

Click the blue "**Next: Permissions**" button. This will take you to the "**Set permissions**" page; click the "**Add user to group**" large menu button, then click "**Create group**" (Figure 4-11).

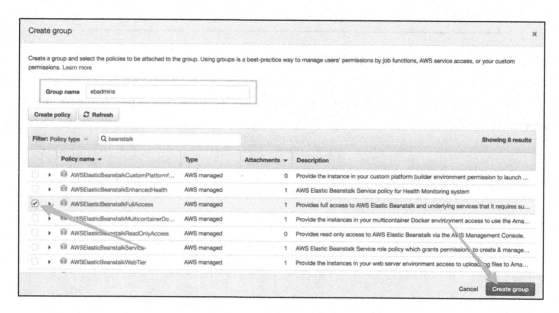

***Figure 4-11.*** *Giving WSElasticBeanstalkFullAccess to new user*

Create a group name, "**ebadmins**" in this case, and assign it the policy name "**WSElasticBeanstalkFullAccess**." Then click the "**Create group**" button to finalize the group. Click the "**Next: review**" blue button and, on the following page, click the blue "**Create user**" button (Figure 4-12).

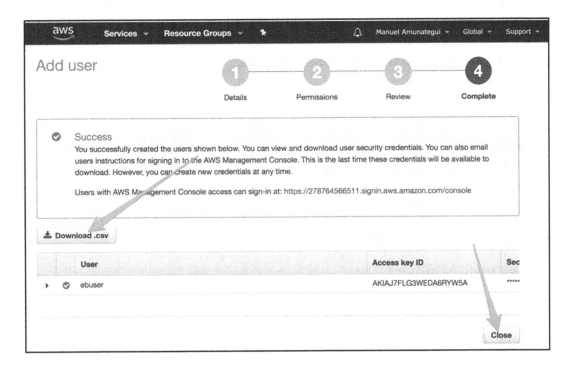

***Figure 4-12.***  *Download access key after successfully creating a user*

Once you see the "**Success**" message, this means you have successfully created the "**ebuser**" account. Make sure you download the "**.csv**" file to your local machine by clicking on the "**Download .csv**" button. This file is important, as it holds your key and secret code. Store it in a known location on your local machine, as you will need that information to connect and Secure Shell (SSH) into your EB.

# Elastic Beanstalk

We'll refer to the Elastic Beanstalk as EB going forward. We need to install the "**awsebcli**" library to interact and manage our EB service on AWS.

For Mac and Linux users (if it complains about the "**user**" parameter, try without it). See Listing 4-25.

***Listing 4-25.*** Installing "**awsebcli**"

```
$ pip install awscli
$ pip install awsebcli
```

For windows (if it complains about the "**user**" parameter, try without it). See Listing 4-26.

***Listing 4-26.*** Installing "**awsebcli**" on Windows

```
$ pip install awscli --user
$ pip install awsebcli --user
```

For more information on installing and troubleshooting the "**awsebcli**" library, refer to Amazon's help document: https://docs.aws.amazon.com/elasticbeanstalk/ latest/dg/eb-cli3-install.html.

# EB Command Line Interface

From the AWS help files:

> *EB is a command line interface (CLI) tool that asks you a series of questions and uses your answers to deploy and manage Elastic Beanstalk applications. This section provides an end-to-end walkthrough using EB to launch a sample application, view it, update it, and then delete it.*[13]

This is a handy command-line set to commands to initialize, push, control, and terminate our EB instance.

- **eb init**: initializes the EB service[14]

- **eb create**: creates a new EB instance

- **eb open**: opens a web page pointing to your EB instance

- **eb deploy**: deploys any changes to code or configuration

- **eb config**: opens EB instance configuration file for reading and editing

---

[13]https://docs.aws.amazon.com/elasticbeanstalk/latest/dg/command-reference-get-started.html

[14]https://docs.aws.amazon.com/elasticbeanstalk/latest/dg/eb3-init.html

- **eb logs**: pulls various log files for the open EB instance

- **eb terminate**: kills the EB instance (always terminate if you don't
  want to keep accruing charges)

For more complete information on EB commands, see `https://docs.aws.amazon.`
`com/elasticbeanstalk/latest/dg/eb-cli3-getting-started.html`.

We waited until now to install the "**awsebcli**" library, as we didn't want any of it to
make it into our requirements.txt file (Listing 4-27).

***Listing 4-27.*** Initializing "**awsebcli**"

```
$ eb init -i
```

This will ask you a series of questions and you can go with most of the defaults.
Under "**Enter Application name**" enter "**winetest**" (Figure 4-13).

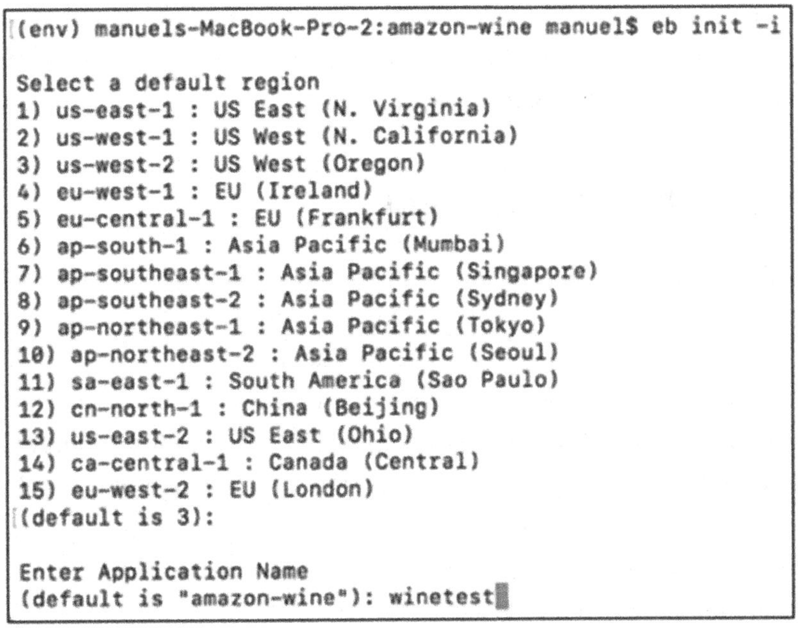

```
(env) manuels-MacBook-Pro-2:amazon-wine manuel$ eb init -i

Select a default region
1) us-east-1 : US East (N. Virginia)
2) us-west-1 : US West (N. California)
3) us-west-2 : US West (Oregon)
4) eu-west-1 : EU (Ireland)
5) eu-central-1 : EU (Frankfurt)
6) ap-south-1 : Asia Pacific (Mumbai)
7) ap-southeast-1 : Asia Pacific (Singapore)
8) ap-southeast-2 : Asia Pacific (Sydney)
9) ap-northeast-1 : Asia Pacific (Tokyo)
10) ap-northeast-2 : Asia Pacific (Seoul)
11) sa-east-1 : South America (Sao Paulo)
12) cn-north-1 : China (Beijing)
13) us-east-2 : US East (Ohio)
14) ca-central-1 : Canada (Central)
15) eu-west-2 : EU (London)
(default is 3):

Enter Application Name
(default is "amazon-wine"): winetest
```

***Figure 4-13.*** *Creating an application name*

If this is your first time running AWS on your computer, it will ask for your
credentials. Open the "**credentials.csv**" that was downloaded on your machine when
you created a user and enter the two fields required (Figure 4-14).

```
● ○ ◉ ▢  serverless-hosting-on-amazon-aws — eb create serverless-hosting-on-amazo...
You have not yet set up your credentials or your credentials are incorrect
You must provide your credentials.
[(aws-access-id): AKIAIMAYE3RBMZ4ALVAQ                                              ]
[(aws-secret-key): We/Ft/WBAEqYAzVFhb6Z4nOtAhztN1w+wrueSffY                         ]
```

***Figure 4-14.*** *Entering your credentials*

Go with the Python defaults (it needs to be a 3.x version) but say yes setting up SSH (Figure 4-15).

```
Do you want to set up SSH for your instances?
[(Y/n): y
```

***Figure 4-15.*** *Creating an SSH key*

Go with the default settings for all the other questions it may ask. Before we create the web application, you need to customize the WSGI configuration file to inform it that you will be requiring the Python sub-interpreter mode.

# Fix the WSGIApplicationGroup

When using Python libraries like NumPy, Pandas, or any other Python-heavy libraries, you need to tell the WSGI to enter a special Python sub-interpreter mode. This is done because these libraries are more complicated to load and require more threading, etc, thus are turned off by default (for more information see Python Simplified GIL State API[15]). The switch entails adding the variable in Listing 4-28 to the configuration file (this is from a Stackoverflow solution[16]). A copy of the file is included in the downloads or you can use the one provided in the folder (this is a hidden folder that you may or may not be able to see—if you aren't sure, try creating the folder as per instructions and if it complains, that means you already have it).

---

[15]http://modwsgi.readthedocs.io/en/develop/user-guides/application-issues.
html#python-simplified-gil-state-api
[16]https://stackoverflow.com/questions/41812497/aws-elastic-beanstalk-script-
timed-out-before-returning-headers-application-p

***Listing 4-28.*** Adding the '**WSGIApplicationGroup**' Variable

```
'WSGIApplicationGroup %{GLOBAL}'
```

To turn this on, you need to create a new folder under the "**web-application**" folder called "**.ebextensions**." Enter the command in Listing 4-29 in your local terminal window.

***Listing 4-29.*** Creating the "**wsgi_fix**" File

```
$ mkdir .ebextensions
```

This will create a new folder called "**.ebextensions**" and open a "**vi**" window, which is a simple text editor. Hit the "**i**" key to switch from read-only to "**insert**" mode and paste the following line at the end of the document (a text file of this fix is also included in the folder with the documents for this chapter). The process reading this file is very finicky; if there are added spaces or tabs, it will fail. Keep a close eye for any errors during the deployment process relating to the file and address accordingly.

Open "**vi**" from your local terminal window (Listing 4-30).

***Listing 4-30.*** Open a "**vi**" Session and Hit "**i**" to Enter Insert Mode

```
$ vi .ebextensions/wsgi_fix.config
```

Paste the following code into it (the code can be copied from file "**ebextensions_fix. txt**" in the downloads for this chapter). This is very finicky; a misplaced tab will break this process, so I recommend getting the content from the downloads or using the default file already provided (Listing 4-31).

***Listing 4-31.*** Paste the Following Code into Your "**vi**" Session

```
#add the following to wsgi_fix.config
files:
  "/etc/httpd/conf.d/wsgi_custom.conf":
    mode: "000644"
    owner: root
    group: root
    content: |
      WSGIApplicationGroup %{GLOBAL}
```

Now hit "**escape**" to exit "**insert**" mode and enter read-only mode, and type "**:wq**" to write and quit "**vi**" (Listing 4-32).

***Listing 4-32.*** Save and Quit "**vi**" Session

```
:wq
```

Next you need to create your EB.

# Creating the EB

Now we are ready to create our web application. Run the "**eb create**" command with the name of the application created earlier (Listing 4-33).

***Listing 4-33.*** Initializing "**awsebcli**"

```
$ eb create winetest
```

This will take a few minutes, and you should get a success message if all goes well. Then you can simply use the "**eb open**" command to view the web application live.

# Take if for a Spin

It may take a little bit of time to run the application the first time around and it may even timeout. If that is the case, try the "**eb open**" command one more time (Listing 4-34 and Figure 4-16).

***Listing 4-34.*** Open Web Site with the Following Command

```
$ eb open
```

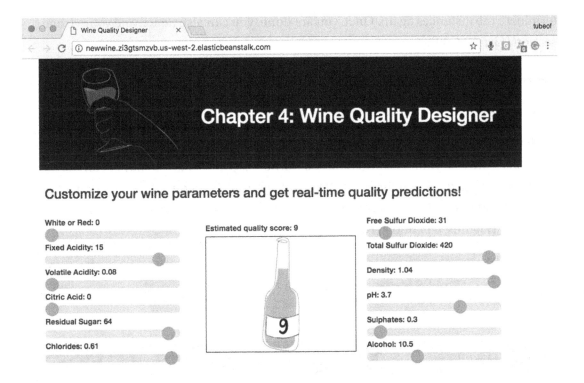

*Figure 4-16.*  *The Wine quality designer running on AWS elastic beanstalk*

# Don't Forget to Turn It Off!

Finally, we need to terminate the Beanstalk instance so as not to incur additional charges. This is an important reminder that most of these cloud services are not free (Listing 4-35).

*Listing 4-35.*  Terminate Your Instance

```
$ eb terminate winetest
```

It does take a few minutes but will take the site down. It is a good idea to double-check on your AWS dashboard that all services are indeed turned off. This is easy to do: simply log into your AWS account at https://aws.amazon.com/ and make sure that your EC2 and Elastic Beanstalk accounts don't have any active services you didn't plan on having. In case you see an instance that seems to keep coming back to life after each time you "**terminate**" it, check under EC2 "**Load Balancers**" and terminate those first, then terminate the instances again. Once you are done, you can also deactivate your virtual session (Listing 4-36).

**Listing 4-36.** Deactivate Your Virtual Environment

```
$ deactivate
```

It is always a good idea (essential idea really) to log into your account in the cloud and make sure everything is turned off (be warned: if you don't, you may get an ugly surprise at the end of the billing cycle). Log into your AWS account and make sure that your EC2 and Elastic Beanstalk accounts don't have any active services you didn't plan on having (Figures 4-17 and 4-18).

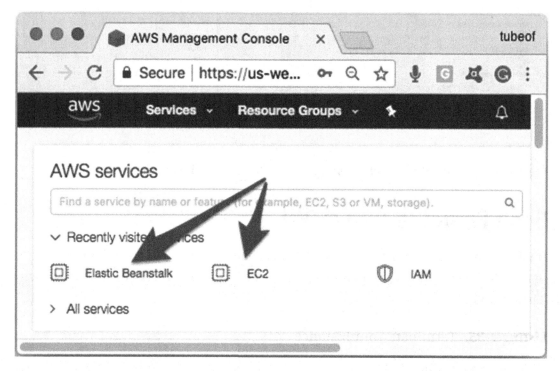

**Figure 4-17.** *Checking for any active and unwanted instances on the AWS dashboard*

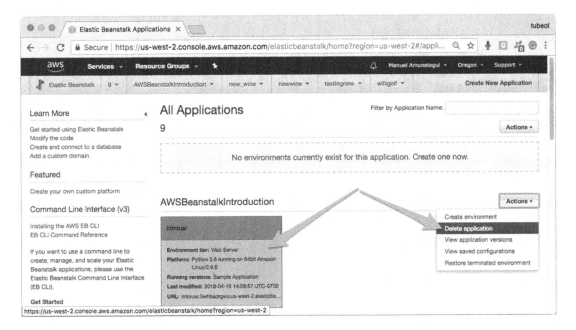

***Figure 4-18.*** *Locate the instance you want to terminate or delete and select your choice using the "**Actions**" dropdown button*

In case you see an instance that seems to keep coming back to life after each time you "**Delete application**," check under EC2 "**Load Balancers**" and terminate those first, then go back and terminate the rogue instance again (Figure 4-19).

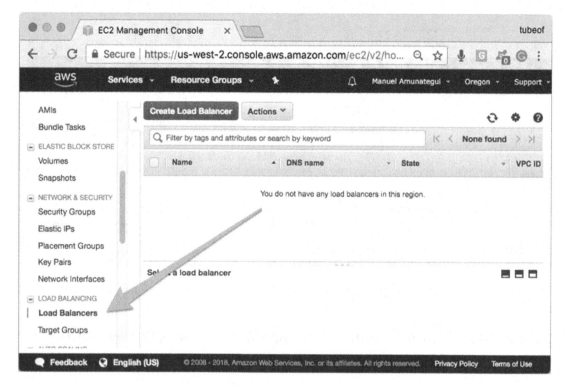

*Figure 4-19. "Load Balancers" can prevent an application from terminating; this can kick in if you inadvertently start multiple instances with the same name*

# Steps Recap

Let's power through the steps to get the Wine Quality Designer web application deployed on Amazon Web Services.

### Step 1: Start virtual environment

```
$ python3 -m venv wineenv
$ source wineenv/bin/activate
```

### Step 2: Install Python libraries

```
$ pip3 install -r requirements.txt
```

### Step 3: Test web application locally

```
$ python3 application.py
```

**Step 4: Create the wsgi_fix.config file**

```
$ mkdir .ebextensions
$ vi .ebextensions/wsgi_fix.config

#add the following to wsgi_fix.config
files:
  "/etc/httpd/conf.d/wsgi_custom.conf":
    mode: "000644"
    owner: root
    group: root
    content: |
      WSGIApplicationGroup %{GLOBAL}
```

**Step 5: Elastic Beanstalk**

```
$ eb init -i
$ eb create winetest
$ eb open
```

**Step 6: Terminate web application**

```
$ eb terminate
```

**Step 7: Exit virtual environment**

```
$ deactivate
```

# Troubleshooting

There are all sorts of things that can go wrong between your working local web application and a working one in the cloud. The first stop if you are not seeing what you were expecting is the logs!

# Access the Logs

If you are having issues, check the logs and look for any errors. Logs can be accessed directly in the terminal window with the following command in Listing 4-37.

***Listing 4-37.*** Accessing the Logs

```
$ eb logs
```

Logs also can be accessed on the Amazon Elastic Beanstalk dashboard page (Figure 4-20).

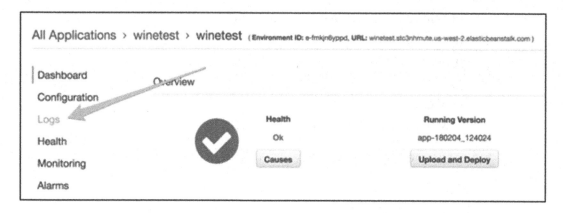

***Figure 4-20.*** *Accessing the logs*

Select the last 100 lines of logs (Figure 4-21).

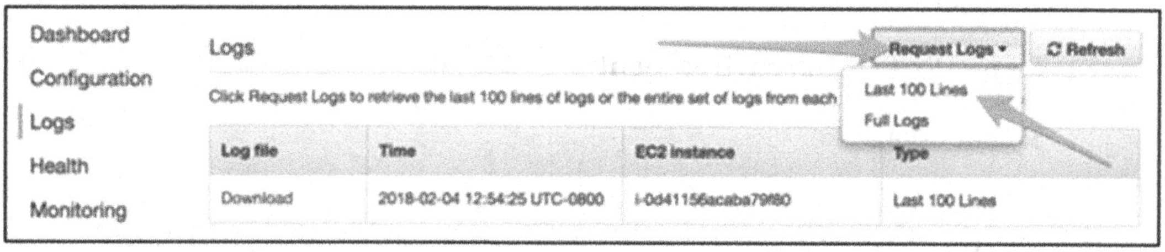

***Figure 4-21.*** *Requesting the last 100 lines*

# SSH into your Instance

If you want to SSH directly into your new instance, see Listing 4-38 and Figure 4-22.

***Listing 4-38.*** SSH into Your Instance

```
$ eb ssh
```

```
INFO: Attempting to open port 22.
INFO: SSH port 22 open.
INFO: Running ssh -i /Users/manuel/.ssh/flask_test ec2-user@54.188.179.155
Are you sure you want to continue connecting (yes/no)? y
Please type 'yes' or 'no': yes
```

***Figure 4-22.*** *SSH'ing directly into your Elastic Beanstalk instance*

It will ask you if you want to SSH into your instance; type "**yes**" (Figure 4-23).

You're in! This is the AWS Beanstalk instance that you just created with the

"**eb create**" command.

```
Warning: Permanently added '54.188.179.155' (ECDSA) to the list of known hosts.
```

```
                         Amazon Linux AMI
```

```
This EC2 instance is managed by AWS Elastic Beanstalk. Changes made via SSH
WILL BE LOST if the instance is replaced by auto-scaling. For more information
on customizing your Elastic Beanstalk environment, see our documentation here:
http://docs.aws.amazon.com/elasticbeanstalk/latest/dg/customize-containers-ec2.html
[ec2-user@ip-10-36-21-84 ~]$
```

***Figure 4-23.*** *Instance ready*

# Conclusion

In this chapter we created a web application around wine quality using colorful images and inviting sliders (who can resist a slider?). The backend processing relied on real-time processing using Ajax for instant feedback without having to refresh the whole page–pretty cool.

There is plenty of great free material on the web regarding AWS Elastic Beanstalk and EB with Flask.

- Getting started with EB: `https://docs.aws.amazon.com/ elasticbeanstalk/latest/dg/command-reference-get-started.html`

- Deploying a Flask Application to AWS Elastic Beanstalk: simple
  Flask example on AWS Beanstalk: `https://docs.aws.amazon.com/`
  `elasticbeanstalk/latest/dg/create-deploy-python-flask.`
  `html#python-flask-`

- Managing Elastic Beanstalk Environments with the EB CLI:
  `https://docs.aws.amazon.com/elasticbeanstalk/latest/dg/`
  `eb-cli3-getting-started.html`

Amazon Web Services is the leader in the cloud space today. It has over a decade of market dominance and that is one of its strengths. Some reasons users may look elsewhere are for diversity, redundancy, and cost savings.

# CHAPTER 5

# Case Study Part 1: Supporting Both Web and Mobile Browsers

Predicting the stock market with web and mobile platforms support on PythonAnywhere.com.

For the first part of our case study, we are going to create a simple trade alerting system. The tool will scan a number of stocks and alert the viewer of any interesting trade setups. The design will be kept simple to work well on both regular and mobile web pages (Figure 5-1).

***Figure 5-1.*** *The final web application for this chapter*

© Manuel Amunategui, Mehdi Roopaei 2018

M. Amunategui and M. Roopaei, *Monetizing Machine Learning*, https://doi.org/10.1007/978-1-4842-3873-8_5

Machine learning and quantitative trading go hand in hand. This shouldn't come as a surprise, as the premise of machine learning is about unearthing patterns, and who doesn't want to find patterns in the stock market? In this chapter we start working on the book's case study, where we develop the core trading signal strategy and dissemination application (keep in mind that this isn't a real trading system, just another interesting applied example of a web application). We will continue to improve this application in subsequent chapters. We'll base our fictional trading system on a popular pair-trading approach.

---

**Note**    Download the files for Chapter 5 by going to www.apress.com/9781484238721 and clicking the source code button. Open Jupyter notebook "**chapter5.ipynb**" to follow along with this chapter's content.

---

# The Pair-Trading Strategy

The idea behind pair trading is that related stocks tend to move together, so when we find stocks behaving abnormally by moving away from each other, we short the highest and buy the lowest in hopes that they revert to the mean. Another advantage to pairs trading is that you are removing the market's volatility by being both long and short in the market at the same time. Obviously, a real pair-trading strategy would have a lot more checks and safeguards before considering a trade, and there is never any guarantee that two values will move in the intended direction. For example, if the company is being investigated for fraud or a particular business model just doesn't make sense anymore, then buying it in hope that it increases in value may be a wishful proposition.

In our case and in a nutshell, we will use the last 90 trading periods to track an index-member stock against the index itself. Then we will consider the stock with the widest positive spread for a "**short**" trade. This is when we borrow a stock and sell it on the market. If it goes down, we repurchase it, return it to its original owner, and pocket the difference. Of course, the borrowed stock can rise and you will lose money when you have to repurchase it to return it to its owner.

And we will consider the stock with the widest negative spread as a "**long**" trade. Let's take a look at an example where Boeing Co ("**BA**") is strong and above the Dow Jones Index ("**^DJI**") and 3M Co ("**MMM**") is weak and below the Dow Jones Index ("**^DJI**"). So the trade would be to buy "**MMM**" and short "**BA**" (Figure 5-2).

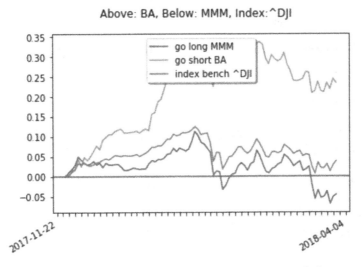

*Figure 5-2.* *An example of a spread between BA, MMM, and the Dow Jones Index*

# Downloading and Preparing the Data

We will compare a subset of stocks that are part of the **Dow Jones Industrial Average Index (^DJI)** against the index itself, and buy the lowest stock and short the highest one. The DJI, is an index based on 30 large publicly traded stocks in the US. The weighting is calculated on the sum price of the share of each company.

For our case study, we are going to keep things simple by focusing only on the top ten, highest weighted stocks contained in the Dow Jones Industrial Average Index, shown in Table 5-1. Go ahead and download the files for this chapter into a folder called "**chapter-5**." Open up the Jupyter notebook to follow along.

**Table 5-1.** *Top 10 Highest Weighted Stocks Contained in the Dow Jones Industrial Average Index*

| Company Name | Stock Symbol |
| --- | --- |
| Boeing | BA |
| Goldman Sachs | GS |
| UnitedHealth Group | UNH |
| 3M | MMM |
| Home Depot | HD |
| Apple | AAPL |
| McDonalds | MCD |
| IBM | IBM |
| Caterpillar | CAT |
| Travelers | TRV |

You can access a snapshot of this data in the repository for this chapter. And if you want to access current data, you can manually download the files from Yahoo Finance. For example, if you wanted to get the latest historical prices for Apple, simply enter the following link:

`https://finance.yahoo.com/quote/AAPL/history?p=AAPL.`

Select the "**Time Period**" desired (one year's worth should do the trick), click "**Apply**," and finally, click the "**Download Data**" link. This will download a CSV file onto your machine with the requested data. This is a great free service currently offered by Yahoo Finance. You can also use other financial data services if this one doesn't work for you, as stock data is pretty much universal.

Let's take a look at one the CSV files included for this project. We'll load "**^DJI**," which is the Dow Jones index and the benchmark we'll use to understand the movement of our ten stocks (Listing 5-1 and Figure 5-3).

*Listing 5-1.* Top Rows of the DJI CSV

```
DJI = pd.read_csv('^DJI.csv')
DJI.head()
```

| | Date | Open | High | Low | Close | Adj Close | Volume |
|---|---|---|---|---|---|---|---|
| 0 | 2017-04-05 | 20745.060547 | 20887.500000 | 20639.550781 | 20648.150391 | 20648.150391 | 284980000 |
| 1 | 2017-04-06 | 20653.769531 | 20746.460938 | 20612.169922 | 20662.949219 | 20662.949219 | 251720000 |
| 2 | 2017-04-07 | 20647.810547 | 20726.070313 | 20606.949219 | 20656.099609 | 20656.099609 | 219730000 |
| 3 | 2017-04-10 | 20668.220703 | 20750.330078 | 20614.859375 | 20658.019531 | 20658.019531 | 230480000 |
| 4 | 2017-04-11 | 20644.320313 | 20660.029297 | 20512.560547 | 20651.300781 | 20651.300781 | 255120000 |

*Figure 5-3.* *First five rows from the DJI CSV*

# Preparing the Data

We automate the loading of all stocks and index CSV files. This should allow you to add and remove stock files with ease (Listing 5-2).

*Listing 5-2.* Loop Through Each CSV and Create on Data Frame

```
stock_data_list = []
for stock in index_symbol + stock_symbols:
    tmp = pd.read_csv(stock + '.csv')
    tmp['Symbol'] = stock
    tmp = tmp[['Symbol', 'Date', 'Adj Close']]
    stock_data_list.append(tmp)

stock_data = pd.concat(stock_data_list)
```

The code snippet loops through each stock "**CSV**" file, pulls the data into a Pandas dataframe, adds a new column to hold the stock-symbol name, and appends it to the "**stock_data_list**" list. Once it has collected all the files, it uses the Pandas "**concat()**" function to create a single data frame containing all ten stocks and the index (Listing 5-3 and Figure 5-4).

*Listing 5-3.* Top Rows of the "**stock_data**" Data Frame

```
stock_data.head()
```

| | Symbol | Date | Adj Close |
|---|---|---|---|
| **0** | ^DJI | 2017-04-05 | 20648.150391 |
| **1** | ^DJI | 2017-04-06 | 20662.949219 |
| **2** | ^DJI | 2017-04-07 | 20656.099609 |
| **3** | ^DJI | 2017-04-10 | 20658.019531 |
| **4** | ^DJI | 2017-04-11 | 20651.300781 |

*Figure 5-4.* *Concatenated data frame with symbol feature added*

## Pivoting by Symbol

To make working with multiple stocks a bit easier, we are going to drop all fields except for "**Date**" and "**Adj Close**." We'll then pivot them all into one big table (see the Jupyter notebook for this chapter for more details on this process). This approach scales well, as I've successfully concatenated over 3,000 stocks with ten years of data for analysis (Listing 5-4 and Figure 5-5).

*Listing 5-4.* Pivot and Make Symbol Column Header and Date Row Index

```
stock_data = stock_data.pivot('Date','Symbol')
stock_data.columns = stock_data.columns.droplevel()
stock_data.head()
```

| Symbol<br>Date | AAPL | BA | CAT | GS | HD | IBM | MCD | MMM | TRV | UNH | ^DJI |
|---|---|---|---|---|---|---|---|---|---|---|---|
| **2017-04-05** | 141.777161 | 172.905151 | 91.817474 | 224.784515 | 143.392029 | 166.105453 | 127.264633 | 185.988541 | 118.215958 | 162.975571 | 20648.150391 |
| **2017-04-06** | 141.422775 | 173.188293 | 93.356865 | 225.752121 | 143.978745 | 165.692307 | 127.010887 | 185.646362 | 117.854195 | 162.945999 | 20662.949219 |
| **2017-04-07** | 141.107742 | 174.633438 | 93.064575 | 225.001724 | 143.871185 | 165.394455 | 126.825478 | 185.744141 | 117.472885 | 163.616150 | 20656.099609 |
| **2017-04-10** | 140.940399 | 173.373825 | 94.642937 | 225.998962 | 144.927246 | 164.491302 | 126.844994 | 185.470398 | 118.411491 | 163.083954 | 20658.019531 |
| **2017-04-11** | 139.424393 | 174.360031 | 94.603958 | 224.863480 | 144.917465 | 163.895584 | 128.035568 | 185.822357 | 118.489716 | 163.428894 | 20651.300781 |

*Figure 5-5.* *Final stock data frame with each stock pivoted into its own column*

This is going to make working with this data and comparing each symbol against the index a lot easier to automate.

# Scaling the Price Market Data

In order to compare moves between differently priced assets, we need to normalize the data. In essence, we are going to rescale all the prices into a common scale. There are many ways of achieving this, and in this web application we will use the percent change and cumulative sum.

# Percent Change and Cumulative Sum

A very simple way to do this is to transform our price data into percentage changes and apply a rolling sum, known as a cumulative sum (Listing 5-5 and Figure 5-6).

*Listing 5-5.* Applying Percent Change and Cumulative Sum to APPL

```
pd.DataFrame({"Price":stock_data['AAPL'], "PercentChange":stock_
data['AAPL'].pct_change().cumsum()}).head()
```

| Date | PercentChange | PercentChangeCumSum | Price |
|---|---|---|---|
| 2017-11-22 | NaN | NaN | 174.249573 |
| 2017-11-24 | 0.000057 | 0.000057 | 174.259521 |
| 2017-11-27 | -0.005029 | -0.004972 | 173.383087 |
| 2017-11-28 | -0.005859 | -0.010831 | 172.367249 |
| 2017-11-29 | -0.020743 | -0.031574 | 168.791809 |
| 2017-11-30 | 0.013984 | -0.017590 | 171.152191 |
| 2017-12-01 | -0.004655 | -0.022246 | 170.355438 |
| 2017-12-04 | -0.007308 | -0.029553 | 169.110519 |
| 2017-12-05 | -0.000942 | -0.030496 | 168.951172 |
| 2017-12-06 | -0.003714 | -0.034210 | 168.323715 |

*Figure 5-6.* *Price, percentage change, and percent change cumulative sum*

The first price in the transformation at date "**2017-11-22**" is lost to a non-number value, or "**NaN**," as we have nothing prior to it to compare against. The second price of 174.259521 has a positive percentage change, as it is up from the previous price of 174.249573. The cumulative change column sums all these changes in temporal order. The "**PercentChangeCumSum**" column becomes a series of data that we can compare against any other stock series, regardless of their values and ranges–a very useful transformation for comparative analysis.

## Plotting the Spread

Now that we have the percentage change cumulative sum, we can easily plot the difference between the index and one of its stocks.

We'll start with the cumulative sum percentage difference between the "**DJI**" and "**BA**" (Boeing) and the "**DJI**" and "**AAPL**" (Apple). "**BA**" is showing a lot of positive strength against the "**DJI**" index for the period tracked (90 trading days between 11/22/1017 and 4/4/2018). The way to read the chart is to think of the green "**0**" line as the stabilized index and the blue squiggly line as the difference in the stock member (Figure 5-7).

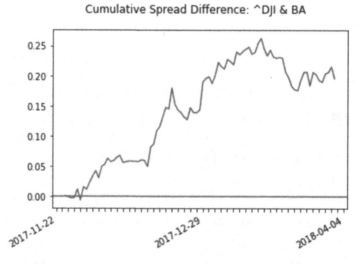

**Figure 5-7.** *Cumulative spread between the Dow Jones Index and Boeing, where the stock is showing a lot of strength compared with the index*

The opposite is happening for "**AAPL**"; it is showing strong negative weakness compared with the "**DJI**" index for the same period (Figure 5-8).

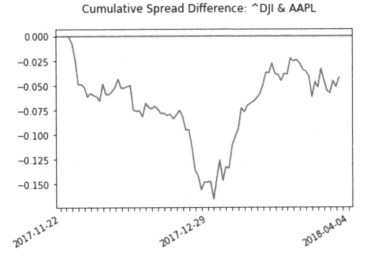

**Figure 5-8.** *Cumulative spread between the Dow Jones Index and Apple, where the stock is showing weakness compared with the index*

And now let's imagine a fictional trade. What if we went short on "**BA**" around 12/29/2017 and went long on "**AAPL**" around the same time. Obviously this is picking stocks in hindsight, but you would have lost money on "**BA**" and made a lot more on "**AAPL**", making this pair trade profitable. Of course, you will have to make sure you are invested in equivalent dollar quantities on both sides for this to work (more on this shortly).

# Serving up Trading Ideas

Now that we have a basic idea of the type of trades we're after, let's find some active setups to offer to our readers.

## Finding Extreme Cases

Let's assume that the financial data we're holding is up to date. In order to find the strongest positive and strongest negative stocks against the "**DJI**," we simply need to create a data frame to hold all our stocks for the desired 90 day look-back period, apply percentage and cumulative sum to all of them, then look for the largest and smallest last-day trading price. Let's walk through this in more detail.

Let's apply this to all of our stocks. As we will track differences only over the last 90 trading days (this can be adjusted for experimentation), we will drop older data (Listing 5-6).

***Listing 5-6.*** Only Using the Last 90 Trading Days of Data

```
stock_data = stock_data.tail(90)
```

Let's loop through each symbol and compare the last percentage change cumulative sum against the DJI and store them for comparison in the dictionary "**last_distance_from_index**" (Listing 5-7).

***Listing 5-7.*** Getting the Distance Between Stocks and the Index

```
stock1 = '^DJI'
last_distance_from_index = {}
temp_series1 = stock_data[stock1].pct_change().cumsum()
for stock2 in list(stock_data):
    # no need to process itself
    if (stock2 != stock1):
        temp_series2 = stock_data[stock2].pct_change().cumsum()
        # we subtract the stock minus the index, if stock is strong
          compared
        # to index, it will show a positive value
        diff = list(temp_series2 - temp_series1)
        last_distance_from_index[stock2] = diff[-1]
```

Let's see what we caught in our dictionary (Listing 5-8).

***Listing 5-8.*** Analyzing the Distances Between Our Stocks and the Index

**Input:**

```
print(last_distance_from_index)
```

**Output:**

```
{'AAPL': -0.042309986580456815,
 'BA': 0.1960194615751124,
 'CAT': 0.03379845694757866,
 'GS': 0.047454711281622486,
 'HD': 0.014178592951754165,
 'IBM': -0.003376107031365594,
 'MCD': -0.06239853566933862,
 'MMM': -0.08366228603707737,
 'TRV': 0.04601871807501723,
 'UNH': 0.060732956928879145}
```

Just by looking at the dictionary, we see that "**BA**" has the highest value and that "**MMM**" has the lowest.

# Making Recommendations

We need to pull these values programmatically if we want to automate the process or scale this up to thousands of stocks. Let's use the convenient lambda functions. This is a style of function that can easily be nested into dictionaries and apply transformations or, in our case, finding the minimum and maximum values in a dictionary (Listings 5-9 and 5-10).

***Listing 5-9.*** Applying a Lambda Function to Find the Minimum Value in Our Dictionary

**Input:**

```
weakest_symbol = min(last_distance_from_index.items(), key=lambda x: x[1])
print('Weakest symbol: %s' % weakest_symbol[0])
```

**Output:**

```
Weakest symbol: MMM
```

***Listing 5-10.*** Applying a Lambda Function to Find the Maximum Value in Our Dictionary

**Input:**

```
strongest_symbol = max(last_distance_from_index.items(), key=lambda x: x[1])
print('Strongest symbol: %s' % strongest_symbol[0])
```

**Output:**

```
Strongest symbol: BA
```

Let's visualize the two extreme cases: "**BA**" and "**MMM**" (Figures 5-9 and 5-10).

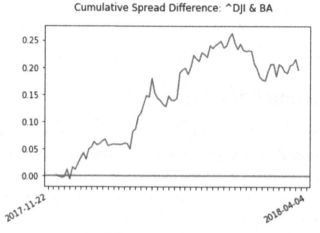

***Figure 5-9.*** *The strongest stock of the set vs. the Dow Jones Index*

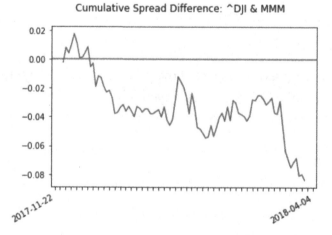

***Figure 5-10.*** *The weakest stock of the set vs. the Dow Jones Index*

178

We now have our two recommendations and what they look like plotted on a graph (Figure 5-11).

1. Buy "**MMM**"

2. Short "**BA**"

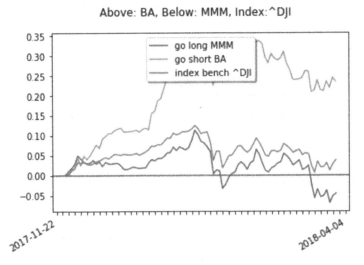

***Figure 5-11.***  *Recommendation from our algorithm: short BA, buy MMM*

# Calculating the Number of Shares to Trade

You want to get into the market with equal dollar amounts on your long and short positions. In other words, you want to be dollar neutral (granted this is harder with smaller budgets or stocks that trade at big values).

You need to set a budget value, the total number of dollars that you are willing to use for the trade (though I refer to dollars for this use-case, any other currency can be substituted). For this example, we'll go with $10,000 (Listing 5-11).

***Listing 5-11.***  Setting Our Trading Budget

```
trading_budget = 10000
```

As we are dealing with dollar amounts, we need to use the actual stock price, not our percent change and cumulative sums. We will continue working with "**BA**" and "**MMM**" (Listing 5-12).

***Listing 5-12.*** Getting the Last Trading Price for Both Recommended Stocks

**Input:**

```
short_symbol = strongest_symbol[0]
short_last_close = stock_data[strongest_symbol[0]][-1]
print('Strongest symbol %s, last price: $%f' % (strongest_symbol[0],
short_last_close))

long_symbol = weakest_symbol[0]
long_last_close = stock_data[weakest_symbol[0]][-1]
print('Weakest symbol %s, last price: $%f' % (weakest_symbol[0],
long_last_close))
```

**Output:**

```
Strongest symbol BA, last price: $327.440002
Weakest symbol MMM, last price: $217.559998
```

Let's apply the trading budget and figure out how many shares of each we can trade. The formula is simply to divide half the budget against the price of the stock (Listing 5-13 and Listing 5-14).

***Listing 5-13.*** Getting the Last Trading Price for the First Recommended Stock

**Input:**

```
print('For %s, at $%f, you need to short %i shares' %
      (short_symbol, short_last_close, (trading_budget * 0.5) /
      short_last_close ))
```

**Output:**

```
For BA, at $327.440002, you need to short 15 shares
```

***Listing 5-14.*** Getting the Last Trading Price for the Second Recommended Stock

**Input:**

```
print('For %s, at $%f, you need to buy %i shares' %
      (long_symbol, long_last_close, (trading_budget * 0.5) /
       long_last_clse ))
```

180

**Output:**

```
For MMM, at $217.559998, you need to buy 22 shares
```

So, in order to make a dollar-neutral pair trade, we need to short 15 shares of "**BA**" and buy 22 shares of "**MMM**." This should make sense, as 327 times 15 is a around $5,000, and 217 times 22 is around $5,000 as well.

# Designing a Mobile-Friendly Web Application to Offer Trading Ideas

Let's get to work and build our local Flask application that will offer trading recommendations. As usual, we have to ask ourselves what is it that we want our end users to experience.

This is going to be a simple application. In the first part of our case study, we'll simply offer up trading ideas based on the strategy in a visually pleasant manner using large colored arrows and clear instructions. The user will be able to input their total budget for the trade and the application will calculate the quantity of shares to buy and sell in order to remain dollar neutral.

## Fluid Containers

The application also needs to be mobile friendly. This means it needs to make use of "**responsive fluid containers**" from Bootstrap. Bootstrap 3 is mobile friendly out of the box[1] but we can enhance and control specific behavior by using the right tagging.

Most scripts in this book follow some of the "**responsive**" behavior recommended, such as offering proper page rendering and touch zooming that you will see in each project's headers (Listing 5-15).

---

[1]https://getbootstrap.com/docs/3.3/css/

***Listing 5-15.*** Handling Different Web Viewing Devices

```
<meta name="viewport" content="width=device-width, initial-scale=1">
```

We will add fluid containers to create smart rendering, depending on the device or size of web page used to view the site. In normal mode, we see the wide version where the green and red arrows are drawn on the right and left sides of the page (Figure 5-12).

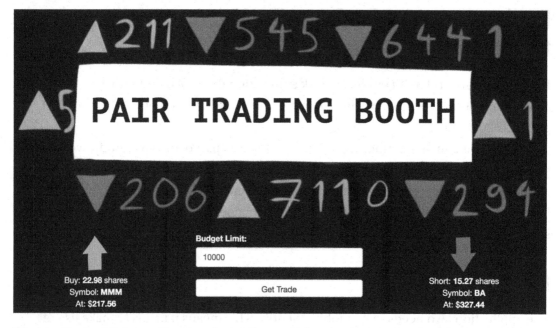

***Figure 5-12.*** *Web application in wide mode for computers and tablets*

And when the application is viewed on a mobile or narrow page, we see a version where the green and red arrows render above and below the "**Budget Limit:**" text box (Figure 5-13).

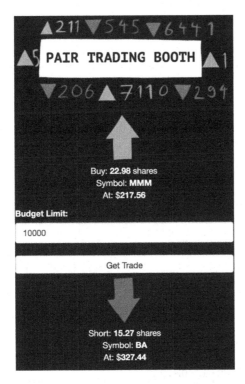

***Figure 5-13.*** *Web application in narrow mode for mobile devices*

This ensures that the application is always easy to use no matter the format. For more on this topic, see "**Bootstrap Grid Examples**" on w3schools at `https://www.w3schools.com/bootstrap/bootstrap_grid_examples.asp`.

# Running the Local Flask Version

Download the files for Chapter 5 to your local machine if you haven't done so already. In a command/terminal window, enter the "**web-application**" folder. Your file structure should look like Listing 5-16.

***Listing 5-16.*** Our Web Application's File Structure

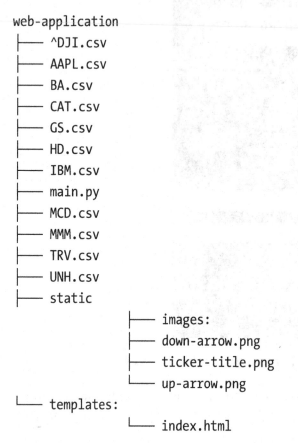

```
web-application
├── ^DJI.csv
├── AAPL.csv
├── BA.csv
├── CAT.csv
├── GS.csv
├── HD.csv
├── IBM.csv
├── main.py
├── MCD.csv
├── MMM.csv
├── TRV.csv
├── UNH.csv
├── static
                ├── images:
                ├── down-arrow.png
                ├── ticker-title.png
                └── up-arrow.png
└── templates:
                └── index.html
```

You can install all the required Python libraries by running the "**pip3 install -r**" command (Listing 5-17).

***Listing 5-17.*** Installing Requirements

```
$ pip3 install -r requirements.txt
```

Go ahead and take if for a spin by typing the "**python3 main.py**" command in your terminal window (Listing 5-18).

***Listing 5-18.*** Our Local Web Application

```
$ python3 main.py
```

You should see the "**Pair Trading Booth**" web application (Figure 5-14).

*Figure 5-14.* *Local rendering of the web application*

# What's Going on Here?

Compared with the other sites we've built so far, this chapter is fairly straightforward. We did implement some mobile-friendly tags that were discussed earlier, and here are two more things worthy of mention.

## Bootstrap Input Field Validation

Bootstrap has great form-validation features that are trivial to implement. By just telling Bootstrap what the data type of an input field is, it can handle it automatically for you and save you a lot of coding and form processing headaches (Listing 5-19).

*Listing 5-19.* Bootstrap Field Validation

```
<input type="number" class="form-control" value="" name=...>
```

To see it in action, try entering a non-numeric character in the "**Budget Limit**" input box and you will see the front end pop up a message box stating that you can only enter numbers. This is a phenomenal feature to leverage, especially in this day-and-age of text injection attacks (Figure 5-15).

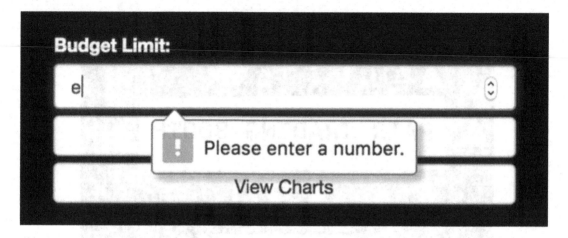

**Figure 5-15.** *Form-validation in action and rejecting a non-numeric character*

For more information on these form validators, check out the Bootstrap documentation at: `https://getbootstrap.com/docs/4.0/components/forms/`.

We will keep building out the "**Pair Trading Booth**" web application in the next sections, so keep the code handy.

# Running on PythonAnywhere

Log into your PythonAnywhere account you used in Chapter 1. Navigate to the "**Files**" button on the top right of the dashboard. Create a new folder on PythonAnywhere called "**pair-trading-booth**." Click on the "**Files**" link in the top menu bar, then enter "**pair-trading-booth**" in the directories text box and hit "**New Directory**" (Figure 5-16).

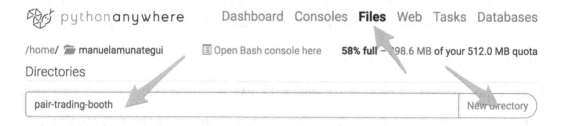

**Figure 5-16.** *Creating our site's root directory*

Enter the "**pair-trading-booth**" folder and create two more folders under it: "**static**" and "**templates**."

Right under the "**pair-trading-booth**" directory, hit the "**Upload a file**" button and upload all "**CSV**" files and the "**main.py**" file under the "**web-application**" folder for this chapter. Unfortunately you will need to do this once for every file, so 12 times (Figure 5-17).

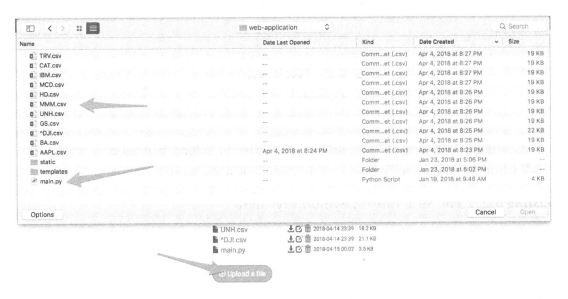

***Figure 5-17.*** *Uploading files manually to the cloud*

Your folder structure should look like Listing 5-20.

***Listing 5-20.*** File Structure on PythonAnywhere

```
^DJI.csv
AAPL.csv
BA.csv
CAT.csv
GS.csv
HD.csv
IBM.csv
main.py
MCD.csv
MMM.csv
TRV.csv
```

```
UNH.csv
├── static
└── templates
```

Next, enter the "**static**" folder and create an "**images**" folder. Enter the "**images**" folder and upload all the "**PNG**" files. Your folder should look like Listing 5-21.

***Listing 5-21.*** File Structure on PythonAnywhere

```
└── static:
        └── images:
                        ├── down-arrow.png
                        ├── ticker-title.png
                        └── up-arrow.png
```

Backtrack to the "**templates**" folder and enter it and upload; you guessed it, the "**index.html**" file. The final full structure should look like Listing 5-22.

***Listing 5-22.*** File Structure on PythonAnywhere

```
^DJI.csv
AAPL.csv
BA.csv
CAT.csv
GS.csv
HD.csv
IBM.csv
main.py
MCD.csv
MMM.csv
TRV.csv
UNH.csv
└── static:
        └── images:
                        ├── down-arrow.png
                        ├── ticker-title.png
                        └── up-arrow.png

└── templates:
                └── index.html
```

The file structure is exactly the way we want it. Now we need to work on the web settings to tell PythonAnywhere what to serve and how. Click the "**Web**" tab.

# Fixing the WSGI File

We need to set up the Web Server Gateway Interface file (WSGI). The WSGI file is a common interface between different web frameworks and the Python programming language.

# Source Code

Click on the "**Web**" menu button on PythonAnywhere (Figure 5-18).

***Figure 5-18.*** *Enter the Web tab on PythonAnywhere*

Scroll down to the "**Code**" section and change the "**Source code**" path to end with "**pair-trading-booth**" (Listing 5-23).

***Listing 5-23.*** Find the "**mysite**" Source Code Link

**Source code:** /home/<YOUR-ACCOUNT-NAME>/mysite

Change "**mysite**" to "**pair-trading-booth**" (Listing 5-24).

***Listing 5-24.*** Update the Source Code Link with Our Web Application Path

**Source code:** /home/<YOUR-ACCOUNT-NAME>/pair-trading-booth

This way, PythonAnywhere knows where to find the source code (Figure 5-19).

Code:

What your site is running.

Source code:  egui/pair-trading-booth ⊗ ✓ ✕                    ➔Go to directory

***Figure 5-19.***   *You should end up with the correct application name in the source code text box*

# WSGI Configuration

Next, and in the same section, update the "**WSGI configuration file**." Click on the link shown in Figure 5-20.

Code:

What your site is running.

Source code:  /home/manuelamunategui/pair-trading-booth       ➔Go to directory
Working directory:  /home/manuelamunategui/                    ➔Go to directory
WSGI configuration file:  /var/www/manuelamunategui_pythonanywhere_com_wsgi.py

***Figure 5-20.***   *Click on the "**WSGI configuration file**" link to update the configuration file*

There are two edits to perform on this file. Update the "**project_home**" variable to include your account name (which it should do automatically), and change the folder name to "**pair-trading-booth**." This informs the web server that the "**pair-trading-booth**" folder is where it will find all the files needed to serve the web application. Next, update the last line of the script to import from "**main**" (short for main.py) as shown in Listing 5-25.

***Listing 5-25.***   Updating the WSGI Configuration File

```
# This file contains the WSGI configuration required to serve up your
# web application at http://<your-username>.pythonanywhere.com/
# It works by setting the variable 'application' to a WSGI handler of some
# description.
#
# The below has been auto-generated for your Flask project
```

```
import sys

# add your project directory to the sys.path
project_home = u'/home/<YOUR-ACCOUNT-NAME>/pair-trading-booth'
if project_home not in sys.path:
    sys.path = [project_home] + sys.path

# import flask app but need to call it "application" for WSGI to work
from main import app as application
```

After editing the file, hit the green "**Save**" button in the upper right corner and click on the "**Web**" tab again (Figure 5-21).

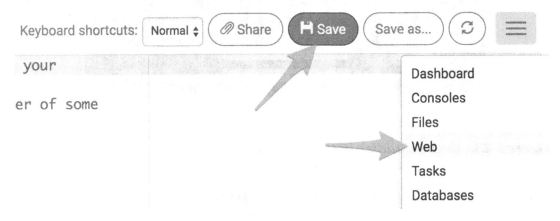

***Figure 5-21.***  *Saving the WSGI configuration file*

# Reload Web Site

There's only one more thing to do and that is the to hit the big green "**Reload <<YOUR ACCOUNT>>.pythonanywhere .com**" button. Click on your website URL and you should see the "**Pair Trading Booth**" web application (Figure 5-22).

Configuration for
manuelamunategui.pythonanywhere.com

Reload:

⟳ Reload manuelamunategui.pythonanywhere.com

***Figure 5-22.***  *The big green reload button to update the web server after any changes*

Go ahead, change the "**Budget Limit**" size and get trades. Also resize the web site's window to see the wide vs. mobile display change (Figure 5-23).

*Figure 5-23.* *The pair trading booth web application is live!*

# Troubleshooting PythonAnywhere

If you see the "**Something went wrong :-(**" page (Figure 5-24) instead of the web application, click on the error log. This will help you pinpoint what went wrong and how to fix it. Anytime you make a change to server-side scripts, you will need to hit the big green "**Reload <<YOUR ACCOUNT>>.pythonanywhere .com**" button.

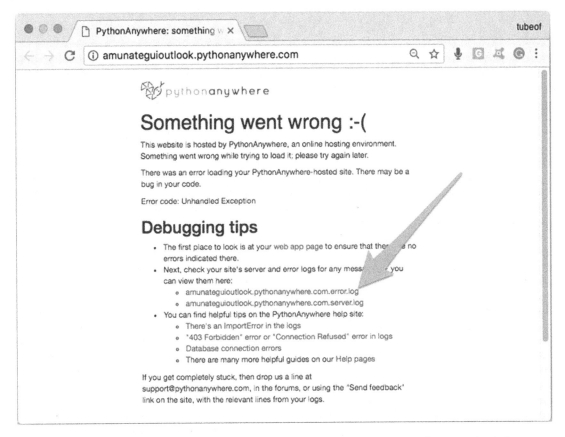

*Figure 5-24.* *Link to the error and server log files in case something isn't right*

# Conclusion

PythonAnywhere.com is not only extremely simple to use, it is intuitive and, in some instances, free! This makes it a strong contender to extend simple Python ideas on the World Wide Web and reach anybody with access to the Internet!

One more thing: because we are only using a free instance on PythonAnywhere, there is no urgency to terminate the service whenever not in use. You can let this run and show your friends and family. This won't be the case for some of the bigger providers–if you leave your page up and running on some of them, you will incur charges.

And I'll relentlessly keep reminding you to turn the services off after each chapter.

# CHAPTER 6

# Displaying Predictions with Google Maps on Azure

Where will crime happen next in San Francisco? Let's build an interactive predictive mapping dashboard using Google Maps and Microsoft Azure.

We step up our game once more by displaying information using the powerful Google Maps API. We will build a web dashboard to predict, map, and visualize crime in San Francisco (Figure 6-1). We will use a dataset from "**DataSF | San Francisco Open Data**" derived from the real and regularly updated SFPD Crime Incident Reporting System. With minimal work, we can benefit from the world's most popular mapping framework.

© Manuel Amunategui, Mehdi Roopaei 2018
M. Amunategui and M. Roopaei, *Monetizing Machine Learning*, https://doi.org/10.1007/978-1-4842-3873-8_6

*Figure 6-1.*  *The final web application for this chapter*

**Note**    Download the files for Chapter 6 by going to www.apress.com/
9781484238721 and clicking the source code button. Open the Jupyter notebook
"**chapter6.ipynb**" to follow along with this chapter's content.

# Planning our Web Application

This is an ambitious project, as we're going to build an application that will predict and visualize where crime will happen in the future. Not only does it require a modeling layer around crime data and time, but also a visualization layer that needs to be intuitive, inviting, and appealing.

# Exploring the Dataset on SF Crime Heat Map on DataSF

If you want to access some up-to-date crime data for a big city, look no further than DataSF.[1] It actually offers a lot more than just crime data; as of the last time I checked, it had over 462 published datasets available on a variety of topics around the city of San Francisco.

We will be using the SF Crime Heat Map; it has fairly current data, usually updated at the end of each month. The DataSF also has a dashboard where you can visualize the data on a map–cool stuff (see it at `https://data.sfgov.org/Public-Safety/SF-Crime-Heat-Map/q6gg-sa2p/data`). Go ahead and download the files for this chapter into a folder called "**chapter-6**" and open up the Jupyter notebook to follow along.

This is a fairly large dataset (over 250 MB); so let's download the data only once and run off a local copy during subsequent runs. In the Jupyter notebook, run the data-downloading code and then turn the flag "**already_have_the_data**" to "**True**" once the code has been successfully downloaded and saved locally. This will ensure that any subsequent run of the notebook will pull the data from your local machine.

The dataset may be different when you download it, as they keep adding data to it. We use the Pandas "**read_csv()**" to download it and save it locally. From a cursory look at the data, we see that it contains 14 columns and over 1.7 million rows (Listing 6-1).

*Listing 6-1.* Get Dataset Shape

**Input:**

```
crime_df.shape
```

**Output:**

```
(2192062, 12)
```

---

[1]`https://datasf.org/about/`

We see that we have data from 2003 to 2018 (Listing 6-2).

***Listing 6-2.*** Years Covered by Data

**Input:**

```
years = [int(dte.split("/")[2]) for dte in crime_df['Date']]
print('Max year %i, min year %i' % (max(years), min(years)))
```

**Output:**

```
Max year 2018, min year 2003
```

We also notice that it contains both interesting and not-so-interesting fields (Listing 6-3).

***Listing 6-3.*** Feature Names (before any cleanup)

**Input:**

```
list(crime_df)
```

**Output:**

```
['IncidntNum',
 'Category',
 'Descript',
 'DayOfWeek',
 'Date',
 'Time',
 'PdDistrict',
 'Resolution',
 'Address',
 'X',
 'Y',
 'Location']
```

# Data Cleanup

To make the data easier to work with and more intuitive, we'll only keep "**Category**," "**DayOfWeek**," "**Date**," "**Time**," "**X**," "**Y**," and we will rename the "**X**" and "**Y**" to "**Longitude**" and "**Latitude**" (Listing 6-4 and Figure 6-2).

*Listing 6-4.* Drop Unwanted Features and Rename Columns

```
crime_df = crime_df[['Category', 'DayOfWeek', 'Date', 'Time', 'X', 'Y']]
crime_df.columns = ['Category', 'DayOfWeek', 'Date', 'Time', 'Latitude',
'Longitude']
crime_df.head()
```

| | Category | DayOfWeek | Date | Time | Longitude | Latitude |
|---|---|---|---|---|---|---|
| 0 | NON-CRIMINAL | Monday | 01/19/2015 | 14:00 | -122.421582 | 37.761701 |
| 1 | ROBBERY | Sunday | 02/01/2015 | 15:45 | -122.414406 | 37.784191 |
| 2 | ASSAULT | Sunday | 02/01/2015 | 15:45 | -122.414406 | 37.784191 |
| 3 | SECONDARY CODES | Sunday | 02/01/2015 | 15:45 | -122.414406 | 37.784191 |
| 4 | VANDALISM | Tuesday | 01/27/2015 | 19:00 | -122.431119 | 37.800469 |

*Figure 6-2.* *Keeping only useful features with better column header names*

# Rebalancing the Dataset

Let's take an exploratory detour and learn more about this data. We'll start by looking at the categories and how many reports each contains (Listing 6-5).

*Listing 6-5.* Categories in Dataset

**Input:**

```
crime_df['Category'].value_counts()
```

**Output:**

| | |
|---|---:|
| LARCENY/THEFT | 473842 |
| OTHER OFFENSES | 306575 |
| NON-CRIMINAL | 235669 |
| ASSAULT | 192459 |
| VEHICLE THEFT | 125983 |
| DRUG/NARCOTIC | 118911 |
| VANDALISM | 114688 |
| WARRANTS | 100512 |
| BURGLARY | 90495 |
| SUSPICIOUS OCC | 79618 |
| MISSING PERSON | 64332 |
| ROBBERY | 55332 |
| FRAUD | 41104 |
| SECONDARY CODES | 25495 |
| FORGERY/COUNTERFEITING | 22938 |
| WEAPON LAWS | 21991 |
| TRESPASS | 19195 |
| PROSTITUTION | 16669 |
| STOLEN PROPERTY | 11771 |
| SEX OFFENSES, FORCIBLE | 11554 |
| DISORDERLY CONDUCT | 9988 |
| DRUNKENNESS | 9781 |
| RECOVERED VEHICLE | 8716 |
| DRIVING UNDER THE INFLUENCE | 5629 |
| KIDNAPPING | 5307 |
| RUNAWAY | 4403 |
| LIQUOR LAWS | 4078 |
| ARSON | 3887 |
| EMBEZZLEMENT | 2961 |
| LOITERING | 2420 |
| SUICIDE | 1285 |
| FAMILY OFFENSES | 1177 |
| BAD CHECKS | 921 |
| BRIBERY | 804 |

| | |
|---|---|
| EXTORTION | 733 |
| SEX OFFENSES, NON FORCIBLE | 425 |
| GAMBLING | 343 |
| PORNOGRAPHY/OBSCENE MAT | 57 |
| TREA | 14 |

Beyond the fact that these are scary categories, you will also notice that the top ones have the majority of counts. We may need to reorder things into new categories, smaller and more balanced categories. This is easy to do with Python Pandas. Let's create the following logical groups (at least they made sense to me; Listing 6-6).

***Listing 6-6.*** Higher Level Groupings

```
THEFT = ["LARCENY/THEFT", "VEHICLE THEFT", "BURGLARY", "ROBBERY", "STOLEN
PROPERTY"]

IMPAIRED = ["DRUNKENNESS", "DRIVING UNDER THE INFLUENCE", "LIQUOR LAWS",
"DISORDERLY CONDUCT", "DRUG/NARCOTIC", "LOITERING"]

VIOLENCE = ["ASSAULT", "VANDALISM", "SUSPICIOUS OCC", "TRESPASS",
"SEX OFFENSES,
FORCIBLE" , "SEX OFFENSES, NON FORCIBLE"]

OTHER = ["OTHER OFFENSES", "NON-CRIMINAL"]
```

Using NumPy's handy "**select()**" function, we can create new super categories (Listing 6-7).

***Listing 6-7.*** Using NumPy's "**select()**" Feature to Create New Categories

```
selections = [
    (crime_df['Category'].isin(THEFT)),
    (crime_df['Category'].isin(IMPAIRED)),
    (crime_df['Category'].isin(VIOLENCE)),
    (crime_df['Category'].isin(OTHER))]

new_categories = ['THEFT', 'IMPAIRED', 'VIOLENCE', 'OTHER']
crime_df['CAT'] = np.select(selections, new_categories, default='OTHER')
```

If we look at our categories by counts, we see that we have a somewhat better balanced, and more importantly, much shorter set to deal with (Listing 6-8).

***Listing 6-8.*** Counts of New Categories

**Input:**

```
crime_df_tmp['CAT'].value_counts()
```

**Output:**

```
OTHER        865893
THEFT        757423
VIOLENCE     417939
IMPAIRED     150807
```

# Exploring by Day-of-the-Week

Let's break down some of these categories by day-of-the-week. Theft happens most prevalently on Fridays, followed by Saturdays (Listing 6-9 and Figure 6-3).

***Listing 6-9.*** Theft by Day-of-the-Week

```
crime_df_tmp = crime_df[crime_df['CAT'] == 'THEFT']
crime_df_tmp['DayOfWeek'].value_counts().plot(kind='bar')
plt.suptitle('Category: THEFT')
```

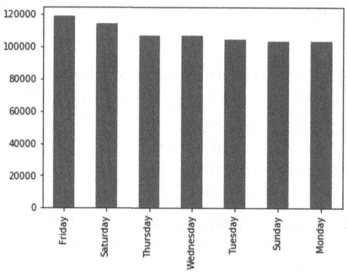

***Figure 6-3.*** *Theft by day-of-the-week*

Let's break down some of these categories by day-of-the-week. Being **"Impaired"** surprisingly happens the most on week days (Figure 6-4).

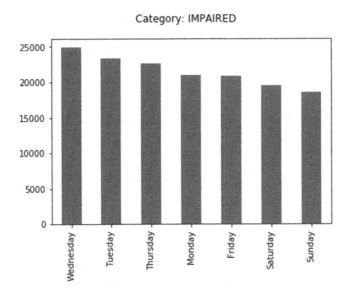

Category: IMPAIRED

*Figure 6-4.* *Impaired by day-of-the-week*

# Feature Engineering

Let's transform some of this data into simpler groups that will help us model this in a more generalizable way.

# Creating a Month-of-the-Year Feature

We'll start with an easy one and create a month-of-the-year feature. We aren't interested in the year, just the month; we're making the assumption that most Januarys over the years are similar. This simply entails pulling the first two digits from the date field. Had it not been in such a clean format, we would have had to cast it as a date field and use some specialized function (Listing 6-10 and Figure 6-5).

*Listing 6-10.* Aggregating by Months

```
crime_df["Month_of_year"] = [int(dte.split("/")[0]) for dte in crime_
df['Date']]
crime_df["Month_of_year"].value_counts()
```

| | Month_of_year |
|---|---|
| 1 | 197255 |
| 2 | 173798 |
| 3 | 193254 |
| 4 | 180322 |
| 5 | 184082 |
| 6 | 174736 |
| 7 | 181810 |
| 8 | 187368 |
| 9 | 183389 |
| 10 | 190431 |
| 11 | 174967 |
| 12 | 170650 |

**Figure 6-5.** *Month-of-the-year total reported crime counts table*

As expected, we have 12 groups. Let's plot them and see what month has the most reported crime (Listing 6-11 and Figure 6-6).

**Listing 6-11.** Reported Crime Counts by Month

```
plt.barh(crime_by_month.index, crime_by_month['Month_of_year'],
align='center', alpha=0.5)
objects = ['Jan','Feb','Mar','Apr','Ma','Jun','Jul','Aug','Sept','Oct',
'Nov','Dec']
plt.yticks(crime_by_month.index, objects)
plt.xlabel('Crime Reports')
plt.title('Total Crime By Month')
plt.show()
```

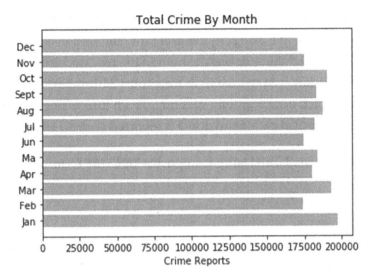

***Figure 6-6.*** *Month-of-the-year total reported crime counts*

We can dig deeper by segmenting the data by crime category (Figure 6-7).

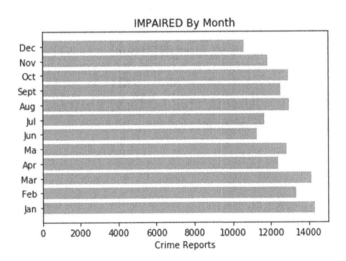

***Figure 6-7.*** *Month-of-the-year reported crime counts for category "**IMPAIRED**"*

# Creating Time Segments

To simplify our time data, we're going to segment it into three categories:

- Morning

- Afternoon

- Night

We'll reuse our "**np.select()**" function but there are plenty of other ways of achieving the same results, like using "**np.where()**" or comprehensions. We first create a new feature called "**Hour**" that extracts the hour of the reported crime from the full timestamp. We then simply create three time zones on a 24-hour timeline and filter out which hour fits into which time segment. We end up with three categories that are fairly well balanced, with a slight skew toward non-AM reported crimes (Listing 6-12).

*Listing 6-12.* Transform Time Using Categories "**AM**," "**AFT**," and "**NT**"

**Input:**

```
# create AM, AFT, NT
crime_df["Hour"] = [int(hr.split(":")[0]) for hr in crime_df['Time']]
crime_df["Hour"]

# create new groups
selections = [
    (crime_df['Hour'] > 5) & (crime_df['Hour'] <=13),
    (crime_df['Hour']  > 13) & (crime_df['Hour'] <= 19),
    (crime_df['Hour']  > 18) & (crime_df['Hour'] <= 5)]

new_categories = [0, 1, 2] # ['AM', 'AFT', 'NIT']
crime_df['Day_Segment'] = np.select(selections, choices, default=2)
crime_df['Day_Segment'].value_counts()
```

**Output:**

```
2    683804
1    674292
0    620236
```

# Exploring by Time Segment

Reported "**IMPAIRED**" crimes seem to be more prevalent in the afternoons, while reported "**VIOLENCE**" crime seem more prevalent at night (Figures 6-8 and 6-9).

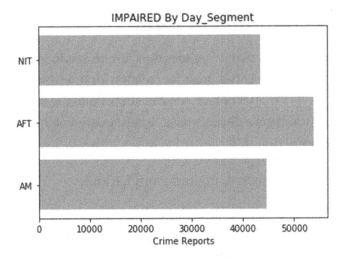

***Figure 6-8.*** *Breakdown of "**IMPAIRED**" Reported Crimes by Time Segment*

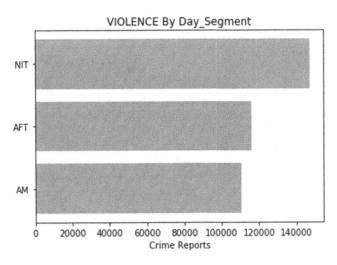

***Figure 6-9.*** *Breakdown of "**VIOLENCE**" Reported Crimes by Time Segment*

---

**Note**   Refer to the Jupyter notebook for this chapter for more graphs.

---

# Visualizing Geographical Data

A great feature of the dataset is that it includes the latitude and longitude of the location where the crime was reported. This is great for creating location-based models but also great for visualization. As a matter of fact, you can simply plot the latitude and longitude in Matplotlib as x and y and get a decent visual representation of the data. Let's try it (Listing 6-13 and Figure 6-10).

***Listing 6-13.*** Plotting Crime by Longitude and Latitude

```
plt.plot(crime_df['Longitude'].head(50000),
        crime_df['Latitude'].head(50000),
        linestyle='none', marker='.')
plt.show()
```

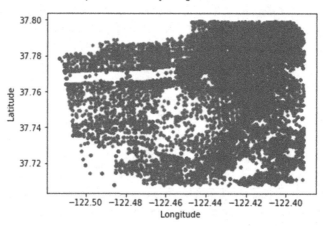

***Figure 6-10.*** *Raw plotting of the longitude and latitude of where the crime was reported*

As you can see from the plot, we can easily make out the Golden Gate Park, Presidio, and Lake Merced Park by looking at the space areas (Figure 6-11).

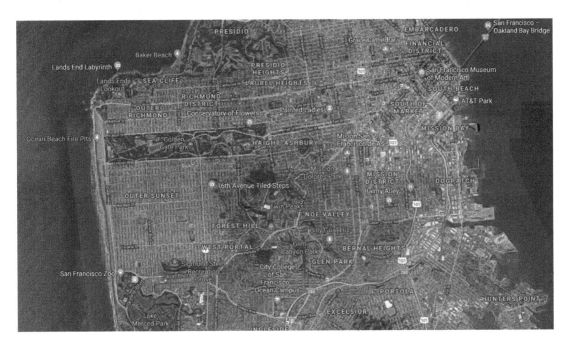

***Figure 6-11.*** *Comparative satellite plot for the same area from Google Maps (source Google Maps)*

# Rounding Geocoordinates to Create Zone Buckets

A trick to working with geocoordinates is that you can round them easily and they will create a comprehensive grid–for a tighter grid, round less, for a looser one, round more. We create a simple "**rounding_factor**" variable to help us better find the proper perspective needed. Let's take a look (Listings 6-14 and 6-15; Figures 6-12 and 6-13).

***Listing 6-14.*** Rounding Geocoordinates

```
rounding_factor = 4
plt.plot(np.round(crime_df['Longitude'].head(10000),rounding_factor),
         np.round(crime_df['Latitude'].head(10000),rounding_factor),
         linestyle='none', marker='.')
```

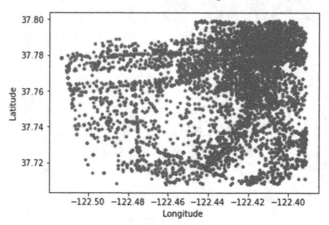

***Figure 6-12.*** *Rounding the longitude and latitude to four numbers after the decimal point*

***Listing 6-15.*** Rounding Geocoordinates

```
rounding_factor = 2
plt.plot(np.round(crime_df['Longitude'].head(10000),rounding_factor),
        np.round(crime_df['Latitude'].head(10000),rounding_factor),
        linestyle='none', marker='.')
```

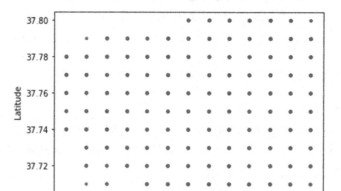

***Figure 6-13.*** *Rounding the longitude and latitude to two numbers after the decimal point*

We will go with the last case, rounding the number down to only two numbers after the decimal point. As you can see from the corresponding figure, it forms a clean and evenly spaced grid perfect for a generalized perspective of reported crime activities. As a final Matplotlib experiment, we can create a heat map of crime (Listing 6-16 and Figure 6-14).

***Listing 6-16.*** Experimenting with Heatmaps

```
from matplotlib.colors import LogNorm
x = np.round(crime_df['Longitude'].head(10000),rounding_factor)
y = np.round(crime_df['Latitude'].head(10000),rounding_factor)
fig = plt.figure()
plt.suptitle('Reported Crime Heatmap')
plt.xlabel('Latitude')
plt.ylabel('Longitude')
H, xedges, yedges, img = plt.hist2d(x, y, norm=LogNorm())
extent = [yedges[0], yedges[-1], xedges[0], xedges[-1]]
```

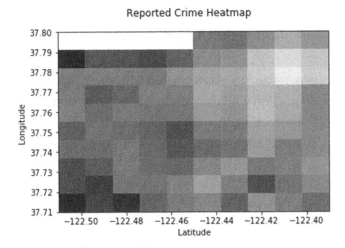

***Figure 6-14.*** *Heat map of reported crime; clearly, the brightest area is in the north-east of San Francisco*

# Using the Past to Predict the Future

We are going to aggregate the essential features we've created so far into a generalized, time-based representation. Let's take a look (Listing 6-17 and Figure 6-15).

***Listing 6-17.*** Capping the Data Down to Only Essential Features

```
crime_df = crime_df[['CAT', 'Day_of_month','Month_of_year', 'Day_Segment',
'Longitude', 'Latitude']]
crime_df.head()
```

|   | CAT | Day_of_month | Month_of_year | Day_Segment | Longitude | Latitude |
|---|-----|--------------|---------------|-------------|-----------|----------|
| 0 | OTHER | 19 | 1 | 1 | -122.422 | 37.762 |
| 1 | THEFT | 1 | 2 | 1 | -122.414 | 37.784 |
| 2 | VIOLENCE | 1 | 2 | 1 | -122.414 | 37.784 |
| 5 | OTHER | 1 | 2 | 1 | -122.452 | 37.787 |
| 8 | THEFT | 31 | 1 | 1 | -122.407 | 37.788 |

***Figure 6-15.*** *A look at the data that we will feed into our web application*

The data holds a generalized location portion with the longitude and latitude, and a generalized time-based portion with the month-of-year, day-of-month, day-segment. We need to aggregate this information down to the time and location level. This will allow us to sum up reports and build intensity maps depending on quantity of reports for a period.

We start by adding a "**Count**" feature and apply the handy Pandas "**groupby()**" function (Listing 6-18 and Figure 6-16).

***Listing 6-18.*** Aggregating Information by Time and Location

```
crime_df['Count'] = 0
crime_df_agg = crime_df.groupby(['CAT', 'Day_of_month', 'Month_of_
year', 'Day_Segment', 'Longitude', 'Latitude',]).count().reset_index()
crime_df_agg.tail()
```

| | CAT | Day_of_month | Month_of_year | Day_Segment | Longitude | Latitude | Count |
|---|---|---|---|---|---|---|---|
| 1257202 | VIOLENCE | 31 | 12 | 2 | -122.392 | 37.730 | 1 |
| 1257203 | VIOLENCE | 31 | 12 | 2 | -122.392 | 37.732 | 1 |
| 1257204 | VIOLENCE | 31 | 12 | 2 | -122.392 | 37.758 | 1 |
| 1257205 | VIOLENCE | 31 | 12 | 2 | -122.392 | 37.789 | 1 |
| 1257206 | VIOLENCE | 31 | 12 | 2 | -122.391 | 37.719 | 1 |

*Figure 6-16.* *Information aggregated down to time and location level*

Now we can ask for a particular date signature and get all the reported crimes per location (Listing 6-19 and Figure 6-17).

*Listing 6-19.* Information Aggregated by Date Signature

```
Day_of_month = 1
Month_of_year = 1
Day_Segment = 1
crime_df_agg_tmp = crime_df_agg[(crime_df_agg['Day_of_month'] ==
Day_of_month) &
                              (crime_df_agg['Month_of_year'] ==
                              Month_of_year) &
                                (crime_df_agg['Day_Segment'] ==
                                Day_Segment)]
crime_df_agg_tmp.head()
```

| | CAT | Day_of_month | Month_of_year | Day_Segment | Longitude | Latitude | Count |
|---|---|---|---|---|---|---|---|
| 16463 | IMPAIRED | 6 | 6 | 0 | -122.508 | 37.754 | 1 |
| 16464 | IMPAIRED | 6 | 6 | 0 | -122.505 | 37.745 | 2 |
| 16465 | IMPAIRED | 6 | 6 | 0 | -122.503 | 37.747 | 1 |
| 16466 | IMPAIRED | 6 | 6 | 0 | -122.501 | 37.777 | 1 |
| 16467 | IMPAIRED | 6 | 6 | 0 | -122.466 | 37.773 | 5 |

*Figure 6-17.* *Information aggregated by date signature*

And we can plot this information as well (Figure 6-18).

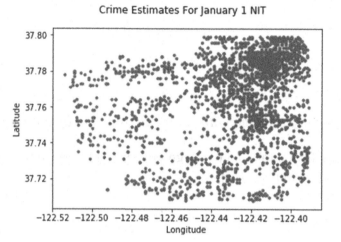

*Figure 6-18.* *Crime estimates for January 1st at night by longitude and latitude*

This chart doesn't discriminate on count intensity; let's fix that. We'll switch from Matplotlib's "**plot()**" to "**scatter()**" and use the "**s**" or size parameter (Listing 6-20 and Figure 6-19).

*Listing 6-20.* Scatter Plot of Crime Data Using Dot Sizing

```
plt.scatter(crime_df_agg_tmp['Longitude'],
        crime_df_agg_tmp['Latitude'], s=crime_df_agg_tmp['Count'])
plt.suptitle(title)
plt.xlabel('Longitude')
plt.ylabel('Latitude')
plt.show()
```

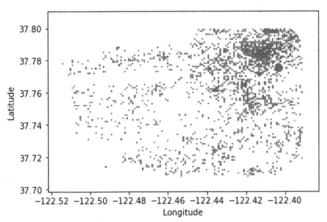

***Figure 6-19.*** *Plotting with Matplotlib's Scatter function and passing the "**Count**" feature to adjust dot size for January 1s^t at night*

And for good measure, let's dial up a completely different time (Figure 6-20).

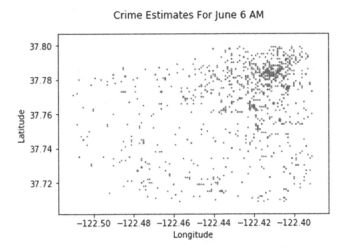

***Figure 6-20.*** *Looking at reported crime aggregates for June 6th in the morning*

To conclude our quick estimates, there is more reported crime on January 1st at night than June 6th in the morning.

# Google Maps Introduction

If there is one great and easily customizable visualization tool, it's got to be Google Maps! Let's change gears and try our data using Google Maps instead of plain old Matplotlib. For this part and to get the web application working, you will need to create a free Google Maps API Key at https://developers.google.com/maps/documentation/javascript/get-api-key.

In the past you could get Google Maps to work without it, but these days you need it, and for moderate use you can get away with the free tier.

So, get that key and try this simple example where you input an address and API key in order to get all sorts of corollary information regarding that location. Enter the following address and URL link into your browser: "**1600 Amphitheatre Parkway, Mountain View, CA**" (Listing 6-21).

***Listing 6-21.*** If Your API Key Is Valid and the URL Is Well Formed, You Should See a Long XML Response with Similar Data

**Input:**

```
https://maps.googleapis.com/maps/api/geocode/xml?address=1600+Amphitheatre+
Parkway,+Mountain+View,+CA&key=<<ADD YOUR GOOGLE MAP API KEY>>
```

**Output:**

```
<GeocodeResponse>
<status>OK</status>
<result>
<type>premise</type>
<formatted_address>
Google Building 41, 1600 Amphitheatre Pkwy, Mountain View, CA 94043, USA
</formatted_address>
...
<geometry>
<location>
<lat>37.4224082</lat>
<lng>-122.0856086</lng>
</location>
...
```

# Heatmap Layer

A heatmap is a way of visualizing information by intensity. This is a very useful tool to overlay on top of maps, as it can relay where things are happening a lot vs. happening only a little–great for reporting crime!

For a great example of creating heatmaps on Google Maps (Figure 6-21), check out the Google Maps API example from Google's documentation at: `https://developers.google.com/maps/documentation/javascript/examples/layer-heatmap`.

***Figure 6-21.*** *Example script from Google Maps API; you will need an API key for this*

# Google Maps with Crime Data

Let's inject our crime data into a Google Map. First, we need to understand the format expected by the Google Maps "**LatLng**" function in JavaScript on an HTML page (Listing 6-22).

*Listing 6-22.* The "**getPoints()**" JavaScript Function

```
<script>
      function getPoints() {
      return [
        new google.maps.LatLng(37.782551, -122.445368),
        new google.maps.LatLng(37.782745, -122.444586),
            ...
  ];}
</script>
```

Therefore, we need to extract our latitudes and longitudes from our "**crime_df_agg_tmp**" and format them into the correct format expected by Google Maps. We then concatenate each into a long string that we can pass using Flask into the HTML script (Listing 6-23).

*Listing 6-23.* Creating New Google Maps "**LatLng()**" Objects in a Loop

```
LatLngString = "
for index, row in crime_df_agg_tmp.iterrows():
    LatLngString += "new google.maps.LatLng(" + str(row['Latitude']) + ","
    + str(row['Longitude']) + "),"
```

As an example, I manually pasted the "**LatLngString**" output into a sample HTML page and this is the result (Figure 6-22).

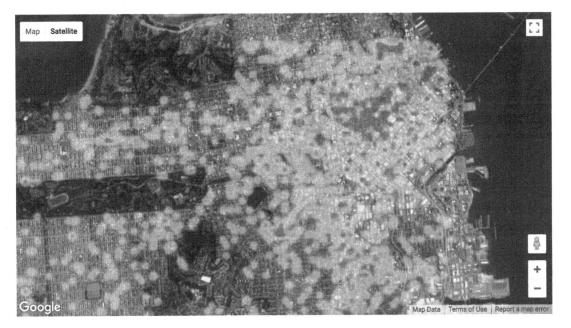

***Figure 6-22.*** *Some San Francisco crime data plotted into Google Maps*

# Abstracting Our Crime Estimator

As usual, we need to organize our web application's engine in a clean and simple manner, so we can drop it into our main Flask script. We'll create the **"GetCrimeEstimates()"** function that will take in a date and **"time_segment"** (whether it is in the morning, afternoon, or night; Listing 6-24).

***Listing 6-24.*** The **"GetCrimeEstimates()"** Function

```
def GetCrimeEstimates(horizon_date, horizon_time_segment):
    Day_of_month = int(horizon_date.split('/')[1])
    Month_of_year = int(horizon_date.split('/')[0])
    Day_Segment = int(horizon_time_segment) # 0,1,2

        crime_horizon_df_tmp = crime_horizon_df[
                    (crime_horizon_df['Day_of_month'] == Day_of_month) &
            (crime_horizon_df['Month_of_year'] == Month_of_year) &
            (crime_horizon_df['Day_Segment'] == Day_Segment)]
```

```
# build latlng string for google maps
LatLngString = "
for index, row in crime_horizon_df_tmp.iterrows():
    LatLngString += "new google.maps.LatLng(" + str(row['Latitude'])
                        + "," + str(row['Longitude']) + "),"

return (LatLngString)
```

The "GetCrimeEstimates()" is the brains of the application. It takes in a date that the user selects via the slider on the web application along with a time segment, and returns all the aggregated crime for that date.

For example, when calling the "**GetCrimeEstimates()**," we get back a string of concatenated "**google.maps.LatLng**" coordinates ready to be fed into Google Maps (Listing 6-25).

*Listing 6-25.* Calling "**GetCrimeEstimates()**"

**Input:**

```
GetCrime('10/10/2018', 0)
```

**Output:**

```
new google.maps.LatLng(37.764,-122.508),
new google.maps.LatLng(37.781,-122.49),
new google.maps.LatLng(37.711,-122.469),
new google.maps.LatLng(37.764,-122.46700000000001),
new google.maps.LatLng(37.763000000000005,-122.464),
...
```

# Designing a Web Application to Enable Viewers to Enter a Future Date and Visualize Crime Hotspots

Go ahead and download the code for the web application for this chapter. Open a command line window and change the drive to the "**web-application**" folder. It should contain the following files (Listing 6-26).

***Listing 6-26.*** Web Application Files

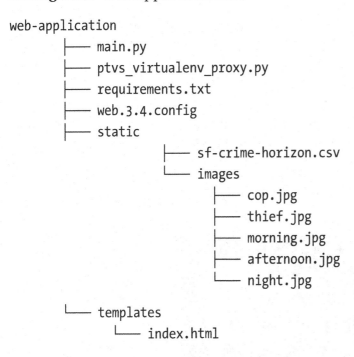

```
web-application
        ├── main.py
        ├── ptvs_virtualenv_proxy.py
        ├── requirements.txt
        ├── web.3.4.config
        ├── static
                    ├── sf-crime-horizon.csv
                    └── images
                              ├── cop.jpg
                              ├── thief.jpg
                              ├── morning.jpg
                              ├── afternoon.jpg
                              └── night.jpg

        └── templates
                └── index.html
```

As usual, we'll start a virtual environment to segregate our Python library installs (Listing 6-27).

***Listing 6-27.*** Starting a Virtual Environment

```
$ python3 -m venv predictingcrimeinsanfrancisco
$ source predictingcrimeinsanfrancisco/bin/activate
```

Then install all the required Python libraries by running the "**pip3 install -r**" command (Listing 6-28).

***Listing 6-28.*** Code Input

```
$ pip3 install -r requirements.txt
```

# Add Your Google API Key

In an editor, open up the file "**index.html**" and add in your own API key where it says "**ADD_YOUR_API_KEY_HERE**". You will need to update the code in order to see Google Maps, otherwise you will get an error message.

# Take It for a Spin

As usual, take the site for a spin on a local Flask instance (Listing 6-29).

*Listing 6-29.* Code Input

```
$ python3 main.py
```

You should see the web application with a working Google Map if all went well. Go ahead and take it for a spin and look at crime predictions for future dates (Figure 6-23).

*Figure 6-23.* *Running the local version of our web application*

# Git for Azure

Initialize a Git session (Listing 6-30).

***Listing 6-30.*** Initialize Git

```
$ git init
```

It is a great idea to run "**git status**" a couple times throughout to make sure you are tracking the correct files (Listing 6-31).

***Listing 6-31.*** Running "**git status**"

```
Input:

$ git status

Output:

Untracked files:
  (use "git add <file>..." to include in what will be committed)

        main.py
        predictingcrimeinsanfrancisco/
        ptvs_virtualenv_proxy.py
        requirements.txt
        sf-crime-horizon.csv
        static/
        templates/
        web.3.4.config
```

Add all the web-application files from the "**web-application**" file using the "**git add .**" command and check "**git status**" again (Listing 6-32).

***Listing 6-32.*** Adding to Git

```
Input:

$ git add .
$ git status
```

Output:

```
Changes to be committed:
  (use "git rm --cached <file>..." to unstage)

        new file:    main.py
        new file:    predictingcrimeinsanfrancisco/bin/activate
        new file:    predictingcrimeinsanfrancisco/bin/activate.csh
        new file:    predictingcrimeinsanfrancisco/bin/activate.fish
        new file:    predictingcrimeinsanfrancisco/bin/easy_install
        new file:    predictingcrimeinsanfrancisco/bin/easy_install-3.6
...
```

You may have noticed that we have added a lot of files to our "**git add .**" command. As per instructions from "**git status**," it tells us how to remove files that we don't want to commit to Git with the "**rm**" command. Let's remove all files and folders from the virtual environment "**predictingcrimeinsanfrancisco**" that aren't needed for the project (Listing 6-33).

***Listing 6-33.*** Removing "**predictingcrimeinsanfrancisco**" from Git

Input:

```
$ git rm -r --cached predictingcrimeinsanfrancisco
$ git status
```

Output:

```
Changes to be committed:
  (use "git rm --cached <file>..." to unstage)

        new file:    main.py
        new file:    ptvs_virtualenv_proxy.py
        new file:    requirements.txt
        new file:    sf-crime-horizon.csv
        new file:    static/images/afternoon.jpg
        new file:    static/images/cop.jpg
        new file:    static/images/morning.jpg
        new file:    static/images/night.jpg
        new file:    static/images/thief.jpg
```

```
new file:    templates/index.html
new file:    web.3.4.config
```

We now only have the files we need. So, do a local "**git commit**" and add a comment that makes sense in case you need to revisit past actions in the future (Listing 6-34).

***Listing 6-34.*** Git Commit

Input:

```
$ git commit -am 'where crime happens'
```

Output:

```
[master (root-commit) 1b87606] where will crime happen next
 11 files changed, 120065 insertions(+)
 create mode 100644 main.py
 create mode 100644 ptvs_virtualenv_proxy.py
 create mode 100644 requirements.txt
 create mode 100644 sf-crime-horizon.csv
 create mode 100644 static/images/afternoon.jpg
 create mode 100644 static/images/cop.jpg
 create mode 100644 static/images/morning.jpg
 create mode 100644 static/images/night.jpg
 create mode 100644 static/images/thief.jpg
 create mode 100644 templates/index.html
 create mode 100644 web.3.4.config
```

For more information on the Git Deployment to Azure App Service, see https://docs.microsoft.com/en-us/azure/app-service/app-service-deploy-local-git.

# The azure-cli Command Line Interface Tool

We will use the "**azure-cli**" tool to deploy our web application on Microsoft Azure (if you don't already have it installed, refer back to Chapter 2).

# Step 1: Logging In

Create an "**az**" session (Listing 6-35 and Figure 6-24).

*Listing 6-35.* Code Input

```
$ az login
```

```
[manuels-MacBook-Pro-2:web-application manuel$ az login
 To sign in, use a web browser to open the page https://microsoft.com/devicelogin
 and enter the code B54YSXKF2 to authenticate.
```

*Figure 6-24.* *Logging into Azure from azure-cli*

Follow the instructions, point a browser to the given URL address, and enter the code accordingly (Figure 6-25).

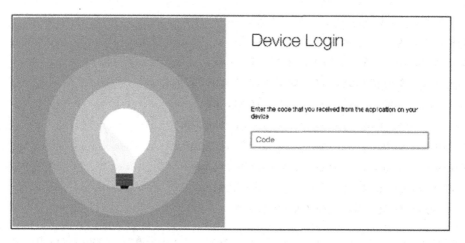

*Figure 6-25.* *Authenticating session*

If all goes well (i.e., you have an Azure account in good standing), it will connect the azure-cli terminal to the cloud server. Also, once you are authorized, you can safely close the browser window.

Make sure your command-line tool is pointing inside this chapter's "**web-application**" folder.

# Step 2: Create Credentials for Your Deployment User

This user will have appropriate rights for FTP and local Git use. Here I set the user-name to "**flaskuser10**" and password to "**flask123**." You should only have to do this once, then you can reuse the same account. In case it gives you trouble, simply create a different user name (or add a number at the end of the user name and keep incrementing it like I usually do; Listing 6-36).

*Listing 6-36.* Code Input

```
$ az webapp deployment user set --user-name flaskuser10 --password flask123
```

As you proceed through each "**azure-cli**" steps, you will get back JSON replies confirming your settings. In the case of the "**az webapp deployment**" most should have a null value and no error messages. If you have an error message, then you have a permission issue that needs to be addressed ("**conflict**" means that name is already taken so try another, and "**bad requests**" means the password is too weak).

# Step 3: Create Your Resource Group

This is going to be your logical container. Here you need to enter the region closest to your location (see https://azure.microsoft.com/en-us/regions/). Going with "**West US**" for this example isn't a big deal even if you're worlds away, but it will make a difference in a production setting where you want the server to be as close as possible to your viewership for best performance.

Here I set the name to "**myResourceGroup**" (Listing 6-37).

*Listing 6-37.* Code Input

```
$ az group create --name myResourceGroup --location "West US"
```

## Step 4: Create your Azure App Service Plan

Here I set the name to "**myAppServicePlan**" and select a free instance (sku) (Listing 6-38).

*Listing 6-38.*  Code Input

```
$ az appservice plan create --name myAppServicePlan --resource-group
myResourceGroup --sku FREE
```

## Step 5: Create your Web App

Your "**webapp**" name needs to be unique, and make sure your "**resource-group**" and "**plan**" names are the same as what you set in the earlier steps. In this case I am going with "**amunateguicrime.**" For a full list of supported runtimes, run the "**list-runtimes**" command (Listing 6-39).

*Listing 6-39.*  Code Input

```
$ az webapp list-runtimes
```

To create the web application, use the "**create**" command (Listing 6-40).

*Listing 6-40.*  Code Input

```
$ az webapp create --resource-group myResourceGroup --plan myAppServicePlan
--name amunateguicrime --runtime "python|3.4" --deployment-local-git
```

The output of "**az webapp create**" will contain an important piece of information that you will need for subsequent steps. Look for the line "**deploymentLocalGitUrl**" (Figure 6-26).

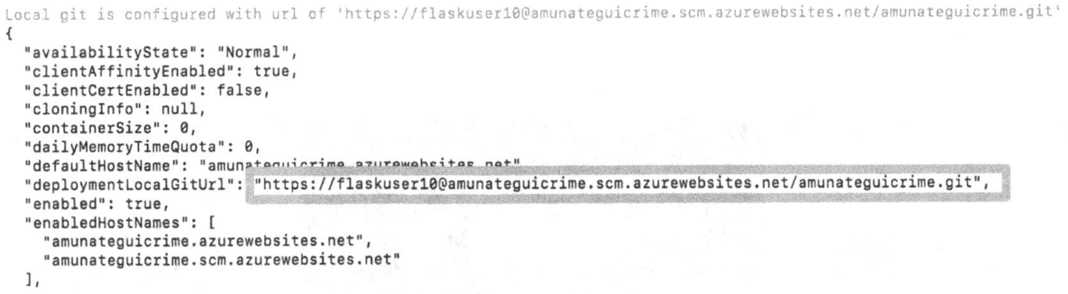

```
Local git is configured with url of 'https://flaskuser10@amunateguicrime.scm.azurewebsites.net/amunateguicrime.git'
{
  "availabilityState": "Normal",
  "clientAffinityEnabled": true,
  "clientCertEnabled": false,
  "cloningInfo": null,
  "containerSize": 0,
  "dailyMemoryTimeQuota": 0,
  "defaultHostName": "amunateguicrime.azurewebsites.net"
  "deploymentLocalGitUrl": "https://flaskuser10@amunateguicrime.scm.azurewebsites.net/amunateguicrime.git",
  "enabled": true,
  "enabledHostNames": [
    "amunateguicrime.azurewebsites.net",
    "amunateguicrime.scm.azurewebsites.net"
  ],
```

*Figure 6-26.* *"webapp create" command and resulting "deployment LocalGitUrl" value*

## Step 6: Push Git Code to Azure

Now that you have a placeholder web site, you need to push out your Git code to Azure (Listing 6-41).

*Listing 6-41.* Code Input

```
# if git remote is say already exits, run 'git remote remove azure'
$ git remote add azure https://flaskuser10@amunateguicrime.scm.
azurewebsites.net/amunateguicrime.git
```

Finally, push it out to Azure (Listing 6-42).

*Listing 6-42.* Code Input

```
$ git push azure master
```

It will prompt you for the "**webapp deployment user**" password you set up earlier. If all goes well, you should be able to enjoy the fruits of your labor. Open a web browser and enter your new URL that is made of your "**webapp**" name followed by "**.azurewebsites.net**" (Figure 6-27).

***Figure 6-27.*** *Enjoy the fruits of your hard work!*

On the other hand, if the Azure-cli returns error messages, you will have to address them (see the troubleshooting section).

Anytime you update your code and want to redeploy it, see Listing 6-43.

***Listing 6-43.*** For Code Updates

```
$ git commit -am "updated output"
$ git push azure master
```

You can also manage your application directly on Azure's web dashboard. Log into Azure and go to App Services (Figure 6-28).

***Figure 6-28.*** *Microsoft Azure dashboard*

# Troubleshooting

It can get convoluted to debug web application errors. One thing to do is to turn on logging through Azure's dashboard (Figure 6-29).

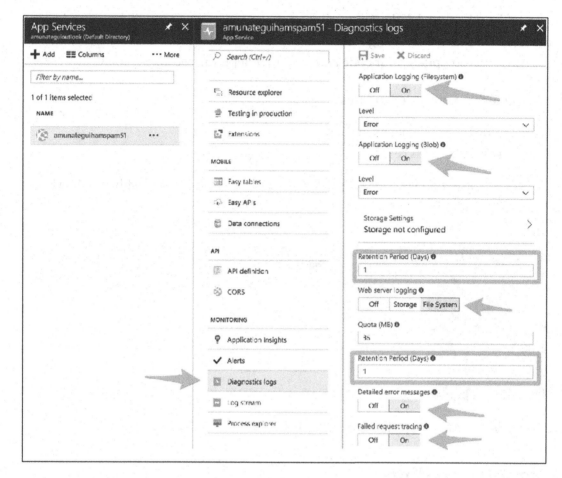

***Figure 6-29.*** *Turning on Azure's Ddiagnostics logs*

Then you turn the logging stream on to start capturing activity (Figure 6-30).

***Figure 6-30.*** *Capturing log information*

You can also check your file structure using the handy Console tool built into the Azure dashboard (Figure 6-31).

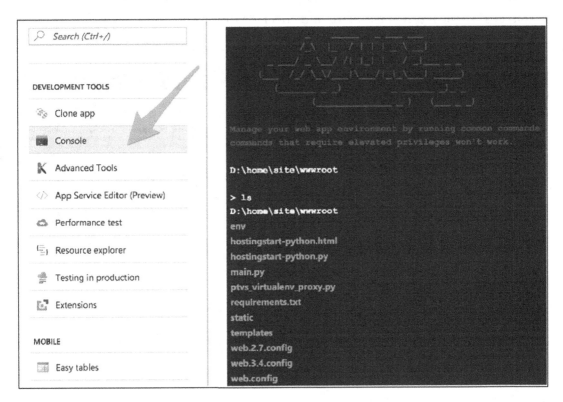

**Figure 6-31.** *Azure's built-in command line tool*

You can also access the tail of the log in your command window (Listing 6-44).

**Listing 6-44.** Code Input

```
$ az webapp log tail --resource-group myResourceGroup --name
amunateguicrime
```

You can even check if your "**requirement.txt**" file works by calling the "**env\scripts\ pip**" function (Listing 6-45).

**Listing 6-45.** Code Input

```
$ env\scripts\pip install -r requirements.txt
```

# Don't Forget to Turn It Off!

As usual, stop and delete anything that you don't need anymore. Go to "**All resources**" in the Azure dashboard and check anything you don't need, and delete away (Figure 6-32).

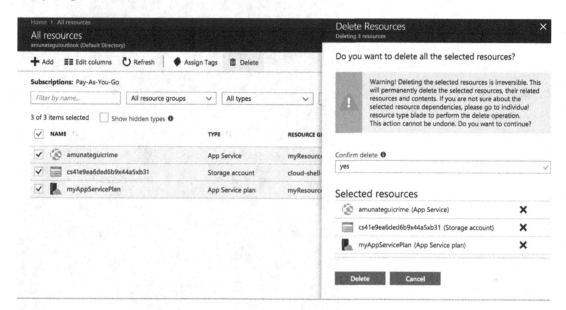

*Figure 6-32.* *Turning everything off once finished*

And finally, deactivate your virtual environment (Listing 6-46).

*Listing 6-46.* Code Input

```
$ deactivate predictingcrimeinsanfrancisco
```

# Conclusion

This chapter introduces an obvious great piece of technology, Google Maps. There is so much that can be done with this front-end dashboard. You can get user's location (with their consent), you can visualize geographical data in many different ways from satellite to street views, and the list keeps going on. The best part is that very little is required in terms of programming. The Google Maps API is mature and abstracts a lot of the heavy lifting for you. If you end up building a high-traffic site using Google Maps, you will most likely need a paid account.

Those with a sharp eye may have noticed that the Jupyter code uses data frames to analyze the crime data, while the Flask application uses a NumPy array. This isn't the only time we will use this trick. The Panda library is such a large and complex library that it is sometimes hard to get it to play nice with serverless instances that don't like libraries with deep tentacles into the OS and file system. As a rule of thumb, the least amount of imports you need to declare at the top of your Flask application, the better.

# CHAPTER 7

# Forecasting with Naive Bayes and OpenWeather on AWS

Will I golf tomorrow? Find out using naive Bayes and real-time weather forecasts on Amazon Web Services.

In this chapter, we will take a look at the famed "**Golf|Weather Dataset**" from Gerardnico's blog.[1] I say "**famed**" because it seems that whenever somebody does an introductory piece on the Bayes, they use this dataset. It makes sense, as it is a very simple and intuitive collection of environmental readings, and whether or not a player ends up playing golf. It is to the point and very amiable to modeling with the Bayes Theorem without a computer or even a calculator. But don't fear, we'll be using the sklearn library as usual. We will model what it takes to go golfing, incorporating the OpenWeatherMap[2] to pull real forecast based on user-selected locations (Figure 7-1).

---

[1]https://gerardnico.com/data_mining/weather
[2]https://openweathermap.org/

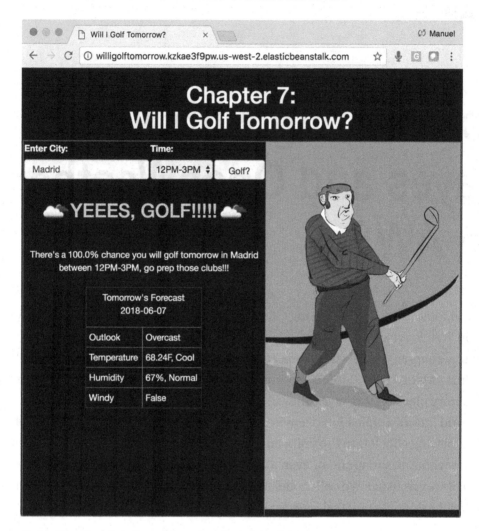

***Figure 7-1.*** *The final web application for this chapter*

---

**Note**    Download the files for Chapter 7 by going to `www.apress.com/`
`9781484238721` and clicking the source code button. Open Jupyter notebook
"**chapter7.ipynb**" to follow along with this chapter's content.

---

# Exploring the Dataset

Go ahead and download the files for this chapter into a folder called "**chapter-7.**" Open
up the Jupyter notebook to follow along.

As it is a very small dataset, there is nothing to download and we will recreate it manually in Python. It contains only 14 rows (Listing 7-1 and Figure 7-2).

*Listing 7-1.* Let's Load Our Dataset into a Pandas Data Frame, Create Column Names and Cast the Boolean Fields into Integers

```
golf_data_header = ['Outlook', 'Temperature_Numeric', 'Temperature_
Nominal', 'Humidity_Numeric', 'Humidity_Nominal', 'Windy', 'Play']

golf_data_set = [['overcast',83,'hot',86,'high',False,True],
['overcast',64,'cool',65,'normal',True,True],
['overcast',72,'mild',90,'high',True,True],
['overcast',81,'hot',75,'normal',False,True],
['rainy',70,'mild',96,'high',False,True],
['rainy',68,'cool',80,'normal',False,True],
['rainy',65,'cool',70,'normal',True,False],
['rainy',75,'mild',80,'normal',False,True],
['rainy',71,'mild',91,'high',True,False],
['sunny',85,'hot',85,'high',False,False],
['sunny',80,'hot',90,'high',True,False],
['sunny',72,'mild',95,'high',False,False],
['sunny',69,'cool',70,'normal',False,True],
['sunny',75,'mild',70,'normal',True,True]]

golf_df = pd.DataFrame(golf_data_set, columns=golf_data_header)
golf_df[['Windy','Play']] = golf_df[['Windy','Play']].astype(int)

golf_df.head()
```

| | Outlook | Temperature_Numeric | Temperature_Nominal | Humidity_Numeric | Humidity_Nominal | Windy | Play |
|---|---|---|---|---|---|---|---|
| 0 | overcast | 83 | hot | 86 | high | 0 | 1 |
| 1 | overcast | 64 | cool | 65 | normal | 1 | 1 |
| 2 | overcast | 72 | mild | 90 | high | 1 | 1 |
| 3 | overcast | 81 | hot | 75 | normal | 0 | 1 |
| 4 | rainy | 70 | mild | 96 | high | 0 | 1 |

*Figure 7-2.* First few rows of the "*Golf\Weather*" Dataset

The "**Outlook**" field is an overall take on the weather, like a super category. It is made up of three values: "**Overcast**," "**Rainy**," and "**Sunny**." The temperature reading comes in two flavors: "**Temperature_Numeric**," which is the numeric temperature in Fahrenheit, and "**Temperature_Nominal**," which is a categorical representation broken into three values: "**hot**," "**mild**," and "**cold**." Humidity also comes in two flavors: "**Humidity_Numeric**," which is the percent humidity reading, and "**Humidity_Nominal**," which is a categorical variable with two values: "**high**" and "**normal**." "**Windy**" is a Boolean variable that states whether it is windy or not. Finally, "**Play**" is the outcome variable and the resulting truth whether the player did or didn't play golf according to said conditions.

# Naive Bayes

Naive Bayes is a group of algorithms based on Bayes Theorem and conditional probabilities (Figure 7-3). It considers predictors independently to determine the probability of an outcome. It is called "**naive**" because it assumes independence between the predictors, but short of each predictor happening on a different planet, it is hard to know for sure. That said, such assumption simplifies the model tremendously; it makes the model simple, fast, and transparent. It is perfect for working with large datasets in distributed environments.

$$P(A \mid B) = \frac{P(B \mid A)\, P(A)}{P(B)},$$

where $A$ and $B$ are events and $P(B) \neq 0$

**Figure 7-3.** *Bayesian probabilistic formula*

In very simple terms, Naive Bayes classification creates a frequency table cataloging every possible value combination from some historical dataset, including both positive and negative outcomes. Its simpler to visualize by thinking of simple categorical features, but it can handle any data type. The Bayes theorem can then use the collected frequencies to yield new probabilities.

Which states that:

- **P(A|B)** is the probability of the outcome happening given certain equal values

- **P(B|A)** is the probability for those values for that outcome multiplied

- **P(A)** is the the probability for that outcome regardless of the values divided

- **P(B)** is the probability for those values regardless of the outcome

- If this isn't clear, check out a brief and funny video from the good folks at RapidMiner at `https://www.youtube.com/watch?v=IlVINQDk4o8`.

# Sklearn's GaussianNB

As mentioned in the introduction to this chapter, we're going to use the "**sklearn. naive_bayes**" "**GaussianNB**" library. This is a simple model that does offer a few tunable parameters: see `http://scikit-learn.org/stable/modules/generated/sklearn. naive_bayes.GaussianNB.html` for more information.

It is straightforward to use, and here is an example on calling the Naive Bayes model for classification and how to extract probabilities and predictions. The "**predict()**" function returns a true/false prediction based on what it trained on. A "**1**" means the model predicts that golfing will happen (Listing 7-2).

*Listing 7-2.*  Calling predict() on the GaussianNB Model

**Input:**

```
from sklearn.naive_bayes import GaussianNB
naive_bayes = GaussianNB()
naive_bayes.fit(X_train[features],  y_train))
print(naive_bayes.predict(X_test))
```

**Output:**

```
[0 0 1 0 0 0 1]
```

The "**predict_proba()**" function returns a pair of values. The first value represents the probability of being false, while the second value is the probability of being true (Listing 7-3).

*Listing 7-3.*  Getting Probabilities out of the GaussianNB Model

**Input:**

```
print(naive_bayes.predict_proba(X_test))
```

**Output:**

```
array([[  9.99994910e-01,    5.09005696e-06],
       [  9.99968916e-01,    3.10842486e-05],
       [  0.00000000e+00,    1.00000000e+00],
       [  8.84570501e-01,    1.15429499e-01],
       [  8.00907988e-01,    1.99092012e-01],
       [  9.99932094e-01,    6.79055800e-05],
       [  0.00000000e+00,    1.00000000e+00]])
```

Obviously, you can use either value; just remember which means what. In our case we'll use the second value, as we aren't that interested in the probability of not golfing vs. the probability of golfing. Both numbers add up to 1.

# Realtime OpenWeatherMap

We're going use real weather forecasts into our "**Will I Golf Tomorrow**" web application. Go ahead and sign up for an API from openweathermap.org; it's free (Figure 7-4)! A big thanks to the folks over at Open Weather–love the service!

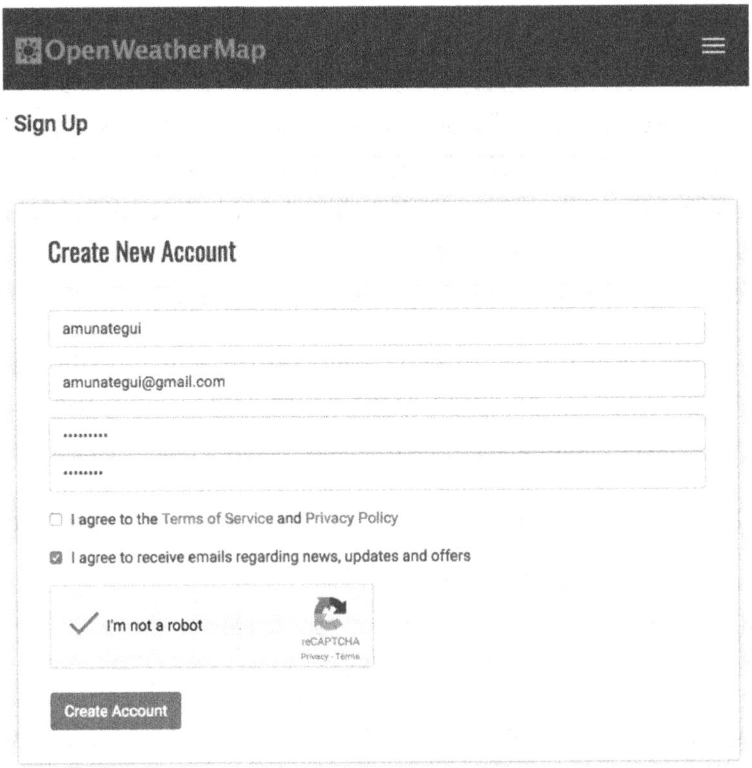

**Figure 7-4.** *OpenWeatherMap.org sign up screen*

They will send you an email confirmation containing your API key, along with an example. It states that it can take up to ten minutes to authorize the new key. For me, it took more like 30 minutes. Then run the example with your new key to double-check that your account is working (Listing 7-4).

**Listing 7-4.** URL for Weather–Add Your API Key

```
http://api.openweathermap.org/data/2.5/weather?q=London,uk&APPID=<<YOU
R_API_KEY>>
```

It does indeed take a little while to propagate on their system, but it does work, and the example returns the following JSON string (of course yours will have different weather data; Listing 7-5).

***Listing 7-5.*** Raw JSON String

```
{"coord":{"lon":-0.13,"lat":51.51},"weather":[{"id":803,"main":"Clouds",
"description":"broken clouds","icon":"04n"}],"base":"stations","main":{"temp":
284.37,"pressure":1014,"humidity":76,"temp_min":283.15,"temp_max":285.15},
"visibility":10000,"wind":{"speed":6.7,"deg":240},"clouds":{"all":75},"dt":
1524531000,"sys":{"type":1,"id":5091,"message":0.0065,"country":"GB","sunrise":
1524545173,"sunset":1524597131},"id":2643743,"name":"London","cod":200}
```

But a better way to access REST API JSON data is to do it all in Python. This allows you to make the call to the API and process the return data in a fully programmatic manner. Let's take a look (Listing 7-6).

***Listing 7-6.*** Bringing in Real Weather Data Using "**api.openweathermap.org**"

```
from urllib.request import urlopen
weather_json = json.load(urlopen("http://api.openweathermap.org/data/2.5/
weather?q=Barcelona&appid=<<YOUR_API_KEY>>"))
```

In return, we get a JSON object that can be easily accessed via key pair calls (Listing 7-7).

***Listing 7-7.*** JSON Content

**Input:**

```
print(weather_json)
```

**Output:**

```
{'base': 'stations',
 'clouds': {'all': 0},
 'cod': 200,
 'coord': {'lat': 41.38, 'lon': 2.18},
 'dt': 1524538800,
 'id': 3128760,
 'main': {'humidity': 72,
  'pressure': 1018,
  'temp': 287.15,
  'temp_max': 288.15,
```

```
 'temp_min': 286.15},
'name': 'Barcelona',
'sys': {'country': 'ES',
 'id': 5470,
 'message': 0.0034,
 'sunrise': 1524545894,
 'sunset': 1524595276,
 'type': 1},
'visibility': 10000,
'weather': [{'description': 'clear sky',
 'icon': '01n',
 'id': 800,
 'main': 'Clear'}],
'wind': {'deg': 330, 'speed': 2.6}}
```

Individual elements can easily be access by appending key names, just like you would with a Pandas object (Listing 7-8).

*Listing 7-8.* JSON "**main**" Content

**Input:**

```
weather_json['main']
```

**Output:**

```
{'humidity': 72,
 'pressure': 1018,
 'temp': 287.15,
 'temp_max': 288.15,
 'temp_min': 286.15}
```

# Forecasts vs. Current Weather Data

We want to use forecasts for the following day, and "**OpenWeatherMap**" does offer a five-day forecast API service. It returns data in three-hour increments. Let's see how this works (Listing 7-9).

***Listing 7-9.*** URL for Forecast–Add Your API Key. This Will Return a Large Amount of Text with Five-Days' Worth of Three-Hour Increment Weather Forecasts

**Input:**

```
http://api.openweathermap.org/data/2.5/forecast?q=Barcelona&APPID=<<YOUR
API KEY>>
```

**Output:**

```
{"dt":1524679200,"main":{"temp":293.08,"temp_min":291.228,"temp_max":293.08
,"pressure":1021.76,"sea_level":1030.11,"grnd_level":1021.76,"humidity":83,
"temp_kf":1.85},"weather":[{"id":802,"main":"Clouds","description":"scatter
ed clouds","icon":"03d"}],"clouds":{"all":48},"wind":{"speed":0.98,"deg":31
.502},"sys":{"pod":"d"},"dt_txt":"2018-04-25 18:00:00"}
```

The key pair "**dt_txt**" is the start time for the contained weather forecast. So, in this example, for April the 24th in Barcelona between 6 PM and 9 PM, there will be scattered clouds. Being able to access three-hour forecasts offers a great level of granularity for our golfing predictions. See the corresponding Jupyter notebook for ways of pulling specific dates.

# Translating OpenWeatherMap to "Golf|Weather Data"

There are a couple of data transformations needed to get the "**OpenWeatherMap**" data into the correct "**Golf|Weather Dataset**" format. Let's go ahead, change some scales, and fix some categorical data.

### Outlook

The "**outlook**" categorical feature in the golf set has three possible values: "**Overcast**," "**Rainy**," and "**Sunny**." A close equivalent in the "**OpenWeatherMap**" is the "**weather.main**," variable, which offers nine possible values:[3]

- Clear Sky
- Few Clouds
- Scattered Clouds

---

[3]https://openweathermap.org/weather-conditions

- Broken Clouds

- Shower Rain

- Rain

- Thunderstorm

- Snow

- Mist

Though this a subjective endeavor, we need to make a decision as to what goes where. Let's group these and build a function to handle equivalencies (and please change them around if you don't like mine).

**Sunny**

- Clear Sky

- Few Clouds

**Overcast**

- Scattered Clouds

- Broken Clouds

- Mist

**Rainy**

- Shower Rain

- Rain

- Thunderstorm

- Snow

We package our groupings into a clean function that can handle the equivalencies between "**OpenWeatherMap**" and "**Golf|Weather Data**" (Listing 7-10). We also leverage a neat offering by "**OpenWeatherMap**" to supply graphic icons of the weather that we will display on our web application (see the complete list of icons at https://openweathermap.org/weather-conditions).

**Listing 7-10.** Function "**GetWeatherOutlookAndWeatherIcon**"

```
def GetWeatherOutlookAndWeatherIcon(main_weather_icon):
      # truncate third char - day or night not needed
      main_weather_icon = main_weather_icon[0:2]

      # return "Golf|Weather Data" variable and daytime icon
      if (main_weather_icon in ["01", "02"]):
            return("sunny", main_weather_icon + "d.png")
      elif (main_weather_icon in ["03", "04", 50]):
            return("overcast", main_weather_icon + "d.png")
      else:
            return("rain", main_weather_icon + "d.png")
```

### Numeric Temperature

You may have noticed that temperature isn't in Fahrenheit or Celsius but in Kelvins. So, we need to filter it through the following formula for Fahrenheit (though you can have the API do this for you, we will do it ourselves):

```
Fahrenheit = T × 1.8 - 459.67
```

And for Celsius:

```
Celsius = K - 273.15
```

### Nominal Temperature

Nominal temperature is a categorical variable made up of three values "**cool**," "**mild**," and "**hot**." As these are subjective groupings, we're going to infer the ranges so we can create new ones based on the live forecast from "**OpenWeatherMap**" (Listing 7-11).

**Listing 7-11.** Nominal Temperatures

**Input:**

```
golf_df[['Temperature_Numeric', 'Temperature_Nominal']].
groupby('Temperature_Nominal').agg({'Temperature_Numeric' : [np.min,
np.max]})
```

**Output:**

```
                    Temperature_Numeric
                            amin amax
Temperature_Nominal
cool                          64   69
hot                           80   85
mild                          70   75
```

"**Cool**" ranges from 64 to 69 degrees Fahrenheit while "**mild**" ranges from 70 to 75. This is easy, as there is no gap between both values. "**Hot**," on the other hand, starts at 80. So, we have a gap between 75 and 80 to account for. To keep things simple, we'll extend the "**mild**" range to 80. And we end up with the following function (Listing 7-12).

***Listing 7-12.*** Nominal Temperatures

```
def GetNominalTemparature(temp_fahrenheit):
        if (temp_fahrenheit < 70):
                return "cool"
        elif (temp_fahrenheit < 80):
                return "mild"
        else:
                return "hot"
```

**Humidity Numeric**

Humidity is given in percentages on "**OpenWeatherMap**" so we'll use it in its exact numerical form.[4]

**Humidity Nominal**

Just like we did with the categorical nominal temperature, we will have to apply the same logic on the nominal humidity. There are definitely different ways to slice this one, but a choice has to be made to translate a percentage into a category that exists in the current "**Golf|Weather Data**" dataset (Listing 7-13).

---

[4]https://openweathermap.org/current

***Listing 7-13.*** Humidity

**Input:**

```
golf_df[['Humidity_Numeric', 'Humidity_Nominal']].groupby('Humidity_
Nominal').agg({'Humidity_Numeric' : [np.min, np.max]})
```

**Output:**

```
                    Humidity_Numeric
                          amin amax
Humidity_Nominal
high                        85    96
normal                      65    80
```

According to our historical data, we only have two choices: "**normal**" or "**high**." We'll take the easy route and consider 81% and higher as high, and everything else as normal (Listing 7-14).

***Listing 7-14.*** Function "**GetNominalHumidity**"

```
def GetNominalHumidity(humidity_percent):
        if (humidity_percent > 80):
                return "high"
        else:
                return "normal"
```

**Windy**

"**OpenWeatherMap**" states that wind speeds are in meters per second.[5] We'll use the Beaufort scale, a scale that relates wind speeds to different land and sea conditions, and its definition of a "**strong breeze**" category to determine what is and what isn't windy (Figure 7-5) and abstract a function (Listing 7-15). The midpoint of the scale is at wind speeds above 10.8 meters per second, considered "**strong breeze**."

---

[5]https://openweathermap.org/current

***Listing 7-15.*** Function "**GetWindyBoolean**"

```
def GetWindyBoolean(wind_meter_second):
        if (wind_meter_second > 10.8):
                return(True)
        else:
                return(False)
```

**Beaufort Wind Scale**

| 0 | 1 | 2 | 3 | 4 | 5 | 6 | 7 | 8 | 9 | 10 | 11 | 12 |
|---|---|---|---|---|---|---|---|---|---|----|----|----|
| Calm | Light Air | Light Breeze | Gentle Breeze | Moderate Breeze | Fresh Breeze | Strong Breeze | Near Gale | Gale | Strong Gale | Storm | Violent Storm | Hurricane Force |
| Light Winds | | | | | | High Winds | | Gale-force | | Storm-force | | Hurricane-force |
| <1 mph | 1–3 mph | 4–7 mph | 8–12 mph | 13–18 mph | 18–24 mph | 25–31 mph | 31–38 mph | 39–46 mph | 47-54 mph | 55–63 mph | 64–72 mph | ≥73 mph |
| <1 knot | 1–3 knots | 4–6 knots | 7–10 knots | 11–16 knots | 17–21 knots | 22–27 knots | 28–33 knots | 34–40 knots | 41–47 knots | 48–55 knots | 56–63 knots | ≥63 knots |
| <0.3 m/s | 0.3–1.5 m/s | 1.6–3.3 m/s | 3.4–5.5 m/s | 5.5–7.9 m/s | 8.0–10.7 m/s | 10.8–13.8 m/s | 13.9–17.1 m/s | 17.2–20.7 m/s | 20.8–24.4 m/s | 24.5–28.4 m/s | 28.5–32.6 m/s | ≥32.7 m/s |

***Figure 7-5.*** *Beaufort wind scale (source Wikipedia)*

# Designing a Web Application "Will I Golf Tomorrow?" with Real Forecasted Weather Data

As usual, we want our application to be intuitive, visual, and fun. This will be the go-to application for all golfers around the world (yeah right!). It is also a powerful application that will use real weather forecasts from anywhere around the world while remaining extremely simple to build. This is the beauty of a Bayesian model: it is simple and fast and makes for a great real-time and scalable modeling option for web applications.

Our web page only needs two input boxes so the user can enter his or her location and tomorrow's time they wish to golf. The application will attempt to find a weather forecast for the location and time, translate the "**OpenWeatherMap**" JSON data into the required "**Golf|Weather Dataset**" format, and return a "**yes**" or "**no**" to the question "**will I golf tomorrow.**" Pretty straightforward, right?

## Download the Web Application

Go ahead and download the code for this chapter, open a command line window, and change the drive to the "**web-application**" folder. It should contain the same files as in Listing 7-16. Here we are showing the hidden folder ".**ebextensions**" needed for AWS EB. You can either use it as-is or create your own in the "**Fix the WSGIApplicationGroup**" section (don't worry about this when running the local version of the site, as it isn't affected by this fix).

***Listing 7-16.*** Web Application Files

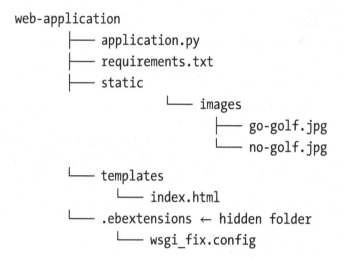

```
web-application
        ├── application.py
        ├── requirements.txt
        ├── static
                    └── images
                            ├── go-golf.jpg
                            └── no-golf.jpg
        └── templates
              └── index.html
        └── .ebextensions ← hidden folder
              └── wsgi_fix.config
```

You should run this application in Python 3.x and, even better, in a virtual environment so you can isolate exactly what is needed to run the web application from what you already have installed on your machine.

Make sure your command window is pointing to the "**web-application**" folder for this chapter and start a virtual environment. Start a virtual environment name "**willigolf-tomorrow**" to insure we're in Python 3 and to install all required libraries (Listing 7-17).

***Listing 7-17.*** Starting a Virtual Environment

```
$ python3 -m venv willigolftomorrow
$ source willigolftomorrow/bin/activate
```

Then install all the required Python libraries by running the "**pip install -r**" command (Listing 7-18).

***Listing 7-18.*** Install Requirements

```
$ pip3 install -r requirements.txt
```

Next you have to open "**application.py**" and add your "**OpenWeatherMap**" API key. Look for "**<<YOUR_API_KEY>>**" in the **PlayGolf()** function (Listing 7-19).

***Listing 7-19.*** Adding Your "**OpenWeatherMap**" API Key

```
openweathermap_url = "http://api.openweathermap.org/data/2.5/forecast?q=" +
selected_location + "&mode=json&APPID=<<YOUR_API_KEY>>"
```

Once you have added your API key, you should be ready to run your local web application as per usual (Listing 7-20).

**Listing 7-20.** Take It for a Spin

```
$ python3 application.py
```

You should see something along the lines of Figure 7-6 (bummer, no golfing tomorrow at the North Pole).

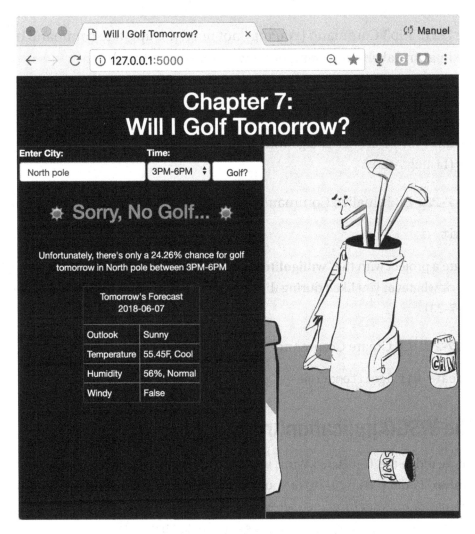

**Figure 7-6.** *Running the Flask application locally*

After you have confirmed that the web application is correctly running on your local machine, "**ctrl-c**" out of it but stay in the virtual environment session.

# Running on AWS Elastic Beanstalk

Still in the "**willigolftomorrow**" virtual environment session, install the latest "**awsebcli**" (Listing 7-21). We are skipping a few steps, as you should already have all the security layers set up by now (if not, refer back to Chapters 1 and 4).

*Listing 7-21.* Pip3 Command (you may not need the upgrade command because you are in a virtual environment, but it won't hurt anything)

```
$ pip3 install awscli --upgrade
$ pip3 install awsebcli --upgrade
```

Initializes the EB service and go with your usual settings as per previous AWS projects (Listing 7-22).

*Listing 7-22.* EB Initialize Command

```
$ eb init -i
```

Create a project with the "**willigolftomorrow**" name, say yes to "**SSH**," and go with defaults or whatever you liked during the AWS EB runs we did in the previous chapters (Listing 7-23).

*Listing 7-23.* EB Create Command

```
$ eb create willigolftomorrow
```

# Fix the WSGIApplicationGroup

Just like we did in the Top-Rated Wine, you need to create a new folder under the "**web-application**" folder (Listing 7-24) or you can use the one provided (this is a hidden folder that you may or may not be able to see–if you aren't sure, try creating the folder as per instructions and if it complains, that means you already have it).

***Listing 7-24.*** Create wsgi_fix File

```
$ mkdir .ebextensions
$ vi .ebextensions/wsgi_fix.config
```

This will create a new folder called "**.ebextensions**" and open a VI window (known in Unix speak as visual instrument), which is a simple text editor. Hit the "**i**" key to switch from read-only to "**insert**" mode and paste the following line at the end of the document (a text file of this fix is also included in the folder with the documents for this chapter). The process reading this file is very finicky; if there are added spaces or tabs, it will fail. Keep a close eye for any errors during the deployment process relating to the file and address accordingly (Listing 7-25).

***Listing 7-25.*** Add Fix

```
#add the following to wsgi_fix.config
files:
  "/etc/httpd/conf.d/wsgi_custom.conf":
    mode: "000644"
    owner: root
    group: root
    content: |
      WSGIApplicationGroup %{GLOBAL}
```

Now hit "**escape**" to exit "**insert**" mode and enter read-only mode, and type "**:wq**" to write and quit "**vi**" (Listing 7-26).

***Listing 7-26.*** Quit "**vi**"

```
:wq
```

# Take It for a Spin

Open web site with the "**open**" command (Listing 7-27).

***Listing 7-27.*** Taking the Cloud Version of the Web Application for a Spin

```
$ eb open willigolftomorrow
```

It may take a little bit of time to run the application the first time around and may even timeout. If that is the case, try "**eb open**" one more time (see Figure 7-7).

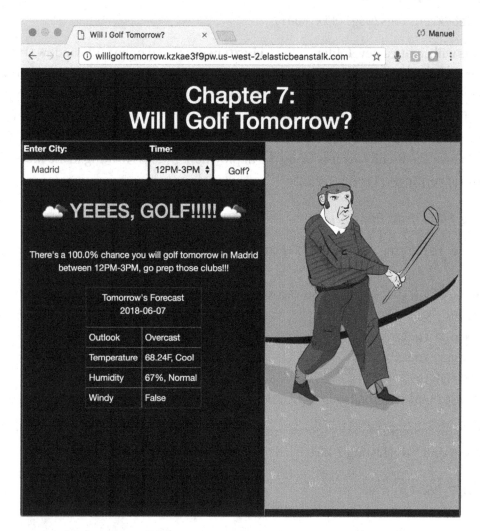

***Figure 7-7.*** *The "**Will I Golf Tomorrow?**" web application running on Elastic Beanstalk*

# Don't Forget to Turn It Off!

Finally, we need to terminate the Beanstalk instance as not to incur additional charges. This is an important reminder that most of these cloud services are not free (if it states that names do not match, try it again). It will ask you to confirm your decision (Listing 7-28).

***Listing 7-28.*** Terminate EB

```
$ eb terminate willigolftomorrow
```

In case you need to do any edits to the code, you simply perform them in your local directory and call the "**eb deploy**" function (Listing 7-29).

***Listing 7-29.*** To Deploy Fixes or Updates

```
$ eb deploy willigolftomorrow
```

Finally, once you've confirmed that your instance is terminated, you can get out of your virtual environment by calling the command (Listing 7-30).

***Listing 7-30.*** Kill the Virtual Environment

```
$ deactivate
```

It is always a good idea (essential idea really) to log into your account in the cloud and make sure everything is turned off (be warned: if you don't, you may get an ugly surprise at the end of the billing cycle). Log into your AWS account and make sure that your EC2 and Elastic Beanstalk accounts don't have any active services you didn't plan on having (Figures 7-8 and 7-9).

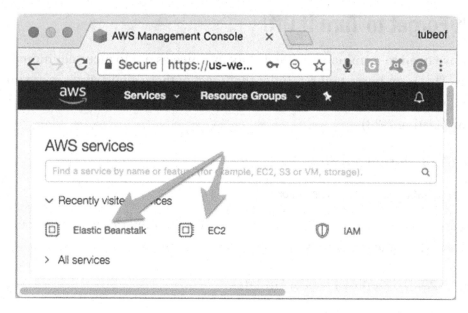

***Figure 7-8.*** *Checking for any active and unwanted instances on the AWS dashboard*

In case you see an instance that seems to keep coming back to life after each time you "**Delete application**," check under EC2 "**Load Balancers**" and terminate those first, then go back and terminate the rogue instance again (Figure 7-10).

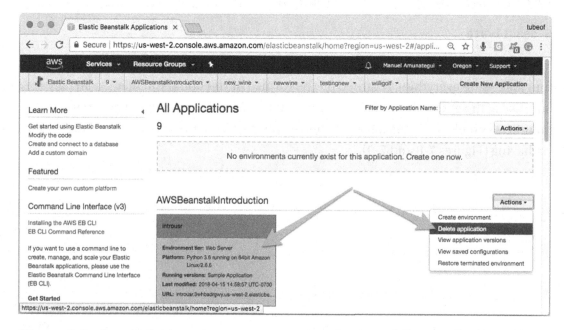

***Figure 7-9.*** *Locate the instance you want to terminate or delete, and select your choice using the "**Actions**" dropdown button*

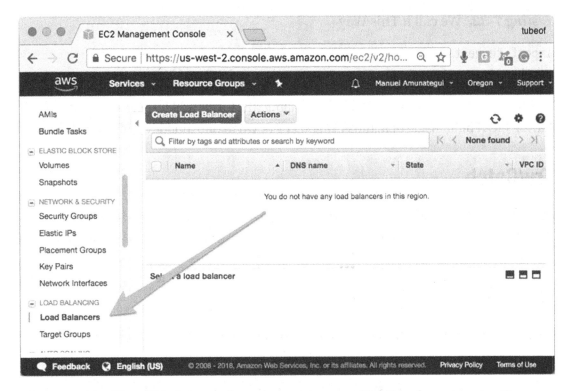

*Figure 7-10.* *"Load Balancers"* *can prevent an application from terminating; this can kick in if you inadvertently start multiple instances with the same name*

# Conclusion

On the surface, this may seem like a slight variation from what we've built in the past, with the exception of the dataset and model, but that really isn't the case. Let's take a look at some of the highlights.

## Accessing OpenWeatherMap Data

Unlike how we called the REST API service in Jupyter using urllib.request's "**urlopen**," in Flask we use the "**requests**" library (Listings 7-31 and 7-32).

*Listing 7-31.* Instead of

```
from urllib.request import urlopen
import json
weather_json = json.load(urlopen(openweathermap_url))
```

***Listing 7-32.*** We call it This Way

```
import requests
weather_json = requests.get(openweathermap_url).json()
```

This is a slightly more popular way of calling REST APIs and has the advantage of having JSON built inside of it, thus bringing us down to one function call instead of two.

# Try/Catch

We also use a try/catch (or in this case a try/error) to capture missing locations. It is critical that the application not crash on a user, and it is also important that an issue, whether an error or not, is handled properly. If you pass an unknown location to "**OpenWeatherMap**," it will return an error. This is easy to catch, leverage, and return an informative message to the user to try something else (Listing 7-33).

***Listing 7-33.*** Try/Catch to Handle Missing Locations

```
try:
    weather_json = requests.get(openweathermap_url).json()
except:
    # couldn't find location
    e = sys.exc_info()[0]
    message = "Cannot find that location, please try again"
```

Even though we capture the error message through the exception variable "**e**," we aren't doing anything with it here. I am leaving it in so you know how to access it, so you can extend it into your own applications via logging or smart displaying.

# Handling User-Entered-Data

This is an important topic that isn't really addressed in this book. Depending on the type of application you are building, you need to make sure that user-entered data won't harm your application, your data, or your hardware. Things like "**SQL injection**"[6] where a user can transit a system command through a text box to delete all files come to mind.

---

[6]https://en.wikipedia.org/wiki/SQL_injection

Instead, here we are making sure that the user-entered text will work with "**OpenWeatherMap**." If you take the raw http string and add spaces into it, it will fail to work (Listing 7-34).

***Listing 7-34.*** Handling Spaces in URLs

```
http://api.openweathermap.org/data/2.5/weather?q=New York City&appid...
```

One easy way of handling these issues is to use the "**quote_plus()**" function from the urllib.parse library. It will take any text input and render it HTML friendly so that it can be added to URLs without interfering with HTML commands. Let's look at an example (Listing 7-35).

***Listing 7-35.*** Handling Spaces in URLs

```
import urllib.parse
urllib.parse.quote_plus('New York City!')

'New+York+City%21'
```

This is easily extended to our Flask script by adding right after the "**request.form**" call and filtering the user data through it before proceeding further (Listing 7-36).

***Listing 7-36.*** Handling Spaces in URLs

```
selected_location = request.form['selected_location']
selected_location = urllib.parse.quote_plus(selected_location)
```

# Interactive Drawing Canvas and Digit Predictions Using TensorFlow on GCP

Let's build an interactive drawing canvas to enable visitors to draw and predict digits using TensorFlow image classification on Google Cloud.

Be forewarned, this is such a fun and interactive chapter that I ended up wasting too much time playing with the final product (Figure 8-1). This is one of the inherent risks of creating web applications using machine learning!

© Manuel Amunategui, Mehdi Roopaei 2018
M. Amunategui and M. Roopaei, *Monetizing Machine Learning*, https://doi.org/10.1007/978-1-4842-3873-8_8

***Figure 8-1.*** *The final web application for this chapter*

Here, we're going to leverage the awesome power of TensorFlow[1] to model the famous MNIST database. Unless you've been living under a rock, you've most likely heard of both (and if you haven't, don't worry, you will by the end of this chapter). The final web application will have a canvas to allow visitors to draw a digit between 0 and 9 with their mouse or finger and have our trained TensorFlow model predict it.

---

[1]https://www.tensorflow.org/

264

**Note** Download the files for Chapter 8 by going to `www.apress.com/`
`9781484238721` and clicking the source code button. Open Jupyter notebook
"**chapter8.ipynb**" to follow along with this chapter's content.

# The MNIST Dataset

The MNIST database contains 60,000 training images and 10,000 testing images. It's
the "**Hello World**" of image recognition classification. It is made up of single digits
between "**0**" and "**9**" written by both high school students and employees from the US
Census Bureau. The best way to understand the data is to take a look at a few examples.
Download the files for this chapter into a directory called "**chapter-8**" and open the
Jupyter notebook to follow along. When you install Tensorflow, you will have the
ability to download the MNIST directly from the "**input_data()**" function within the
"**tensorflow.examples.tutorials.mnist**" library. This will make training our model that
much easier, as they have already split the data into training and testing sets. Let's load
MNIST in memory and pull out a few samples (Listing 8-1).

*Listing 8-1.* Loading MNIST

**Input:**

```
mnist = input_data.read_data_sets("MNIST_data/", one_hot=True)
```

**Output:**

```
Extracting MNIST_data/train-images-idx3-ubyte.gz
Extracting MNIST_data/train-labels-idx1-ubyte.gz
Extracting MNIST_data/t10k-images-idx3-ubyte.gz
Extracting MNIST_data/t10k-labels-idx1-ubyte.gz
```

This will automatically download and unpack four files: two sets of images and
two sets of corresponding labels. Let's open a couple of digits and labels. We'll start by
looking at the very first image in the training dataset and pull the corresponding label
(Listing 8-2 and Figure 8-2).

***Listing 8-2.*** Viewing Digit

```
import matplotlib.pyplot as plt
first_digit = mnist.train.images[0]
first_digit = np.array(first_image, dtype='float')
first_digit = first_digit.reshape((28, 28))
plt.imshow(sample_digit)
plt.show()
```

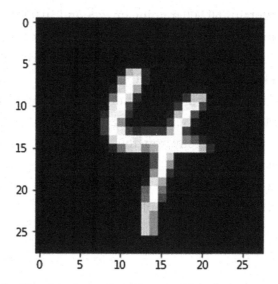

***Figure 8-2.*** *A Matplotlib visual render of one of the digits in MNIST*

We can also see the corresponding label; the format is an array of 1ten0 digits that each represents a value from 0 to 9 (Listing 8-3).

***Listing 8-3.*** Viewing Digit

**Input:**

```
mnist.train.labels[0]
```

**Output:**

```
array([ 0.,  0.,  0.,  0.,  1.,  0.,  0.,  0.,  0.,  0.])
```

By using an "**argmax()**" function, we can get the index of the largest value and, as they are conveniently sorted in ascending order, we automatically get the digit in question (Listing 8-4).

**Listing 8-4.** Listing Digit

**Input:**

```
np.argmax(mnist.train.labels[0])
```

**Output:**

4

And just for kicks, we'll use another way of sifting through the data by using the built-in "**next_batch()**" function that we will rely on later to feed the data into our TensorFlow model for training (Listing 8-5 and Figure 8-3).

**Listing 8-5.** Viewing Digit

```
batch = mnist.train.next_batch(1)
sample_digit = batch[0]
sample_digit = sample_digit.reshape(28, 28)
plt.imshow(sample_digit)
plt.show()
```

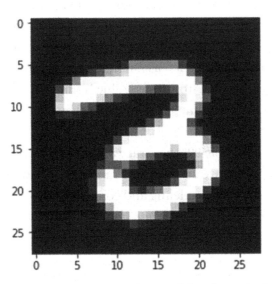

**Figure 8-3.** A Matplotlib visual render of one of the digits in MNIST

The digits are in gray scale and are all 28 by 28 pixels; there really isn't much more to say about the data, as it is self-explanatory.

# TensorFlow

TensorFlow is an open-source library made available by the kind folks at Google and is designed for high-performance numerical computation. It uses a data flow graph to represent mathematical operations that can then easily be computed on local or distributed devices. It has plenty of functionality for computation, number crunching, and normal to deep modeling. It was released under the Apache 2.0 Open Source License in November of 2015.

There is so much material out there on this topic that I will not rehash the subject but instead dive right into our task at hand: modeling handwritten digits! So, go ahead and download the files for this chapter into a folder called "**chapter-8**" if you haven't already done so. Open up the Jupyter notebook to follow along.

# Modeling with TensorFlow and Convolutional Networks

The MNIST dataset has probably been modeled with every single model on earth,[2] but a powerful and relatively easy one to use is convolutional networks known as "**CNN**"s or "**Covnets.**" This is an extremely powerful approach that can be as easy or as complicated as you want it to be. They were originally designed to model images but have proved to be very useful in many other areas such as natural language processing and time-series modeling. We'll leverage the code from TensorFlow's suite of tutorials entitled: "**Deep MNIST for Experts.**"[3] It isn't the simplest model that they offer, but it is still considered an introductory level approach.

The model gets an incredible 99% accuracy at classifying handwritten digits. This is even more interesting when we contrast it to Yann LeCun's journey with this dataset. He is one of the fathers of vision modeling and convolutional neural networks and the Director of AI research at Facebook. He benchmarked this data over a few decades and

---

[2]http://yann.lecun.com/exdb/mnist/
[3]https://tensorflow.org/versions/r1.1/get_started/mnist/pros

worked closely at increasing the modeling recognition accuracy from 12% all the way up to a tiny fraction of a percent. Even more incredible is that today we can open up a tutorial on this topic and get this incredible score with fewer than 50 lines of code.

Let's take a brief look at the model we will use for this web application. Here are some of the highlights from the tutorial (see the full tutorial for more details, at https://www.tensorflow.org/versions/r1.1/get_started/mnist/pros).

# Placeholders (tf.placeholder)

These are conduits for our image and label data streams. This is an important concept in TensorFlow, where you build a functioning graph before you feed any actual data into it (Listing 8-6).

*Listing 8-6.* Code Input

```
x = tf.placeholder(tf.float32, shape=[None, 784])
y_ = tf.placeholder(tf.float32, shape=[None, 10])
```

### Variables (tf.Variable)

Variables are made to hold values and you can initialize them with actual values (Listing 8-7).

*Listing 8-7.* Function "**weight_variable()**"

```
def weight_variable(shape):
  initial = tf.truncated_normal(shape, stddev=0.1)
  return tf.Variable(initial)
```

# Building Modeling Layers

We can define our specialized network layers as functions and be able to reuse them however many times we want, depending on the complexity of the neural network (Listing 8-8).

*Listing 8-8.* Abstracting Functions "**conv2d()**" and "**max_pool_2x2()**".

```
def conv2d(x, W):
  return tf.nn.conv2d(x, W, strides=[1, 1, 1, 1], padding='SAME')

def max_pool_2x2(x):
  return tf.nn.max_pool(x, ksize=[1, 2, 2, 1],
                        strides=[1, 2, 2, 1], padding='SAME')
```

We then can create as many layers as needed by calling the conv2d() and max_pool_2x2() (Listing 8-9).

*Listing 8-9.* Creating Layers

```
h_conv1 = tf.nn.relu(conv2d(x_image, W_conv1) + b_conv1)
h_pool1 = max_pool_2x2(h_conv1)
```

# Loss Function

The original tutorial model uses the "**tf.nn.softmax_cross_entroy_with_logits()**" function, which is a mouth full. Softmax returns a probability over $n$ classes that sums to 1, and cross entropy handles data from different distributions (Listing 8-10).

*Listing 8-10.* Getting the Cross Entropy

```
cross_entropy = tf.reduce_mean(tf.nn.softmax_cross_entropy_with_
logits(labels=y_, logits=y))
```

The documentation states that it will be deprecated in a later version (and get used to that–it happens in TensorFlow and most libraries in Python), so we'll use a similar approach that is more generic (see the Jupyter notebook for details).

# Instantiating the Session

Once we are ready to run our model, we instantiate the session with the "**sess.run**" command. This turns on all the graphs we set up earlier (Listing 8-11).

*Listing 8-11.* Firing-Up the Session

```
sess.run(tf.global_variables_initializer())
```

# Training

We set an arbitrary number of loops, in this case 1,000, and feed the data as batches into our model (Listing 8-12).

*Listing 8-12.* Setting Model Iterations

```
for _ in range(1000):
    batch = mnist.train.next_batch(100)
    train_step.run(feed_dict={x: batch[0], y_: batch[1]})
```

# Accuracy

In order to not fly blind, we add an accuracy measure to monitor how well our model is training that will print out the progress every 100 steps (Listing 8-13).

*Listing 8-13.* Accessing the Accuracy During Training

```
print(step, sess.run(accuracy, feed_dict={x: mnist.test.images, y_: mnist.
test.labels, keep_prob: 1.0}))
```

There's plenty more going on in this script, so please go over to the actual TensorFlow Tutorial, as it's well worth it if you're interested in deep learning (https://www.tensorflow.org/versions/r1.2/get_started/mnist/pros).

# Running the Script

Running it 2,000 times gives us a decent score but leaves plenty of room for improvement (Listing 8-14).

*Listing 8-14.* Accuracy Output During 2,000 Iterations

```
0  0.0997
100  0.8445
200  0.905
300  0.9264
400  0.9399
500  0.9492
```

```
600  0.9509
700  0.9587
800  0.9596
900  0.9623
1000  0.9668
1100  0.9688
1200  0.9706
1300  0.9719
1400  0.9683
1500  0.9708
1600  0.9754
1700  0.9751
1800  0.9753
1900  0.9738
2000  0.9776
```

If you keep modeling over 20,000 steps like the tutorial suggests, you can achieve that elusive 99.2%! But this can take up to 30 minutes depending on your machine (if you have a GPU, you will zip right through it; Listing 8-15).

***Listing 8-15.*** Accuracy Output During 20,000 Iterations

```
      . . .
19300  0.9935
19400  0.9928
19500  0.9926
19600  0.9923
19700  0.9932
19800  0.993
19900  0.9926
20000  0.9927
```

Once the model has finished training, we save it to file so we can run predictions at a later time (and more importantly in our web application; Listing 8-16).

***Listing 8-16.*** Saving the Trained Weights

```
saver = tf.train.Saver()
save_path = saver.save(sess, save_file)
print ("Model saved in file: ", save_path)
```

# Running a Saved TensorFlow Model

This ability of instantiating trained models is an important aspect of applied modeling and building commercial pipelines. The model we are developing here doesn't take that long to train (2 to 30 minutes depending on the number of steps you use), but no user would be willing to wait that long on a web page if you had to train it on each request. The good news is that it is easy to save and reload a trained model. The key is to call the "**save()**" function of "**tf.train.Saver**" before exiting the TensorFlow session (Listing 8-17).

***Listing 8-17.*** Saving Model

**Input**:

```
saver = tf.train.Saver()
save_path = saver.save(sess, save_file)
print ("Model saved in file: ", save_path)
```

**Output**:

```
Model saved in file:   /Users/manuel/apress-book-repository/chapter-8/model.ckpt
```

Next time you want to run the trained model, all you have to do is set up all your graph variables and call the "**restore()**" function of "**tf.train.Saver**" in a TensorFlow session (Listing 8-18).

***Listing 8-18.*** Restoring a Saved Model

**Input**:

```
saver = tf.train.Saver()
with tf.Session() as sess:
        sess.run(tf.global_variables_initializer())
        saver.restore(sess, save_file)
        print("Model restored.")
```

**Output:**

```
INFO:tensorflow:Restoring parameters from /Users/manuel/apress-book-
repository/chapter-8/model.ckpt
Model restored.
```

# Save That Model!

You will find an already trained model ready to go in the downloads for this chapter. If you want to use your own, see the Jupyter notebook and save the trained weight files (that's how I did it). You will end up with three files that represent the saved mode and that are needed in order to load the model. TensorFlow is smart enough to load the latest version from the files you give (you could store multiple checkpoint files for example and it will use the latest one; Listing 8-19).

*Listing 8-19.* Pretrained Model Files in Downloads for This Chapter if You Don't Want to Train It Yourself

```
checkpoint
model.ckpt.data-00000-of-00001
model.ckpt.index
```

# Drawing Canvas

The canvas is a critical part of the application, as it will allow anybody to get a taste, and a fun one at that, in understanding MNSIT, character-recognition, and convolutional modeling. These are usually difficult concepts associated with advanced classes and industrial modeling tools, but they can be fun too! The canvas is part of HTML5 (for more information on this cool feature, see https://www.w3schools.com/html/html5_ canvas.asp) and allows the creation of a space where a user can interact and create drawings on a web page (Figure 8-4).

*Figure 8-4.* *Finger painting with HTML5 and the <canvas> tag*

Using this approach, we can take the content the user drew on the canvas and translate it into an image that our TensorFlow model can ingest and attempt to predict.

# From Canvas to TensorFlow

This part isn't complicated but requires a few transformations, so hang on. When the visitor hits the "**Predict**" button, it calls the "**toDataURL()**" function of the canvas HTML5 control. This translates whatever data is contained within the canvas tags into text representation of the image in PNG format.

This is a concept we've seen before and will see again in this book. Remember Chapter 3? We relied on image data in text representation to easily pass it from server a client. In this case, we're doing it the other way around–client to server (Figure 8-5).

```
<img src="data:image/
png;base64,6BCOHDnyk+pt7/tycHD4Scd5UC+/
/DLeeust1NfX46uvvkJSUhJsbGyQnJzcKe2hJ09
wcDDWrl2rWtfyeba1tf3J9T+MOh4mIyMjnDt3Tl
. . .
nOzc3F22+/jW+++UZZ17V v2oHfv3hg/
fjwGDBiAxMRE3Lhxgz2CEluzZg3Cw8Mxbdo09O/
ARERERJJhACQiIiKSDAMgERERkWT+B5qsMW4gCB
j4AAAAElFTkSuQmCC">
```

Figure 8-5. *Image data represented as text*

---

**Note** Code partially based on a great snippet found at `https://stackoverflow.com/questions/2368784/draw-on-html5-canvas-using-a-mouse`. Whenever you have questions about coding or problems and need a solution, StackOverflow.com should be your first stop!

---

# Testing on New Handwritten Digits

This is a critical part of our pipeline (and web application). We need to be able to pass new handwritten digits to the model for prediction.

### Processing a Real Image

The difference between testing using the MNIST dataset and a real image is that the MNIST data has already been processed for us. We therefore need to apply the same processing on the new image, so it can be compatible with our trained model's. Imagine you create an image file with a digit; this is how you would pass it to the model. We leverage the PIL and NumPy Python libraries to perform most of the image processing (Listing 8-20).

*Listing 8-20.* Importing an Image

```
from PIL import Image
img = Image.open('my-own-4.png')
```

We resize it to the official 28 by 28 required pixel size. As we will be working with transparent images (only the number will show, not the background), we need to add a white background to comply with the trained MNIST data (Listing 8-21).

***Listing 8-21.*** Processing New Image

```
img = img.resize([28,28])
# add white background
corrected_img = Image.new("RGBA", (28, 28), "white")
# paste both images together
corrected_img.paste(img, (0,0), img)
```

Next, we cast the image into arrays, remove the extra color dimensions that we won't need here as we are working with black and white images, and finally invert the whole thing so that the empty pixels are zeros (Listing 8-22).

***Listing 8-22.*** Processing New Image

```
# remove color dimensions
corrected_img = np.asarray(corrected_img)
# remove color layers
corrected_img = corrected_img[:, :, 0]
# invert colors
corrected_img = np.invert(corrected_img)
```

Finally, we flatten the image from a matrix of 28 by 28 to a flat vector of size 784 and center the data between 0 and 1 instead of 0 and 255. That's it; it is now ready to be fed into our TensorFlow model for prediction (Listing 8-23 and Figure 8-6).

***Listing 8-23.*** Flattening the Data

```
corrected_img = corrected_img.reshape([784])
# center around 0-1
img = np.asarray(corrected_img, dtype=np.float32) / 255.
```

```
array([0.          , 0.          , 0.          , 0.          , 0.          ,
       0.          , 0.          , 0.          , 0.          , 0.          ,
       0.          , 0.          , 0.          , 0.          , 0.          ,
       0.          , 0.          , 0.          , 0.          , 0.          ,
       0.          , 0.          , 0.          , 0.          , 0.          ,
       0.          , 0.          , 0.          , 0.          , 0.          ,
       0.          , 0.          , 1.          , 1.          , 1.          ,
       1.          , 0.99607843, 0.          , 0.          , 0.          ,
       0.          , 0.38431373, 1.          , 1.          , 1.          ,
       1.          , 0.          , 0.          , 0.          , 0.          ,
       0.          , 0.          , 0.          , 0.          , 0.          ,
       0.          , 0.          , 0.          , 0.          , 0.          ,
```

*Figure 8-6.   Partial final output of the transformed image data ready for modeling*

# Designing a Web Application

We are now at the fun part of the chapter; we get to design our web application! We are going to keep things extremely simple. This is meant to be fun and intuitive, and by keeping the buttons and options to a minimum, will allow our visitor to immediately understand and interact with the tool. We'll add a central canvas in the middle, so the visitor can draw a digit between "**0**" and "**9**", and two buttons: one to predict the number and the other to clear the canvas. Finally, we'll also add a drop-down menu to control the thickness of the paint brush–that's it!

On the graphical end of things, we are using a large picture that we cut up into different sections: a top portion that contains the head of the fortune teller and two side portions that contain the arms. It is cut up in order to accommodate the drawing canvas in the center of the web application.

# Download the Web Application

Go ahead and download the code for this chapter if you haven't already done so, open a command line window, and change the drive to the "**web-application**" folder. It should contain the usual files along with our saved checkpoint files (Listing 8-24).

***Listing 8-24.*** Web Application Files

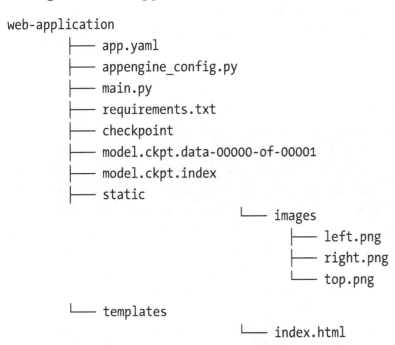

```
web-application
            ├── app.yaml
            ├── appengine_config.py
            ├── main.py
            ├── requirements.txt
            ├── checkpoint
            ├── model.ckpt.data-00000-of-00001
            ├── model.ckpt.index
            ├── static
                              └── images
                                        ├── left.png
                                        ├── right.png
                                        └── top.png

            └── templates
                              └── index.html
```

First, you will need to install TensorFlow on your Python 3.x instance (or install the requirements file in the next step). As usual, we'll start a virtual environment to segregate our Python library installs (Listing 8-25).

***Listing 8-25.*** Starting Virtual Environment and Install TensorFlow

```
$ python3 -m venv whatsmynumber
$ source whatsmynumber/bin/activate
$ pip3 install tensorflow
```

Then install all the required Python libraries by running the "**pip install -r**" command (Listing 8-26).

***Listing 8-26.*** Installing Requirements and Running Local Version

```
$ pip3 install -r requirements.txt
$ python3
```

Run the web application the usual way, and you should see the fortune teller appear. This can take a while to get started, depending on your computing muscle. Go ahead and take it for a spin, and make sure his predictions are worth his salt! (Figure 8-7)

***Figure 8-7.*** *Blank canvas of the **"What's my Number"** web application*

# Google Cloud Flexible App Engine

We'll use the Flexible App Engine in order to run the more demanding Python Libraries like TensorFlow and PIL. We will need to use a slightly more powerful instance in order to handle TensorFlow and our saved model. If you take a peek into the "**app.yaml**" file under the "**web-application**" folder for this chapter, you will see that we upped the memory and disk size (Listings 8-27 and 8-28).

***Listing 8-27.*** Now We're Using the Larger Setup

```
resources:
  cpu: 1
  memory_gb: 3
  disk_size_gb: 20
```

***Listing 8-28.*** Previously We Ran with Fewer Resource Settings

```
resources:
  cpu: 1
  memory_gb: 0.5
  disk_size_gb: 10
```

A word to the wise: the bigger the machine you provision, the bigger the charge. So, make sure you terminate your instance after you're done with it!

# Deploying on Google App Engine

By now you should have some experience with the Google Flexible App Engine, so this will be a quick guide to get this web application up and running.

## Step 1: Fire Up Google Cloud Shell

Log into your instance of Google Cloud and create or select the project in which you want your App Engine to reside (if you don't have one, you will be prompted to create one–see Creating and Managing Projects[4]). Start the cloud shell command line tool by clicking on the upper right caret button. This will open a familiar-looking command line window in the bottom half of the GCP dashboard (Figure 8-8).

---

[4]https://cloud.google.com/resource-manager/docs/creating-managing-projects

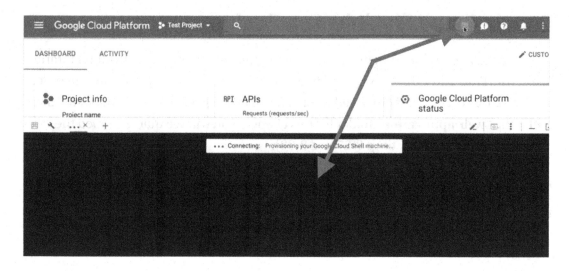

**Figure 8-8.** *Accessing the Google Cloud shell*

## Step 2: Zip and Upload All Files to the Cloud

Zip the files in the "**web-application**" folder but don't zip the virtual environment folder "**whatsmynumber**" as it's not needed (Figure 8-9).

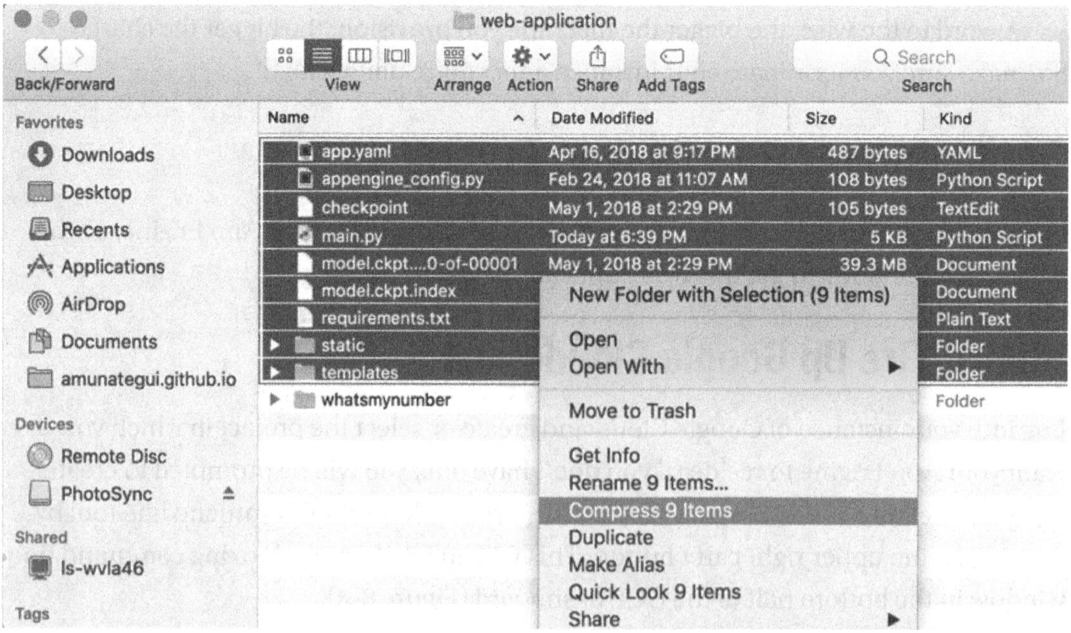

**Figure 8-9.** *Zipping web application files for upload to Google Cloud*

Upload it using the "**Upload file**" option (this is found on the top right side of the shell window under the three vertical dots; Figure 8-10).

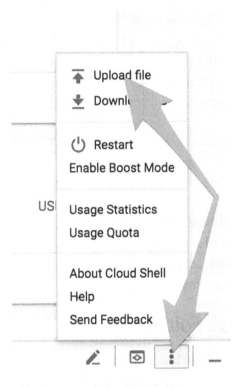

*Figure 8-10.*  *Uploading files via Google Cloud shell*

# Step 3: Create Working Directory on Google Cloud and Unzip Files

Once the file is successfully uploaded, create a new directory named "**chapter-8**" (Listing 8-29).

*Listing 8-29.*  Creating the Cloud Directory

```
$ mkdir chapter-8
$ cd chapter-8
```

Transfer all of the zip Archive into the new folder and unzip it (Listing 8-30).

***Listing 8-30.*** Moving and Unzipping Web Application Files

```
$ mv ../Archive.zip Archive.zip
$ unzip Archive.zip
```

Your folder on Google Cloud should look something like Listing 8-31.

***Listing 8-31.*** Confirming That All Files Are Correctly Uploaded by Running the '**ls**' Command

**Input**:

```
amunategui@cloudshell:~/chapter-8 (apt-memento-192717)$ ls
```

**Output**:

```
appengine_config.py   main.py                        static
app.yaml              model.ckpt.data-00000-of-00001  templates
Archive.zip           model.ckpt.index
checkpoint            requirements.txt
```

# Step 4: Creating Lib Folder

So, run the following command to install all the needed additional libraries to the lib folder. When you deploy your web app, the lib folder will travel along with the needed libraries (Listing 8-32).

***Listing 8-32.*** Installing All Needed Python Libraries into the "**lib**" Folder

```
$ sudo pip3 install -t lib -r requirements.txt
```

# Step 5: Deploying the Web Application

Finally, deploy it to the world with the "**gcloud app deploy**" command (Listing 8-33).

***Listing 8-33.*** Deploying Web Application

```
$ gcloud app deploy app.yaml
```

That's it! Sit back and let the tool deploy the web site. This is the Flexible App Engine, so it can take up to 30 minutes to be fully deployed. Once it is done setting everything up, it will offer a clickable link to jump directly to the deployed web application (Listing 8-34).

**Listing 8-34.** You Can Also Get There with the Following Command

```
$ gcloud app browse
```

Enjoy the fruits of your labor, and make sure to experiment with it by drawing recognizable and nonrecognizable digits (Figure 8-11).

**Figure 8-11.** *The web application on Google Cloud*

# Troubleshooting

There will be cases where you will have issues and the Google Cloud logs will be your best friends. You can easily reach them either directly in the Google Cloud dashboard or by calling the logs URL (Listing 8-35).

***Listing 8-35.***  Logs URL

```
https://console.cloud.google.com/logs
```

Or you can stream the log's tail by entering in the cloud shell the following command in Listing 8-36.

***Listing 8-36.***  Viewing Logs in Terminal Window

```
$ gcloud app logs tail -s default
```

# Closing Up Shop

One last thing before we conclude this chapter: don't forget to stop or delete your App Engine Cloud instance. Even if you are using free credits, the meter is still running and there is no need to waste money or credits.

Things are a little different with the Flexible App Engine over the Standard one, as the Flexible costs more money. So, it is important to stop it if you aren't using it. Also, this can all be conveniently done via the Google Cloud dashboard.

Navigate to App Engine, then Versions. Click on your active version and stop it (Figure 8-12). If you have multiple versions, you can delete the old ones; you won't be able to delete the default one, but stopping it should be enough (if you really don't want any trace of it, just delete the entire project).

***Figure 8-12.***  *Stopping and/or deleting your App Engine version*

That's it! Don't forget to deactivate the virtual environment if you are all done (Listing 8-37).

***Listing 8-37.***  Deactivating the vVirtual Environment

```
$ deactivate
```

# Conclusion

In this chapter, we got to try out some new and old technology.

## HTML5 <canvas> tag

The new "**canvas**" tag in HTML5 is a lot of fun and opens all sorts of new ways of inputting data to devices and into Flask.

## TensorFlow

Working with TensorFlow and having the ability of loading pretrained models into Flask is big. Training these models takes a lot of processing and time, so being able to leverage already trained models allows implementing deep models into web application in a heartbeat. One word of caution here is that you will need a machine commiserate to your TensorFlow needs, the basic simple setup we used in previous chapters won't do the trick (see the app.yaml files for required settings).

## Design

Splitting background image into four sections is a fun an easy way of interlacing large images with input controls, such as the canvas in our case (Figure 8-13).

*Figure 8-13.  Adding extra "**cellpadding**" in the front-end design to see the splits needed to accommodate the drawing canvas*

# CHAPTER 9

# Case Study Part 2: Displaying Dynamic Charts

Displaying dynamic stock charts on PythonAnywhere.

Let's add a few more features to the original case study web application (make sure you are familiar with **"Running on PythonAnywhere"** in Chapter 5). The idiom "a picture is worth a thousand words" is absolutely applicable here and by offering visual chart support of the price action surrounding the recommended pair trade, will go a long way to help the user evaluate things. We will add three charts (Figure 9-1), a chart for each stock in play and the differential showing the percent-change, cumulative sum subtraction between the strong stock minus the weak stock.

© Manuel Amunategui, Mehdi Roopaei 2018
M. Amunategui and M. Roopaei, *Monetizing Machine Learning*, https://doi.org/10.1007/978-1-4842-3873-8_9

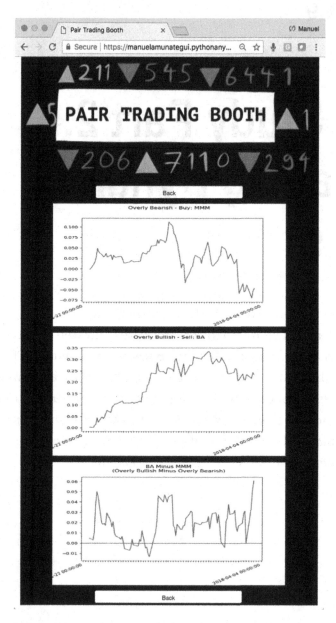

*Figure 9-1.*  *The final web application for this chapter*

The charts are created dynamically in Flask and transformed into PNG files for viewing. This is the same technique we used to build the Titanic web application that allows you to conveniently build dynamic plots using the Matplotlib library and translate them from binary to text for web publishing. This allows the creation of dynamic images at will without having to save anything to file.

---

**Note**    Download the files for Chapter 9 by going to `www.apress.com/`
`9781484238721` and clicking the source code button. Open Jupyter notebook
"**chapter9.ipynb**" to follow along with this chapter's content.

---

# Creating Stock Charts with Matplotlib

Let's get started; download the files for this chapter into a folder called "**chapter-9**" and open up the Jupyter notebook to follow along. This is a method we've used in previous chapters and that we will continue to use. This approach allows us to create images using Matplotlib then translate them into strings, so they can be dynamically fed and understood by an HTML interpreter.

In the following simplified code snippet, we create an image in Python using the "**matplotlib.pyplot**" library (Listing 9-1).

*Listing 9-1.*  Creating Encoded String Images

```
import matplotlib.pyplot as plt
fig = plt.figure()
plt.bar(y_pos, performance, align='center', color = colors, alpha=0.5)
img = io.BytesIO()
plt.savefig(img, format='png')
img.seek(0)
plot_url = base64.b64encode(img.getvalue()).decode()
```

Then the variable "**plot_url**" can be injected into the HTML code using Flask Jinja2 template notation as such (Listing 9-2).

*Listing 9-2.*  Injecting Dynamic Images Using Flask and Jinja2

```
model_plot = Markup('<img src="data:image/png;base64,{}">'.format(plot_url))

...

<div>{{model_plot}}</div>
```

And if you look at the HTML source output, you will see that the HTML image tag is made of an enormous string output (drastically truncated here) that the interpreter knows to translate into an image (Figure 9-2).

```
<img src="data:image/
png;base64,6BCOHDnyk+pt7/tycHD4Scd5UC+/
/DLeeust1NfX46uvvkJSUhJsbGyQnJzcKe2hJ09
wcDDWrl2rWtfyeba1tf3J9T+MOh4mIyMjnDt3Tl
...
nOzc3F22+/jW+++UZZ17V v2oHfv3hg/
fjwGDBiAxMRE3Lhxgz2CEluzZg3Cw8Mxbdo090/
ARERERJJhACQiIiKSDAMgERERkWT+B5qsMW4gCB
j4AAAAAElFTkSuQmCC">
```

***Figure 9-2.*** *Image transformed into string of characters*

# Exploring the Pair-Trading Charts

Go ahead and download the files for this chapter into a folder called "**chapter-9.**" Open up the Jupyter notebook to follow along. You will see a lot of repeated code in this notebook in order to load and process all the financial data needed to get to the charting part.

We're going to offer our visitors three charts, a chart for each stock in play and the differential showing the percent-change, cumulative sum subtraction between the strong stock minus the weak stock. The first half of the Jupyter notebook is a repeat of Chapter 9. We need to keep repeating the code, as we're building these charts onto the previous foundation. Scroll down to "**Part 2.**"

Let's run one chart through the different steps needed to get it into a textual format to be properly served from a web server to a web client. We build it just like we would build any chart in Matplotlib. We create the subplots to initiate a plot object, pass it our financial data, create a title, and rotate the x-axis date field (Listing 9-3 and Figure 9-3).

***Listing 9-3.*** Plotting Price Difference

```
fig, ax = plt.subplots()
ax.plot(temp_series1.index , long_trade_df)
plt.suptitle('Overly Bearish - Buy: ' + weakest_symbol[0])
```

```
# rotate dates
myLocator = mticker.MultipleLocator(2)
ax.xaxis.set_major_locator(myLocator)
fig.autofmt_xdate()
```

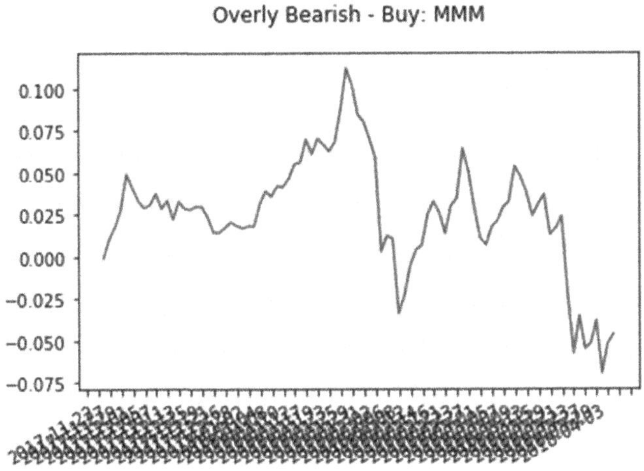

***Figure 9-3.*** *Raw plot with x-axis rotated but still unreadable*

We also do a little extra work to properly format labels on the x-axis. The rotation does help, but we need to prune out some of the dates. There are many ways of approaching this, but we will remove all dates except the first and last ones (to be specific, we are going to keep the second date and the second-to-last date only; Listing 9-4).

***Listing 9-4.*** Fixing Label to Only Show First and Last Date

**Input:**

```
labels = [" for item in ax.get_xticklabels()]
labels[1] = temp_series1.index[0]
labels[-2] = temp_series1.index[-1]
labels = [" for item in ax.get_xticklabels()]
labels[1] = temp_series1.index[0]
labels[-2] = temp_series1.index[-1]
ax.set_xticklabels(labels)
```

**Output:**

```
[Text(0,0,"),
 Text(0,0,'2017-11-22'),
 Text(0,0,"),
 Text(0,0,"),
 Text(0,0,"),
 Text(0,0,"),
 Text(0,0,"),


 ...
 Text(0,0,"),
 Text(0,0,"),
 Text(0,0,"),
 Text(0,0,"),
 Text(0,0,"),
 Text(0,0,"),
 Text(0,0,"),
 Text(0,0,"),
 Text(0,0,"),
 Text(0,0,'2018-04-04'),
 Text(0,0,")]
```

This yields a much more readable and breezy chart showing only two titled dates, the second and second-to-last (Figure 9-4).

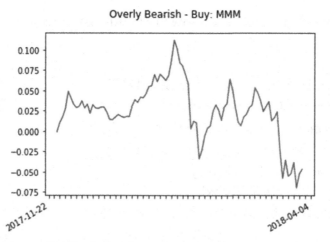

*Figure 9-4.* *A more readable chart showing only the extreme dates*

Finally, we translate the image date into text using the "**BytesIO()**" function from the io library. We call the "**savefig()**" function from Matplotlib to specify the output format (though this isn't being saved to file) and finally call the "**b64encode()**" function of the base64 library. At this point, our "**plot_url**" variable holds a textual representation of our image that we can easily pass using Flask to a web client. This is a very clever way of creating dynamic images in a scalable and session-free manner (Listing 9-5).

***Listing 9-5.*** Sampling Encoded Output

**Input**:

```
img = io.BytesIO()
plt.savefig(img, format='png')
img.seek(0)
plot_url = base64.b64encode(img.getvalue()).decode()
plot_url
```

**Output**:

'iVBORwOKGgoAAAANSUhEUgAAAbAAAAEgCAYAAADVKCZpAAAABHNCSVQICAgIfAhkiAAAAlw
SFlzAAALEgAACxIBOt1+/AAAADlORVhoU29mdHdhcmUAbWFOcGxvdGxpYB2ZXJzaW9u
IDIuMi4yLCBodHRwOi8vbWFOcGxvdGxpYi5vcmcvhp/UCwAAIABJREFUeJzs3Xl4lOXV+PHvJJP
Jvu9MAlkmhBCWAImAK7JFUWMVyiaCVqvaW1vfWrGtYtVX5bWtb1+r/hS1EhSNiEoQZ
ccFRIEgwUBYEkgg+77vs/...'

# Designing a Web Application

Go ahead and download the code for this chapter if you haven't already done so; open a command line window and change the drive to the "**web-application**" folder. It should contain the following files as shown in Listing 9-6.

***Listing 9-6.*** Web Application Files

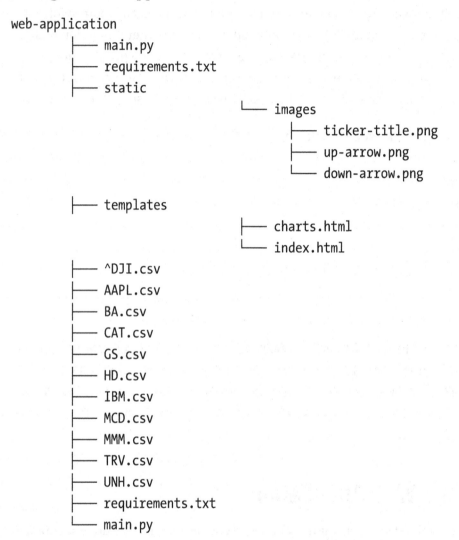

```
web-application
        ├── main.py
        ├── requirements.txt
        ├── static
                        └── images
                                ├── ticker-title.png
                                ├── up-arrow.png
                                └── down-arrow.png

        ├── templates
                        ├── charts.html
                        └── index.html
        ├── ^DJI.csv
        ├── AAPL.csv
        ├── BA.csv
        ├── CAT.csv
        ├── GS.csv
        ├── HD.csv
        ├── IBM.csv
        ├── MCD.csv
        ├── MMM.csv
        ├── TRV.csv
        ├── UNH.csv
        ├── requirements.txt
        └── main.py
```

As usual, we'll start a virtual environment to segregate our Python library installs and create the "**requirements.txt**" file if needed (Listing 9-7).

***Listing 9-7.*** Starting Virtual Environment

```
$ python3 -m venv pairtrading
$ source pairtrading/bin/activate
```

Then install all the required Python libraries by running the "**pip3 install -r**" command (Listing 9-8).

***Listing 9-8.*** Installing Requirements and Run a Local Version of the Web Application

```
$ pip3 install -r requirements.txt
$ python3 main.py
```

Run the web application in the usual manner and make sure it works as advertised. This may be a little slow the first time running but should get more nimble thereafter. Also try the various options on the page to make sure everything works as it should.

# Mobile Friendly with Tables

We are keeping the web application mobile friendly by plotting the charts in table cells to ensure that they resize properly regardless of the screen size. We use percentage sizes in the width and height parameters of the "**<img>**" tag (you can cap the height if you want, but we need the width to be a percentage if we want it to adjust automatically) and wrap each image in a "**<td>**" cell (Listing 9-9).

***Listing 9-9.*** Friendly Tables

**Input**:

```
chart1_plot = Markup('<img style="padding:1px; border:1px solid #021a40;
width: 80%; height: 300px" src="data:image/png;base64,{}">'.format(plot_url))
```

**Output**:

```
<table>
...
    <tr>
        <td align="center">
            {{chart1_plot}}
        </td>
    </tr>
    <tr>
        <td align="center">
            {{chart2_plot}}
```

```
                </td>
        </tr>
        <tr>
                <td align="center">
                        {{chart_diff_plot}}
                </td>
        </tr>
...
</table>
```

This is an easy way of leveraging the flexibility of images in HTML to resize according its holding frame (Figure 9-5).

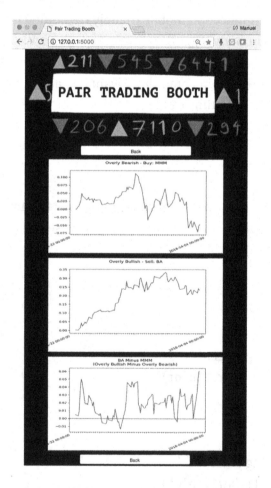

***Figure 9-5.***  *The narrow view of the stock chart automatically resizes according to the client's web page size*

# Uploading our Web Application to PythonAnywhere

Let's upload our updated code to PythonAnywhere. This chapter's project is a continuation of Chapter 5–please tackle the case studies in that order, as we are going to build upon our previous PythonAnywhere work. Log in to your PythonAnywhere account and find the folder "**pair-trading-booth**" that we created previously. Click on the "**Files**" link in the top menu bar and enter the "**pair-trading-booth**" directory (Figure 9-6).

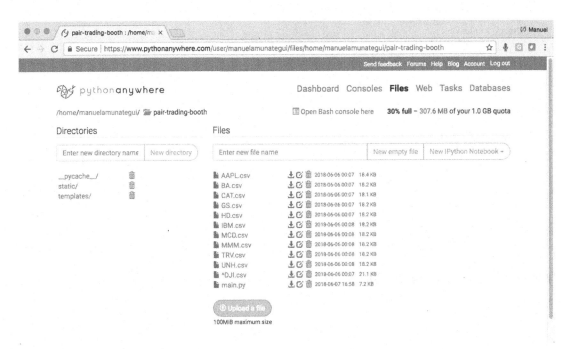

***Figure 9-6.***  *Our pair trading application on PythonAnywhere*

All the financial CSV files needed should already be there (if not, run through Chapter 5 again), and all we need to do is update the "**main.py**" and "**index.html**" files and add a new "**charts.html**" file to display our stock and derived charts. The best way to proceed is to simply open those files in a local editor and copy and paste the content into PythonAnywhere.

For example, let's update "**main.py**", open the file in your local editor and open the file in PythonAnywhere, then copy and paste the new version into your "**main.py**" file on PythonAnywhere. Don't forget to click the green "**Save**" button before moving on to the other files (Figure 9-7).

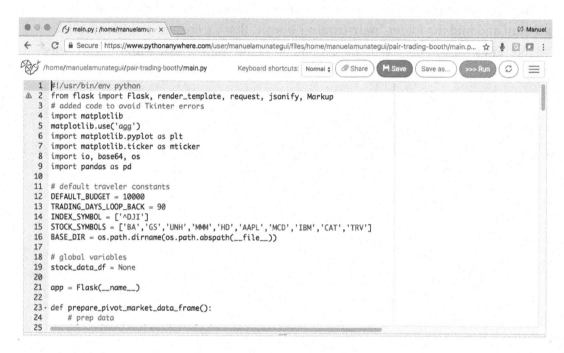

**Figure 9-7.** *Updating the "**main.py**" code base on PythonAnywhere to handle the creation of dynamic charts*

Proceed in the same way in the templates folder for file "**index.html**" and also create a new HTML file called "**charts.html**" (Figure 9-8).

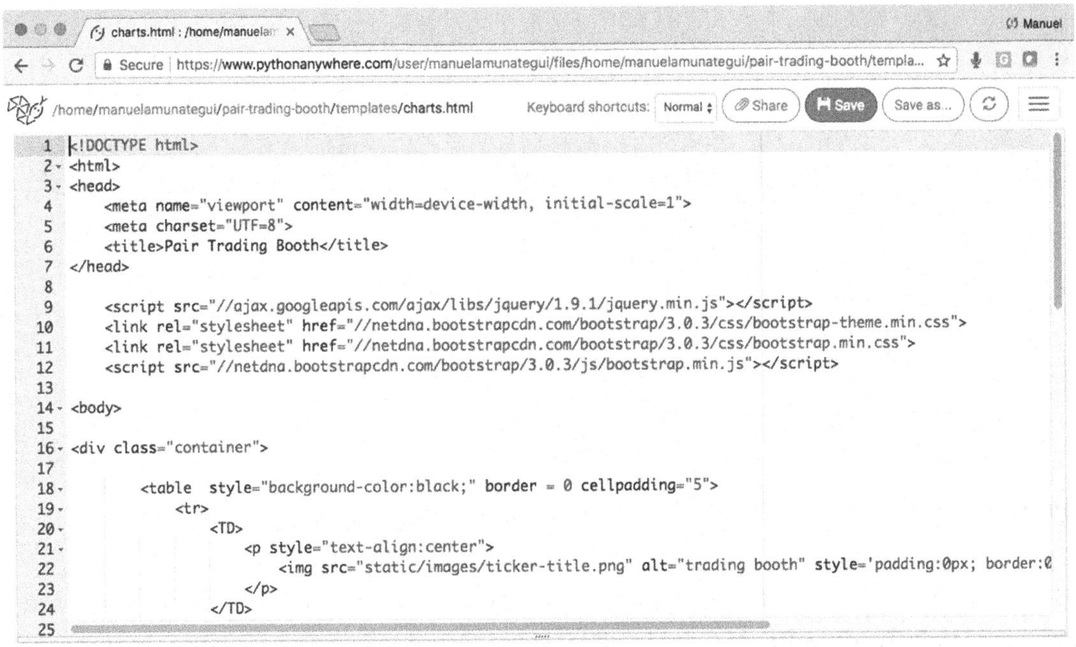

*Figure 9-8.*  *Creating a new file called "**charts.html**" on PythonAnywhere to handle the creation of dynamic charts*

Now you are ready to refresh your web service and fire up the web application. Click on the "**Web**" menu tab and hit the big, green button to reload the application (Figure 9-9).

Figure 9-9.  *Hit the "**Reload**" button to update your web server*

If you enter the URL of your PythonAnywhere site into your browser, you should see the new "**Pair Trading Booth**" site in all its glory. Go ahead and take it through its paces by clicking on the "**Get Trade**" and "**View Charts**" buttons (Figure 9-10).

Figure 9-10.  *The new "**Pair Trading Booth**" site enhanced with charts*

# Conclusion

In this chapter, we took a second pass at the "**Pair Trading Booth**" web application and enhanced it with charting capabilities. Though we haven't introduced any new technology, we successfully enhanced it with extra features while preserving its mobile viewing capabilities. We used table and dynamic image sizing using percentages instead of fixed sizes; sometimes it is the simple things that are the most powerful.

# CHAPTER 10

# Recommending with Singular Value Decomposition on GCP

What to watch next? Let's recommend movie options using SVD and the Wikipedia API on Google Cloud.

In this chapter, we're going to build a movie recommender web application (Figure 10-1) using the MovieLens datasets containing, "**100,000 ratings and 1,300 tag applications applied to 9,000 movies by 700 users**."[1] We will explore different similarity-measurement techniques and design a recommender application using singular value decomposition (SVD) and collaborative filtering to make great movie recommendations.

---

[1]https://grouplens.org/datasets/movielens/

© Manuel Amunategui, Mehdi Roopaei 2018
M. Amunategui and M. Roopaei, *Monetizing Machine Learning*, https://doi.org/10.1007/978-1-4842-3873-8_10

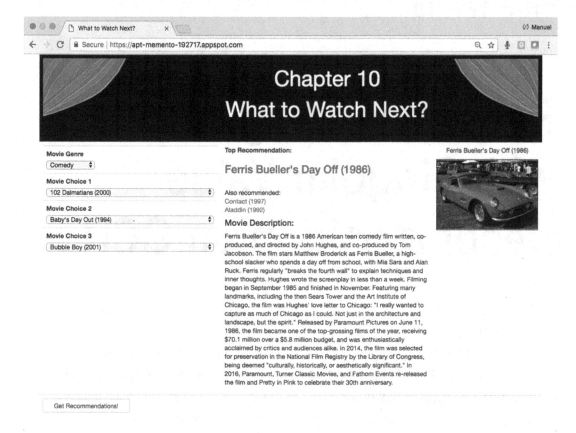

***Figure 10-1.*** *The final web application for this chapter*

---

**Note**    Download the files for Chapter 10 by going to www.apress.com/ 9781484238721 and clicking the source code button. Open Jupyter notebook "**chapter10.ipynb**" to follow along with this chapter's content.

---

# Planning Our Web Application

We will be focusing on collaborative filtering using movie ratings found in the MovieLens dataset.[2] This is probably the most popular dataset to learn about this topic. We're basically going to leverage the wisdom of the crowds to come up with movie recommendations. We'll build a web application where a user can select a couple of

---

[2]https://grouplens.org/datasets/movielens/

movies and the application will return a related movies recommendations. And to make it even more informative and fun, we will also pull related information and images from Wikipedia regarding top movie recommendations.

# A Brief Overview of Recommender Systems

Recommender systems are a big deal on the web and in e-commerce. Anytime a site makes a recommendation based on something you are looking at or on your preferences, it is using some form of recommender model. The types of recommender systems vary widely depending on the tool used and the availability of customized and intelligent data. Two popular areas are "**content based**" and "**collaborative filtering**."

In this chapter, we will focus on collaborative filtering instead of content-based filtering. The data from MovieLens contains user ratings for various movies. The reasoning behind applying CF using this data is if you and a reviewer liked the same movie, then there is a good chance you'll like other movies reviewed by that person.

# Exploring the MovieLens Dataset

Let's take a look at the MovieLens data. According to MovieLens liner notes[3] on the ml-latest-small dataset, the dataset contains 100,004 ratings across 9,125 movies and was created by 671 users between 1995 and 2016.

# More from the MovieLens Dataset's Liner Notes

**Ratings Data File Structure ("ratings.csv"):**

- All ratings are contained in the file "**ratings.csv**." Each line of this file after the header row represents one rating of one movie by one user, and has the following format: userId, movieId, rating, timestamp.

- The lines within this file are ordered first by userId, then, within user, by movieId.

- Ratings are made on a 5-star scale, with half-star increments (0.5 stars–5.0 stars).

---

[3]F. Maxwell Harper and Joseph A. Konstan, 2016. "The MovieLens Datasets: History and Context," *ACM Transactions on Interactive Intelligent Systems (TiiS)* 5 no. 4, Article 19 (January 2016), 19 pages. DOI = https://doi.org/10.1145/2827872

- Timestamps represent seconds since midnight Coordinated Universal Time (UTC) of January 1, 1970.

**Movies Data File Structure ("movies.csv"):**

- Movie information is contained in the file "**movies.csv.**" Each line of this file after the header row represents one movie, and has the following format: movieId, title, genres.

- Movie titles are entered manually or imported from `https://www.themoviedb.org/` and include the year of release in parentheses. Errors and inconsistencies may exist in these titles.

- Genres are a pipe-separated list, and are selected from the following:

  - Action
  - Adventure
  - Animation
  - Children's
  - Comedy
  - Crime
  - Documentary
  - Drama
  - Fantasy
  - Film-Noir
  - Horror
  - Musical
  - Mystery
  - Romance
  - Sci-Fi
  - Thriller
  - War
  - Western
  - (no genres listed)

Go ahead and download the files for this chapter into a folder called "**chapter-10.**" Download the "**ml-latest-small.zip**" dataset (http://files.grouplens.org/datasets/ movielens/ml-latest-small.zip) and unzip it into your "**chapter-10**" folder. We will only use "**ratings.csv**" and "**movies.csv**" datasets. You should have everything you need to follow along in the Jupyter notebook for this chapter.

## Overview of "ratings.csv" and "movies.csv"

Take a quick look at the Pandas "**shape**" and "**tail()**" functions of "**ratings.csv.**" It is a narrow and long table and its timestamp is in Unix time, which is an integer representation of time in the form of seconds from January 1st, 1970 UTC. We're going to fix that. Here we have the CSV files in a folder called "**ml-latest-small**"; adjust yours accordingly (Listing 10-1 and Figure 10-2).

*Listing 10-1.* A look at "**ratings.csv**"

**Input:**

```
pd.read_csv('ml-latest-small/ratings.csv')
print('Shape:', ratings_df.shape)
print('Tail:', ratings_df.tail())
```

**Output:**

Shape: (100004, 4)

|        | userId | movieId | rating | timestamp  |
|--------|--------|---------|--------|------------|
| 99999  | 671    | 6268    | 2.5    | 1065579370 |
| 100000 | 671    | 6269    | 4.0    | 1065149201 |
| 100001 | 671    | 6365    | 4.0    | 1070940363 |
| 100002 | 671    | 6385    | 2.5    | 1070979663 |
| 100003 | 671    | 6565    | 3.5    | 1074784724 |

*Figure 10-2. Raw output of "ratings.csv"*

We'll use the datetime's "**fromtimestamp()**" function to cast Unix time into actual and readable timestamp. Using the "**describe()**" we can confirm the change and see that the ratings range from January 1995 to October 2016 (Listing 10-2).

***Listing 10-2.*** A Look at Timestamps

**Input:**

```
import datetime
ratings_df['timestamp'] = [datetime.datetime.fromtimestamp(dt) for dt in
ratings_df['timestamp'].values]
ratings_df['timestamp'].describe()
```

**Output:**

```
count                    100004
unique                    78141
top        2016-07-23 05:54:42
freq                         87
first      1995-01-09 03:46:49
last       2016-10-16 10:57:24
Name: timestamp, dtype: object
```

When we run Pandas' "**describe()**" on the "**ratings.csv**," we see that the minimum rating is 0.5 and the maximum is 5, with an average of 4 (Listing 10-3 and Figure 10-3).

***Listing 10-3.*** Function "**describe()**" on "**ratings_df**"

```
ratings_df.describe()
```

|       | userId          | movieId         | rating          |
|-------|-----------------|-----------------|-----------------|
| count | 100004.000000   | 100004.000000   | 100004.000000   |
| mean  | 347.011310      | 12548.664363    | 3.543608        |
| std   | 195.163838      | 26369.198969    | 1.058064        |
| min   | 1.000000        | 1.000000        | 0.500000        |
| 25%   | 182.000000      | 1028.000000     | 3.000000        |
| 50%   | 367.000000      | 2406.500000     | 4.000000        |
| 75%   | 520.000000      | 5418.000000     | 4.000000        |
| max   | 671.000000      | 163949.000000   | 5.000000        |

*Figure 10-3. "Describe()" Output of Data Frame "**ratings_df**"*

For our needs, it doesn't really matter how many movies are in the movies data frame; we care about how many movies have been rated in the ratings data (Listing 10-4).

*Listing 10-4.* Count of Unique Movies

**Input:**

```
print('Unique number of rated movies: %i' % len(set(ratings_df['movieId'])))
```

**Output:**

```
Unique number of rated movies: 9066
```

So, we really only have 9,066 movies to work with, not 9,125 as the liner notes mention. And the number of unique reviews is 671 (Listing 10-5).

*Listing 10-5.* Unique User Count with Ratings

**Input:**

```
print('Unique user count with ratings: %i' % len(set(ratings_df['userId'])))
```

**Output:**

```
Unique user count with ratings: 671
```

Let's take a look at "**movies.csv**." Feature "**title**" is straightforward and "**genres**" is an interesting category that we will explore further (Listing 10-6 and Figure 10-4).

*Listing 10-6.*  A Look at "**movies.csv**"

**Input**:

```
movies_df = pd.read_csv('ml-latest-small/movies.csv')
print('Shape:', movies_df.shape)
movies_df.tail()
```

**Output**:

Shape: (9125, 3)

|      | movieId | title | genres |
|------|---------|-------|--------|
| **9120** | 162672 | Mohenjo Daro (2016) | Adventure\|Drama\|Romance |
| **9121** | 163056 | Shin Godzilla (2016) | Action\|Adventure\|Fantasy\|Sci-Fi |
| **9122** | 163949 | The Beatles: Eight Days a Week - The Touring Y... | Documentary |
| **9123** | 164977 | The Gay Desperado (1936) | Comedy |
| **9124** | 164979 | Women of '69, Unboxed | Documentary |

*Figure 10-4.*  *Raw Output of "movies.csv"*

When we run "**describe()**," we notice that feature "**movieId**" doesn't match the data frame's index. The maximum movieId is 164,979, and there are only 9,124 rows. It is most likely a universal MovieLens identifier. This is something we'll need to adjust, to ensure that the movieId follows the table's index; it will make our lives a lot easier once we move from data frames to matrices (Listing 10-7 and Figure 10-5).

*Listing 10-7.*  Running Function "**describe()**" on "**movies_df**"

```
movies_df.describe()
```

| | movieId |
|---|---|
| count | 9125.000000 |
| mean | 31123.291836 |
| std | 40782.633604 |
| min | 1.000000 |
| 25% | 2850.000000 |
| 50% | 6290.000000 |
| 75% | 56274.000000 |
| max | 164979.000000 |

***Figure 10-5.*** *The "**describe()**" output*

# Understanding Reviews and Review Culture

Now that we have a basic understanding of the two data frames, we're going to focus on what they contain. Let's start with the number of reviews per user Ids (Listing 10-8 and Figure 10-6).

***Listing 10-8.*** Plot Reviews per Users

```
plt.plot(sorted(ratings_df['userId'].value_counts(normalize=False)),
marker='o')
plt.suptitle('Number of Reviews per UserId', fontsize=16)
plt.xlabel('Reviewer User ID', fontsize=14)
plt.ylabel('Number of Reviews', fontsize=14)
plt.grid()
plt.show()
```

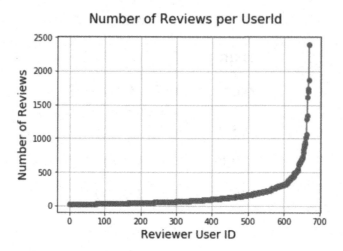

***Figure 10-6.*** *Total number of reviews per user ID*

We see that the majority of the reviewers have fewer than 200 reviews, but one reviewer has almost 2,500! Let's look at the distribution of actual ratings (Listing 10-9 and Figure 10-7).

***Listing 10-9.*** Count of Unique Movies

```
ratings_df['rating'].plot.hist()
plt.suptitle('Rating Histogram', fontsize=16)
plt.xlabel('Rating Category', fontsize=14)
plt.ylabel('Rating Frequency', fontsize=14)
plt.grid()
plt.show()
```

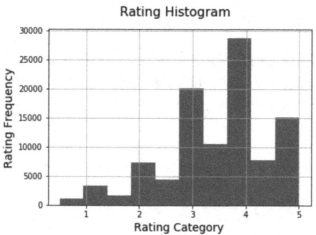

***Figure 10-7.*** *Histogram of total ratings per category*

314

This chart is important, as it shows the type of ratings our reviewers like to use. On a scale of 1 to 5, you would think that an even distribution would make a rating of "**3**" the most used. We clearly see that isn't the case and that a rating of "**4**" is the favorite rating, followed by "**3**" and "**5**." Taking some time to think about this is critical. We are about to build models around user ratings and we need to be able to compare each reviewer on equal footing. How can we generalize each user's ratings to be comparable to all others? One technique is to center each user's reviews around the mean of that user's ratings. This creates a central "**0**" point that will align with all other reviewers' central points–the neutral review level. Of course, this system won't work if a reviewer only has one review.

The categorical "**genres**" field of the movie dataset is interesting to get a quick idea of the type of movies the dataset contains. Drama and comedy seem to be the biggest categories (Figure 10-8).

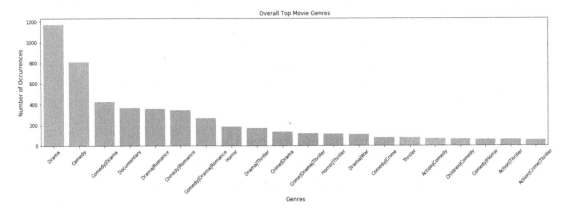

***Figure 10-8.*** *Overall top movie genres*

We can also look at the breakdown of movie "**genres**" over the years they were reviewed, by joining both the ratings and movie data frames. This is easily done with Pandas "**merge()**" function. We join on the common index field "**movieId**" and then pull the year from the "**timestamp**" feature (Listing 10-10).

***Listing 10-10.*** Merging Movies and Reviews by Year

```
reviews_by_genres = pd.merge(movies_df, ratings_df, how = 'inner', on ='movieId')
reviews_by_genres['year'] = reviews_by_genres['timestamp'].dt.year
```

Now we can plot data made up of features from both tables broken down by year, such as the most popular "**genres**" reviewed each year. We have to perform a slightly more involved "**groupby()**" function call to get a "**genres**" total count by year (Listing 10-11 and Figure 10-9).

***Listing 10-11.***  Top Genres by Year Reviewed

```
def top_category_count(x, n=1):
    return x.value_counts().head(n)

reviews_by_genres = reviews_by_genres.groupby(['year']).genres.apply(top_
category_count).reset_index()

plt.figure(figsize=(20,5))
g = sns.barplot(reviews_by_genres['year'], reviews_by_genres['genres'],
alpha=0.8)
plt.title('Top Genres Count by Year Reviewed')
plt.ylabel('Number of Occurrences', fontsize=12)
plt.xlabel('Genres', fontsize=12)
plt.xticks(rotation=45)

for index, row in reviews_by_genres.iterrows():
    g.text(row.name, row.genres, row.level_1, ha="center", rotation=45,
    color='blue', verticalalignment='bottom', fontsize=10)

plt.show()
```

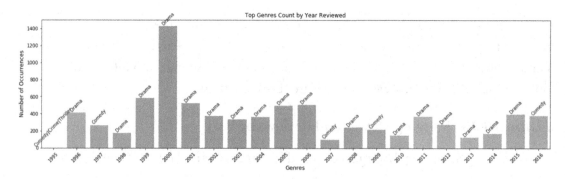

***Figure 10-9.***  *Top genres count by year reviewed*

It shouldn't come as a surprise that "**Drama**" is the most common, with some pockets of "**Comedy**." Year 2000 seems to have been a good year for "**Drama**" movies.

# Getting Recommendations

By now you should have a clear understanding of the data we are using, and this should make the following sections on modeling that much more approachable.

In order to streamline our filtering process, we are going to create a "**matrix**" made with only three fields from the ratings data frame: "**userId**," "**movieId**," and "**ratings**" (Listing 10-12).

***Listing 10-12.*** Creating User by Movie Ratings Matrix

**Input**:

```
ratings_df.set_index(['userId', 'movieId'], inplace=True)
ratings_matrix = sps.csr_matrix((ratings_df.rating,
                    (ratings_df.index.labels[0], ratings_df.index.
                    labels[1]))).todense()
print('shape ratings_matrix:', ratings_matrix.shape)
```

**Output**:

```
shape ratings_matrix: (671, 9066)
```

This is done by setting the index to be both "**userId**" and "**movieId**," then creating a matrix out of it using one index for rows and the other for columns. Because a lot of the ratings will be zeros, we use the "**scipy.sparse.csr_matrix()**" function to create an efficient sparse matrix (for more information, see the official docs at https://docs.scipy.org/doc/scipy-0.15.1/reference/generated/scipy.sparse.csr_matrix.html). The matrix will look like Figure 10-10.

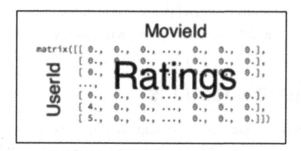

*Figure 10-10.* *The very important "Users" by "Rated Movies" matrix*

Using this matrix will make our querying, measuring similarities, and final SVD model run much faster.

We also need to create a movie list with consecutive indexing sorted in the same order as the original movie Ids from the movies data frame. This can easily be done by dropping the original "**movieId**" field and replacing it with the row index. Then we merge it to the ratings table and adopt the new Id for both ratings and movies (Listing 10-13 and Figure 10-11).

*Listing 10-13.* Fixing Movie Ids

```
movies_df_raw = pd.read_csv('ml-latest-small/movies.csv')
movies_df_raw['movieId_new'] = movies_df_raw.index
movies_df_raw.tail()
```

|  | movieId | title | genres | movieId_new |
|---|---|---|---|---|
| **9061** | 161944 | The Last Brickmaker in America (2001) | Drama | 9061 |
| **9062** | 162376 | Stranger Things | Drama | 9062 |
| **9063** | 162542 | Rustom (2016) | Romance\|Thriller | 9063 |
| **9064** | 162672 | Mohenjo Daro (2016) | Adventure\|Drama\|Romance | 9064 |
| **9065** | 163949 | The Beatles: Eight Days a Week - The Touring Y... | Documentary | 9065 |

*Figure 10-11.* *The old spotty "movieId" versus the new incremental "movieId_new"*

We'll rely on this row indexing to pull similarities. By looking at the tail, we confirm that there is a total of 9,066 unique movies. We need to apply this new movie Id to both the ratings and movies data frames and rename it "**movieId**" for consistency (Listing 10-14).

***Listing 10-14.*** Applying Fixed Movie Id

```
ratings_df_raw = ratings_df_raw.merge(movies_df_raw[['movieId', 'movieId_
new']], on='movieId', how='inner')
ratings_df_raw = ratings_df_raw[['userId',  'movieId_new', 'rating']]
ratings_df_raw.columns = ['userId',  'movieId', 'rating']

movies_df_raw = movies_df_raw[['movieId_new', 'title', 'genres']]
movies_df_raw.columns = ['movieId', 'title', 'genres']
```

One last thing we need to do before diving into similarity metrics is to pull a base movie that we will use in all subsequent similarity algorithms. We'll put the ratings data from the ratings table at index "**0**"; this represents all the ratings for the original Toy Story movie (Listing 10-15).

***Listing 10-15.*** Basing Reviews on Toy Story

**Input:**

```
movie_toy_story = (mat[:,0])
movie_toy_story[0:20]
```

**Output:**

```
matrix([[ 0. ],
        [ 0. ],
        [ 0. ],
        [ 0. ],
        [ 0. ],
        [ 0. ],
        [ 3. ],
        [ 0. ],
        [ 4. ],
        [ 0. ],
        [ 0. ],
        [ 0. ],
        [ 5. ],
        [ 0. ],
        [ 2. ],
```

```
  [ 0. ],
  [ 0. ],
  [ 0. ],
  [ 3. ],
  [ 3.5]])
```

# Collaborative Filtering

Collaborative filtering is a popular type of recommendation system based on leveraging the taste of other users who share similarities.

# Similarity/Distance Measurement Tools

There is a whole slew of distance measuring formulas to measure the similarity or dissimilarity between two lists of numbers. This can be extremely handy when you want to compare various sets of numbers against others. Here we will compare movies by using user recommendations with the Euclidean distance and cosine similarity measures.

We will once again leverage the great Scipy library and use its distance computations "**scipy.spatial.distance()**" function for most of our needs in this chapter.

# Euclidean Distance

Simply put, the Euclidean distance is the distance between two points and probably the most popular distance algorithm. Here we will use Scipy's "**spatial.distance. euclidean()**" function to calculate the Euclidean distance from our seed row of Toy Story against all other rows (Listing 10-16 and Figure 10-12).

*Listing 10-16.* Getting Similar Movies

```
distances_to_movie = []
for other_movies in mat.T:
    distances_to_movie.append(scipy.spatial.distance.euclidean(movie_toy_
    story, other_movies.tolist()))
```

```
# create dataframe of movie and distance scores to Toy Story
distances_to_movie = pd.DataFrame({'movie':movies_df['title'],'distance':
distances_to_movie})
```

```
# sort by ascending distance (i.e. closest to movie_toy_story)
distances_to_movie = distances_to_movie.sort_values('distance')
distances_to_movie.head(10)
```

| | distance | movie |
|---|---|---|
| 0 | 0.000000 | Toy Story (1995) |
| 2506 | 50.882217 | Toy Story 2 (1999) |
| 1866 | 53.849327 | Bug's Life, A (1998) |
| 1019 | 54.904462 | Groundhog Day (1993) |
| 644 | 55.009090 | Independence Day (a.k.a. ID4) (1996) |
| 3803 | 55.056789 | Monsters, Inc. (2001) |
| 4604 | 55.522518 | Finding Nemo (2003) |
| 5611 | 55.709066 | Incredibles, The (2004) |
| 3419 | 56.333826 | Shrek (2001) |
| 866 | 56.643623 | Willy Wonka & the Chocolate Factory (1971) |

*Figure 10-12.* *Movies with shortest Euclidean distance to Toy Story (1995)*

Euclidean distance does a good job linking Toy Story with "**Toy Story 2**" and "**A Bug's Life**"; intuitively, it makes sense.

# Cosine Similarity Distance

The cosine similarity measures the cosine angle between two vectors.[4] This is a more sophisticated form of measurement over the Euclidian distance and one we will be using throughout this chapter.

---

[4]http://mines.humanoriented.com/classes/2010/fall/csci568/portfolio_exports/
sphilip/cos.html

Let's repeat the previous exercise using Scipy's "**spatial.distance.cosine()**" function to calculate the cosine distance from our seed row of Toy Story against all other rows (Listing 10-17 and Figure 10-13).

***Listing 10-17.*** Getting Similar mMovies

```
distances_to_movie = []
for other_movies in mat.T:
    distances_to_movie.append(scipy.spatial.distance.cosine(movie_toy_
    story, other_movies.tolist()))

# create dataframe of movie and distance scores to Toy Story
distances_to_movie = pd.DataFrame({'movie':movies_df['title'],'distance':di
stances_to_movie})

# sort by ascending distance (i.e. closest to movie_toy_story)
distances_to_movie = distances_to_movie.sort_values('distance')
distances_to_movie.head(10)
```

|  | distance | movie |
| ---: | --- | ---: |
| 0 | -2.220446e-16 | Toy Story (1995) |
| 2506 | 4.052902e-01 | Toy Story 2 (1999) |
| 232 | 4.238122e-01 | Star Wars: Episode IV - A New Hope (1977) |
| 321 | 4.354661e-01 | Forrest Gump (1994) |
| 644 | 4.370544e-01 | Independence Day (a.k.a. ID4) (1996) |
| 1019 | 4.519770e-01 | Groundhog Day (1993) |
| 1024 | 4.632997e-01 | Back to the Future (1985) |
| 427 | 4.648029e-01 | Jurassic Park (1993) |
| 3419 | 4.673149e-01 | Shrek (2001) |
| 966 | 4.706660e-01 | Star Wars: Episode VI - Return of the Jedi (1983) |

***Figure 10-13.*** *Movies with shortest cosine distance to Toy Story (1995)*

Cosine distance links Toy Story with "**Toy Story 2**" along with other family movies. It is interesting that cosine distance comes up with a different list than Euclidean distance.

# Singular Value Decomposition

You could easily build a system with collaborative filtering using any of the distance approaches covered, but it won't scale to larger datasets, handle sparse data, or know what to do in cases of cold starts (where there is little or no intersecting user/movie data). A model like SVD is designed to alleviate some of those problems by using lower rank approximation and has the ability of estimating matches on compressed data.[5]

This is definitely not simple stuff, but in a nutshell, it attempts to reduce the data and make the important themes and connections bubble up and lets the noise or the outliers drop to the bottom. It will remove the noisy and irregular information to build a clearer map of what fits where, and what is close and what is far away. SVD will decompose our matrix of movie ratings arranged by genres and users into three parts, a matrix of users, a matrix of movies, and a vector of relationship between both. We can then use the relationship vector to match different users and reviews that are close to each other and collect the surrounding information as potential recommendations. SVD returns a matrix representing the feature space of users and another representing the feature space of movies. We then apply the dot product to find similarities and recommendations.

Also, this isn't meant to be a course on SVD specifics, as it can get complicated. PhDs have been written on this topic and hundreds of tutorials are available on the web for those with a desire to dig deeper. I also recommend the blog post on which this code is based (`https://beckernick.github.io/matrix-factorization-recommender/`), and for a clear and simple example of SVDs in action, Recommendation Engines for Dummies (`http://zwmiller.com/projects/simple_recommender.html`).

# Centering User Ratings Around Zero

As this dataset has been studied for quite a few years now, some interesting tricks have been proved useful and we will apply them here as well. We will subtract each user's recommendations against their mean recommendation. This ensures that all users are scaled accordingly and around zero. If a user's max rating is a 5 and minimum rating is a 3, while other users rate using the whole range, taking each mean and subtracting it against its recommendation will allow both to be comparable (Listing 10-18).

---

[5]`http://web.mit.edu/be.400/www/SVD/Singular_Value_Decomposition.htm`

*Listing 10-18.* Centering All Reviews Around Mean

```
user_ratings_mean = np.mean(ratings_mat, axis = 1)
ratings_mat_centered = ratings_mat - user_ratings_mean.reshape(-1, 1)
```

# A Look at SVD in Action

Let's finish analyzing the SVD code found in the Jupyter notebook (to follow along, refer to e section "**Singular Value Decomposition**"). This is an important piece, as it represents the brains behind our recommender engine.

The first thing we do is build a matrix of users versus rated movies (Figure 10-14). In our case it entails cleaning up a few things like rebuilding the movie Ids to start at zero, making them sequential with no gaps, and resetting the rating Ids to also start at zero.

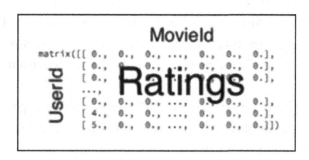

*Figure 10-14.* The very important *"Users"* by *"Rated Movies"* matrix

One way to make our recommender work on new data is to add our visitor's movie taste as a new "**UserId**" to the matrix. We then rebuild the matrix using the "**Compressed Sparse Row matrix**" function "**sps.csr_matrix()**," which will transform the matrix into an SVD-friendly format taking into account the information from our new visitor (Listing 10-19).

*Listing 10-19.* Getting Recommendations for our New User

```
# create a new user id - add 1 to current largest
new_user_id = np.max(ratings_df_cp['userId']) + 1

# add movie preference to matrix and assign 5-star votes
# to all of them
new_user_movie_ids = user_history_movie_ids
```

```
new_user_ratings = [5] * len(new_user_movie_ids)

# fix index to be multilevel with userId and movieId
ratings_df_cp.set_index(['userId', 'movieId'], inplace=True)

# add new movie rating as a pandas series and insert new row
# at end of ratings_df_cp
for idx in range(len(new_user_movie_ids)):
        row_to_append = pd.Series([new_user_ratings[idx]])
        cols = ['rating']
        ratings_df_cp.loc[(new_user_id, new_user_movie_ids[idx]), cols]
=          row_to_append.values

# create new ratings_matrix
ratings_matrix_plus = sps.csr_matrix((ratings_df_cp.rating,
(ratings_df_cp.index.labels[0], ratings_df_cp.index.labels[1]))).todense()
```

We then send our matrix to the "**GetSparseSVD**" function that performs the matrix decomposition, and return the three matrices needed to get the dot product for our new user and other users with similar interests. This uses the Sparse linear algebra SVDS library "**scipy.sparse.linalg.svds**," which does a great job handling large matrices with lots of zeros (most folks have only seen a handful of movies, so when you build a matrix of users to movies, most cells are blanks; Listing 10-20).

**Listing 10-20.**   Getting the SVD

```
Ua, sigma, Vt = GetSparseSVD(ratings_matrix_centered, K=50)
all_user_predicted_ratings = np.dot(np.dot(Ua, sigma), Vt) + user_ratings_
mean.reshape(-1, 1)
```

We run the dot product on the returned matrices (Ua and sigma in our case) and package everything into a data frame called "**predictions_df.**" This data leverages SVD's magic to organize the users and their interests into various dimensions (Listing 10-21 and Figure 10-15).

**Listing 10-21.**   Viewing "**head()**" of Dot Matrix

```
predictions_df = pd.DataFrame(all_user_predicted_ratings, columns = movies_
df.index)
predictions_df.head()
```

```
            0          1          2          3          4          5          6     \
0   -0.054240   0.045128  -0.004833  -0.019825  -0.011278   0.041374  -0.007828
1    0.419849   1.406419  -0.188829   0.156646   0.268030   0.414696   0.052150
2    1.345667   0.266465  -0.012022   0.012361   0.079403   0.090967  -0.122086
3    1.133440   1.047020   0.141292   0.081937  -0.339729  -1.484643  -0.263005
4    1.389618   1.466398   0.605475  -0.029601   0.729323  -0.118494  -0.026019

            7          8          9      ...       9056       9057       9058  \
0   -0.017190   0.012238   0.037665      ...  -0.005258  -0.005453   0.012368
1    0.044731  -0.020230   2.220210      ...  -0.005910  -0.003974  -0.012557
2    0.031366  -0.017969   0.141100      ...  -0.002644  -0.002358  -0.010145
3   -0.169730  -0.021727   1.611773      ...   0.020811   0.000414   0.056051
4    0.065617  -0.156665   0.307791      ...  -0.007421  -0.011804   0.006647

         9059       9060       9061       9062       9063       9064       9065
0   -0.004991  -0.004639  -0.019052   0.021401  -0.006365  -0.006098  -0.004819
1   -0.003555  -0.002712  -0.071607  -0.016215   0.001046  -0.001469  -0.006579
2    0.000278  -0.000116  -0.018086  -0.015750   0.010617   0.006797  -0.006354
3   -0.002815  -0.000765   0.159109   0.087533  -0.030847  -0.021274   0.048537
4   -0.005158  -0.001249  -0.034653   0.016460   0.001714  -0.004163  -0.001864

[5 rows x 9066 columns]
```

*Figure 10-15.* *The dot matrix output organizing our users by similarities on multiple dimensions*

I certainly won't pretend to understand the logic behind these groupings but it works, so I'll trust SVD and I'd recommend you do the same.

The last phase of getting the recommendations for our new user is to extract the SVD's dot product data from "**predictons_df**" for that user only, then append the movie information, and simply pick a handful of new movies that the user hasn't already seen. For example, let's pretend that our new user really likes the following three drama/war movies (Figure 10-16).

| | movieId | title | genres |
|---|---|---|---|
| **188** | 188 | Before the Rain (Pred dozhdot) (1994) | Drama|War |
| **301** | 301 | Walking Dead, The (1995) | Drama|War |
| **472** | 472 | Schindler's List (1993) | Drama|War |

*Figure 10-16.* *Our user's preferences*

We pass this new user's choices as a new user in our "**Users**" by "**Rated Movies**" matrix and add those three choices with high ratings. We then run SVD and pull the predictions for this new user (Figure 10-17).

| | movieId | predictions | title | genres |
|---|---|---|---|---|
| 0 | 472 | 0.683026 | Schindler's List (1993) | Drama\|War |
| 1 | 321 | 0.323625 | Forrest Gump (1994) | Comedy\|Drama\|Romance\|War |
| 2 | 1590 | 0.311840 | Saving Private Ryan (1998) | Action\|Drama\|War |
| 3 | 284 | 0.303246 | Shawshank Redemption, The (1994) | Crime\|Drama |
| 4 | 525 | 0.278348 | Silence of the Lambs, The (1991) | Crime\|Horror\|Thriller |
| 5 | 100 | 0.276620 | Braveheart (1995) | Action\|Drama\|War |
| 6 | 522 | 0.214850 | Terminator 2: Judgment Day (1991) | Action\|Sci-Fi |
| 7 | 2062 | 0.211868 | Matrix, The (1999) | Action\|Sci-Fi\|Thriller |
| 8 | 266 | 0.190663 | Pulp Fiction (1994) | Comedy\|Crime\|Drama\|Thriller |
| 9 | 2288 | 0.181298 | American Beauty (1999) | Drama\|Romance |
| 10 | 1359 | 0.177232 | Titanic (1997) | Drama\|Romance |

***Figure 10-17.*** *The sorted recommendations for our new user*

Obviously, we will not recommend "**Schindler's List**" to this user as it is one they have already seen' instead, we recommend "**Forest Gump**," "**Saving Private Ryan**," and "**The Shawshank Redemption**." Okay, enough on the code; let's build this thing!

# Downloading and Running the "What to Watch Next?" Code Locally

Let's download the files for Chapter 10 and unzip them on your local machine if you haven't already done so. You will need to copy the CSV files "**movies.csv**" and "**ratings. csv**" created earlier to the root directory of the web application files–in the same folder as **main.py**. Your "**web-application**" folder should contain the following files as shown in Listing 10-22.

***Listing 10-22.*** Web Application Files

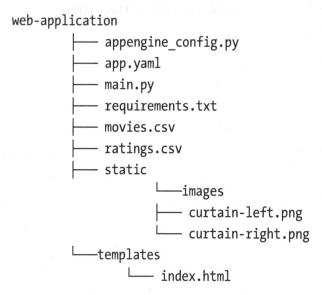

```
web-application
          ├── appengine_config.py
          ├── app.yaml
          ├── main.py
          ├── requirements.txt
          ├── movies.csv
          ├── ratings.csv
          ├── static
                      └──images
                      ├── curtain-left.png
                      └── curtain-right.png
          └──templates
                      └── index.html
```

As is customary, we'll start a virtual environment to segregate our Python library installs (Listing 10-23).

***Listing 10-23.*** Starting Virtual Environment

```
$ python3 -m venv whattowatchnext
$ source whattowatchnext/bin/activate
```

Then install all the required Python libraries by running the "**pip install -r**" command (Listing 10-24).

***Listing 10-24.*** Install Requirements and Take the Site for a Local Spin

```
$ pip3 install -r requirements.txt
$ python3 main.py
```

You can run the application on your local machine just like we did in the previous exercises. Open a command line window, change the drive into the "**web-application**" folder, and run the same commands you ran previous times (such as running "**python3 main.py**"). It should look like the following screen shot in Figure 10-18.

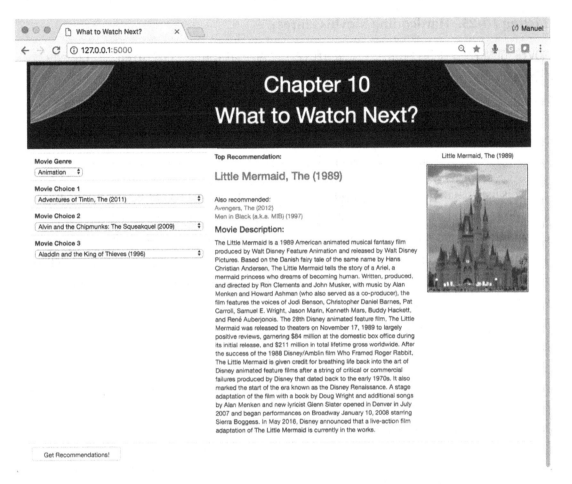

*Figure 10-18.    "What to Watch Next?" running locally*

# What's Going on Here?

Let's take a closer look at some of the interesting things going on in our Flask web application.

# main.py

Let's go over "**main.py**." The movie genres that were extracted during data exploration are copied and hard-coded as a constant list in the script. There is no real value in calculating it dynamically from the data each time, as it tied to the historical data and won't change (Listing 10-25).

***Listing 10-25.*** Hardcoding Movie Genres

```
MOVIE_GENRES = ["Action", "Adventure", "Animation", "Children", "Comedy",
"Crime",
                "Documentary", "Drama", "Fantasy", "Film-Noir", "Horror",
                "IMAX", "Musical", "Mystery", "Romance", "Sci-Fi",
                "Thriller",
                "War", "Western"]
```

This list of genres is used to populate the first drop-down box of the web application. That drop-down box is the only prepopulated field on the page and whenever a user changes it, it automatically populates the other drop downs with movie titles for that genre. This helps focus the application and drastically reduce the number of available choices.

The script contains four convenience functions that are called at different times during a visitor's interaction with the site.

- **GetMoviesByGenres (movies_df, genre)**

  - Is the function that will return a list of all movies and movie IDs for a particular genre. It is called whenever a user changes the "**Movie Genre**" drop-down box.

- **GetSparseSVD (ratings_centered_matrix, K)**

  - Is the SVD algorithm we looked at earlier. It takes a ratings/movieids matrix and "**K**," the number of singular values to compute, and returns a matrix of users, a vector of relationship between both (a diagonal matrix), and a matrix of movies.

- **GetRecommendedMovies (ratings_df, movies_df, user_history_movie_ids)**

  - This is a critical function that will take the ratings and movies data frames, the history of movies the user has watched and liked (the movies selected by the user in the three drop-down boxes). It will append a few new rows on the ratings data frame consisting of the movies our visitor has selected along with a top rating (we assume the user really liked). It will call the "**GetSparseSVD**" function and get an SVD decomposition using the original data along with the append new movies from our visitor. Now that we know the user's movie taste, we can find similar movies from SVD using dot products. The function will return the top three recommended movies.

- **GetWikipediaData (title_name)**

    - This function takes a movie title, will pass it on to the Wikipedia API, and return the first paragraph about the movie along with the first image associated with that movie (usually the movie's poster).

The script also contains three Flask specific functions:

- **startup()**

    - The "**startup()**" function is called whenever the flask server is started–in other words it is a constructor function. This is done by adding the decorator "**@app.before_first_request.**" It loads both datasets: "**movies.csv**" and "**ratings.csv.**" It then cleans them just like we did in the exploration Jupyter notebook by removing unused "**movieId**" and resets all indexes to start at zero.

- **ready()**

    - The "**ready()**" function is called whenever the page is first loaded, refreshed, or when a movie genre is changed. This function sets variable defaults for the "**index.html**" page whenever called for the first time. If it isn't the first time (i.e. it is a form submit via a genre value change and the "**if**" function for "**request.method == 'POST'**" returns true), the function will preserve any value set by a user and get a fresh list of movies choices via the "**GetMoviesByGenres()**" function call. It also checks that the user has at least selected one movie from the three drop downs and passes those Ids to the "**GetRecommendedMovies()**" function. This will return three new movie recommendations. The top movie recommendation is passed to the "**GetWikipediaData()**" function for a description snippet and poster image of said movie.

- **background_process()**

    - This function is tied to the front-end AJAX function in "**index. html**." This is called whenever the user clicks one of the three recommended movie links. It takes the name of the movie and passes it to the "**GetWikipediaData()**" function just like the "**ready()**" function does. This gets a description snippet and poster image for the movie in focus.

# index.html

Let's take a look at some of the interesting things going on in the front-end side in the "**index.html**" script (Listing 10-26).

*Listing 10-26.* On the JavaScript Side

```
<script>

$(document).on("click", "a", function(){
    var move_title = this.innerHTML;
    $(this).text(move_title);
    document.getElementById("movie_poster").innerHTML = move_title;
    fetchdata(move_title);
});

function fetchdata(move_title)
{
    $.ajax({
        type : "GET",
        url:'{{ url_for('background_process') }}',
        data:{ 'movie_title': move_title},
        success: function(data){
            update_dashboard(data.wiki_movie_description, data.wiki_movie_
            poster);
        }

    });
}
```

```
function update_dashboard(wiki_movie_description, wiki_movie_poster){
    document.getElementById('movie_description').innerHTML = wiki_movie_
    description;
    document.getElementById('image_poster').src = wiki_movie_poster;
}

</script>
```

We have three functions:

- **$(document).on("click", "a"...)**

  - This function listens to anchor tags and, when clicked, takes the inner text and passes it to function "**fetchdata()**."

- **fetchdata(move_title)**

  - This function uses AJAX to pass a GET post to the server to get a movie description from Wikipedia and a movie poster, and passes the results to function "**update_dashboard()**."

- **update_dashboard(wiki_movie_description, wiki_movie_poster)**

  - This function gets results from function "**fecthdata()**" and replaces the HTML content description text and the movie poster with whatever is returned.

# Deploying on Google App Engine

By now, you should have some experience with the Google Flexible App Engine, so this will be a quick guide to get this web application up and running.

# Step 1: Fire Up Google Cloud Shell

Log into your instance of Google Cloud and create or select the project in which you want your App Engine to reside (if you don't have one, you will be prompted to create one–see Creating and Managing Projects[6]). Start the cloud shell command line tool by clicking the upper right caret button. This will open a familiar-looking command line window in the bottom half of the GCP dashboard (Figure 10-19).

---

[6]https://cloud.google.com/resource-manager/docs/creating-managing-projects

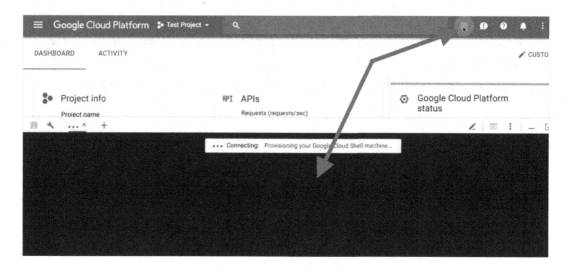

***Figure 10-19.***   *Accessing the Google Cloud shell*

## Step 2: Zip and Upload All Files to The Cloud

Zip the files in the "**web-application**" folder but don't zip the virtual environment folder "**whattowatchnext**" as it's not needed (Figure 10-20).

***Figure 10-20.***   *Zipping web application files for upload to Google Cloud*

Upload it using the "**Upload file**" option (this is found on the top right side of the shell window under the three vertical dots; Figure 10-21).

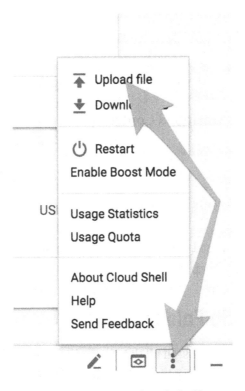

*Figure 10-21.*   *Uploading files via Google Cloud shell*

# Step 3: Create Working Directory on Google Cloud and Unzip Files

Once the file is successfully uploaded, create a new directory, like "**chapter-10**" for example (Listing 10-27).

*Listing 10-27.*  Creating Folder on Cloud

```
$ mkdir chapter-10
$ cd chapter-10
```

Transfer all the zip Archive into the new folder and unzip it (Listing 10-28).

**Listing 10-28.** Loading Needed Files

```
$ mv ../Archive.zip Archive.zip
$ unzip Archive.zip
```

Your folder on Google Cloud should look something like Listing 10-29.

**Listing 10-29.** Checking Unzipped Content

**Input:**

```
$ ls
```

**Output:**

```
appengine_config.py app.yaml Archive.zip lib main.py movies.csv ratings.csv
requirements.txt static templates
```

# Step 4: Creating Lib Folder

Run the following command to install all the needed additional libraries to the lib folder. When you deploy your web app, the lib folder will travel along with the needed libraries (Listing 10-30).

**Listing 10-30.** Installing Required Libraries

```
$ sudo pip3 install -t lib -r requirements.txt
```

# Step 5: Deploying the Web Application

Finally, deploy it to the world with the "**gcloud app deploy**" command (Listing 10-31).

***Listing 10-31.*** Deploying Web Application and Confirming We Want to Deploy Our Application (yes, please)

**Input**:

```
$ gcloud app deploy app.yaml
```

**Output**:

```
Services to deploy:

descriptor: [/home/amunategui/chapter-10/app.yaml]
source: [/home/amunategui/chapter-10]
target project: [apt-memento-192717]target service: [default]
target version: [20180702t150114]
target url: [https://apt-memento-192717.appspot.com]

Do you want to continue (Y/n)?
```

That's it! Sit back and let the tool deploy the web site. This is the Flexible App Engine, so it can take up to 30 minutes to be fully deployed. Once it is done setting everything up, it will offer a clickable link to jump directly to the deployed web application (Listing 10-32).

***Listing 10-32.*** You Can Also Get There with the Following Command:

```
$ gcloud app browse
```

Enjoy the fruits of your labor, and make sure to experiment with it by asking for some movie recommendations! (Figure 10-22).

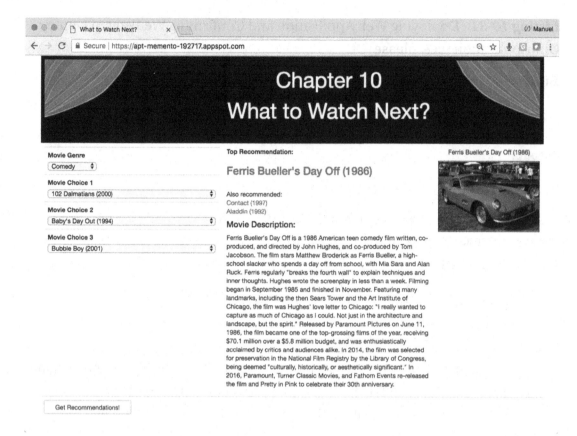

*Figure 10-22.* *The web application on Google Cloud*

# Troubleshooting

There will be cases where you will have issues and the Google Cloud logs will be your best friends. You can easily reach them either directly in the Google Cloud dashboard or by calling the logs URL (Listing 10-33).

***Listing 10-33.*** Logs URL

```
https://console.cloud.google.com/logs
```

Or you can stream the log's tail by entering in the cloud shell the following command in Listing 10-34.

**Listing 10-34.** Viewing Logs in Terminal Window

```
$ gcloud app logs tail -s default
```

# Closing Up Shop

One last thing before we conclude this chapter: don't forget to stop or delete your App Engine Cloud instance. Even if you are using free credits, the meter is still running and there is no need to waste money or credits.

Things are a little different with the Flexible App Engine over the Standard one, as the Flexible costs more money. So, it is important to stop it if you aren't using it. Also, this can all be conveniently done via the Google Cloud dashboard.

Navigate to App Engine, then Versions. Click your active version and stop it (Figure 10-23). If you have multiple versions, you can delete the old ones; you won't be able to delete the default one, but stopping it should be enough (if you really don't want any trace of it just delete the entire project).

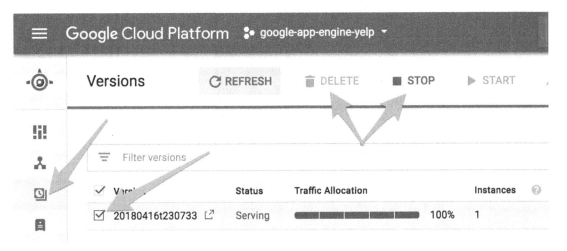

**Figure 10-23.** *Stopping and/or deleting your App Engine version*

That's it! Don't forget to deactivate the virtual environment if you are all done (Listing 10-35).

***Listing 10-35.***  Deactivating the virtual environment

```
$ deactivate
```

# Conclusion

Collaborative filtering for recommender systems are great and really popular in many commercial applications. This was definitely not an easy chapter, as the inner workings of SDVs are rather murky, but the point here is we can build a web application with some serious modeling muscle behind it.

Though SVDs have been around for a while and are still actively in use, interesting advances have been made using convolutional neural networks (CNN).[7]

---

[7]https://medium.com/@libreai/a-glimpse-into-deep-learning-for-recommender-systems-d66ae0681775

# CHAPTER 11

# Simplifying Complex Concepts with NLP and Visualization on Azure

Let's build a simple interactive dashboard to understand the cost of eliminating spam messages using natural language processing on Microsoft Azure.

In this chapter, we will use natural language processing (NLP) on the classic SMS Spam Collection Dataset. We will classify text messages as either ham or spam (i.e., intended messages vs. advertisements) using feature engineering, term frequency–inverse document frequency (TFIDF) and random forests (RF). But the key takeaway will be building a web application to illustrate and learn how to tune a prediction-probability threshold in order to achieve a variety of predictive goals beyond the traditional 0.5 cutoff (Figure 11-1).

© Manuel Amunategui, Mehdi Roopaei 2018
M. Amunategui and M. Roopaei, *Monetizing Machine Learning*, https://doi.org/10.1007/978-1-4842-3873-8_11

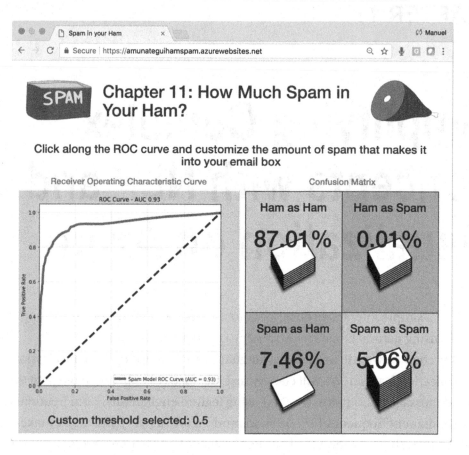

***Figure 11-1.***  *The final web application for this chapter*

---

**Note**    Download the files for Chapter 11 by going to www.apress.com/
9781484238721 and clicking the source code button. Open Jupyter notebook
"**chapter11-ipynb**" to follow along with this chapter's content.

---

# Planning our Web Application—the Cost of Eliminating Spam

Our web application will include an interactive receiver operating characteristic (ROC)
chart where the visitor can click on different thresholds and visualize how many ham
and spam messages get correctly classified. This will help visualize the compromise
between catching all spam messages and the amount of ham messages that get

mislabeled as spam in the process. It will highlight the importance of finding the right threshold to satisfy a particular business requirement. For example, in healthcare, if resources are limited, they may prefer to identify only high-probability patients and accept that lower risk patients may fall through the cracks. While in a ham/spam example, users would rather have some spam and never lose any ham messages. The tolerance for mislabeled predictions needs to be understood through business-domain expertise, but also through modeling know-how and visuals like our web application for this chapter.

# Data Exploration

The classic SMS Spam Collection Dataset is graciously hosted by the University of California's UCI Machine Learning Repository.[1] Go ahead and download the files for this chapter into a folder called "**chapter-11.**" Open up the Jupyter notebook to follow along. According to the UCI Dataset Description, this is a collection of 425 SMS spam messages from the Grumbletext web site and 322 spam messages from the SMS Spam Corpus v.01. Another 4,827 SMS messages from various sources were added as ham messages.[2]

If we call the Pandas "**groupby()**" function on the outcome variable (whether the message is ham or spam), we can understand how balanced the dataset is (Listing 11-1 and Figure 11-2).

*Listing 11-1.* Function "**groupby()**" on the Outcome Variable

```
sms_df.groupby('outcome').describe()
```

| outcome | count | unique | top | freq |
|---|---|---|---|---|
| ham | 4827 | 4518 | Sorry, I'll call later | 30 |
| spam | 747 | 653 | Please call our customer service representativ... | 4 |

*Figure 11-2.* The "*groupby()*" output for variable "*outcome*"

---

[1]https://archive.ics.uci.edu/ml/datasets/sms+spam+collection
[2]https://archive.ics.uci.edu/ml/datasets/sms+spam+collection

This simple command reveals a whole lot of information. We learn that the dataset is comprised of 4,827 ham and 747 spam messages. Some of them are duplicates and will need to be removed. It also shows that data is skewed, with 85% of the messages being ham. This skewness makes sense in the real world but makes modeling more challenging.

# Cleaning Text

Our first step is easy and obvious: we need to remove the duplicate rows identified previously. This can efficiently be done with the Pandas "**drop_duplicates()**" function. By setting the "**keep**" parameter to "**first**", we keep the first occurrence and delete all other subsequent repeats (Listing 11-2).

*Listing 11-2.*  Removing Duplicate Rows

**Input**:

```
print('Duplicates found before clean-up: %i ' % sum(sms_df.duplicated()))
sms_df = sms_df.drop_duplicates(keep='first')
print('Duplicates found after clean-up: %i ' % sum(sms_df.duplicated()))
```

**Output**:

```
Duplicates found before clean-up: 403
Duplicates found after clean-up: 0
```

# Text-Based Feature Engineering

It is important to remember that the majority of models out there, including NLP models, can only work with quantitative data. This means that we need to transform this textual SMS data into numbers. We are going to use various known tricks such as counting words and characters.

Let's start by counting the number of words in each SMS text message. We can use a simple "**comprehension**," which is a fancy Python term for a one-liner loop. We then use those counts as a new feature in our SMS data frame called "**word_count**" (Listing 11-3 and Figure 11-3).

***Listing 11-3.*** Plot Word-Count per SMS

```
sms_df['word_count'] = [len(x.split()) for x in sms_df['sms']]
sms_df['word_count'].hist().plot()
```

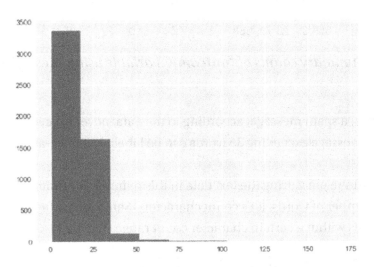

***Figure 11-3.*** *Histogram output of word counts in dataset*

Panda has a handy "**hist().plot**" function that plots the histogram of a data frame series; here we apply it to our new "**word_count**" feature to get a quick feel of the word-count distribution in the dataset. It is clear that the majority of messages contain between 0 and 20 words.

Why would we want to use a word count as a feature? Well, we are hoping that there is an apparent pattern between real messages and spam messages. Maybe real messages range between 5 and 30 words, while spam messages only range between 10 and 20 words. Whatever the case, any differentiating pattern will help our model. Let's find out (Listing 11-4 and Figure 11-4).

***Listing 11-4.*** Differences Between Real and Spam Messages

```
sms_df[['outcome', 'word_count']].groupby('outcome').describe()
```

| word_count | | | | | | | | |
| --- | --- | --- | --- | --- | --- | --- | --- | --- |
| | count | mean | std | min | 25% | 50% | 75% | max |
| outcome | | | | | | | | |
| ham | 4518.0 | 14.233289 | 11.161623 | 1.0 | 7.0 | 11.0 | 19.0 | 171.0 |
| spam | 653.0 | 23.739663 | 5.931064 | 2.0 | 22.0 | 25.0 | 28.0 | 35.0 |

*Figure 11-4.* Summary counts of "**outcome**" variable using "**describe()**" function

So, there it is; a spam message, according to the data, never exceeds 35 words! Right off the bat, any message exceeding 35 words can be labeled as ham–an easy win for the good guys!

We can continue measuring the text data in this manner; for example, instead of counting the number of words, let's count characters. Same idea: Maybe spam messages tend to use words within a certain character count range (Listing 11-5 and Figure 11-5).

*Listing 11-5.* Character cCounts per SMS

```
sms_df['character_count'] = [len(x) for x in sms_df['sms']]
sms_df['character_count'].hist().plot()
```

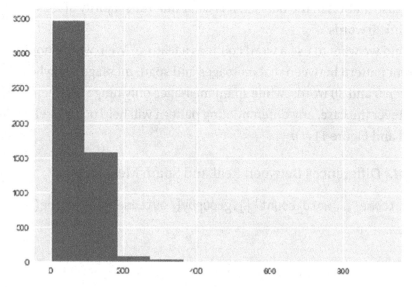

*Figure 11-5.* Character count range histogram

346

Here we see that most messages are less than 100 characters long. And how does this break down between ham and spam messages? (Figure 11-6)

| | character_count | | | | | | | |
| | count | mean | std | min | 25% | 50% | 75% | max |
|---|---|---|---|---|---|---|---|---|
| outcome | | | | | | | | |
| ham | 4518.0 | 70.894865 | 56.590179 | 2.0 | 34.0 | 53.0 | 91.0 | 910.0 |
| spam | 653.0 | 137.710567 | 29.818940 | 13.0 | 132.0 | 148.0 | 157.0 | 223.0 |

*Figure 11-6. The "**character_count**" summary of "**outcome**" variable*

Interestingly, spam messages, according to our historical data, are never shorter than 13 characters–kind of hard selling something when using fewer than 13 characters...

See the corresponding Jupyter notebook for more measurements applied to the SMS data, such as counting punctuation and capital letters. If we can keep highlighting differentiating behavior between both types of messages, our model will keep getting better.

# Text Wrangling for TFIDF

A popular and powerful technique for modeling text is term frequency–inverse document frequency (TFIDF). This is a calculation of the frequency of a word within a document and also within all documents in the corpus. In our case, TFIDF will count the frequency of words in each SMS message and rank them by importance, but will also penalize that importance if it finds that the word is overall too common and not useful in pattern discovery. Before being able to feed our data into TFIDF, we need to wrangle (i.e., prepare) the data a bit more.

A question we need to ask ourselves, and this is relevant to any NLP project, is how much data wrangling is required. When you have a lot of text data and use word vectorization tools such as word2vec,[3] it is recommended to not do any cleaning at all. This is because the model will learn more using raw data than any watered-down version weakened by human assumptions. In those cases, the model will learn best by having

---

[3]https://radimrehurek.com/gensim/models/word2vec.html

access to misspellings or considering words starting with a capital letter and one without as being different, etc. Unfortunately, in order for such approach to be successful, you need lots and lots of data; think Wikipedia-size.

In the case of the SMS dataset, we just don't have enough of it and need to squeeze as much mileage out of it as possible. Take the words "**Won**," "**won**," and "**won!**"; computationally, these are different, but in the hunt for spam, they are the same. Therefore, if we flatten everything down to lower case and remove anything that isn't one of the 26 words of the alphabet (i.e., special characters, numbers, punctuation, etc.), we help the model understand them better by seeing more instances of that word in different situations.

# NLP and Regular Expressions

There are many ways of reducing text data but a popular one is RegEx or Regular expressions. RegEx is a language all unto itself but has been incorporated into many other languages including Python. It uses clever character expression groups to find matches in bodies of text.

Here we will use the Pandas "**str.replace**" function with the regular expression pattern "**[^\w\s]**," which translates to find any ('[') non ('^') word character ('\w') followed by a space character ('\s') and replace them with (""") nothing. This finds anything that isn't a word and removes it. Pretty simple and efficient (for more on this, see JavaScript RegExp Reference on w3schools.com[4]). Then we use the Pandas "**str. lower()**" function to force all the remaining words to lower case (Listing 11-6).

*Listing 11-6.* Remove All Special Characters, Numbers, Punctuation and Force to Lower Case

**Input:**

```
sms_df["sms_clean"] = sms_df['sms'].str.replace('[^\w\s]',")
sms_df["sms_clean"] = sms_df['sms_clean'].str.lower()
sms_df["sms_clean"].head()
```

---

[4]https://www.w3schools.com/jsref/jsref_obj_regexp.asp

**Output**:

```
0    go until jurong point crazy available only in ...
1                              ok lar joking wif u oni
2    free entry in 2 a wkly comp to win fa cup fina...
3           u dun say so early hor u c already then say
4    nah i dont think he goes to usf he lives aroun...
Name: sms_clean, dtype: object
```

We end up with funny looking sentences that aren't very easy to read, at least not for humans.

The list of additional things we could do in the wrangling department is long. Even though we won't use them here, good next steps are "**stemming**" and "**lemmatization**" to reduce words even further down to a common root. For example, if you feed it the words "**organize**," "**organizes**," and "**organizing**," it will reduce them all down to "**organi**."[5] We are getting close to feeding our cleaned data into TFIDF and RF, but there is one easier win for us to claim before modeling.

## Using an External List of Typical Spam Words

Spam has been around for a long time and examples are plentiful. Many services and amateurs collect and curate lists of words, sentences, and even full messages deemed as spam-like. I have curated a simple list of my own, containing words such as **baldness**, **cash**, **cure**, **guaranteed**, **lifetime**, **opportunity**, **wealth**, **winning**, etc.

I included this simple list in this chapter's downloads, and we will use it to compare against each SMS message and tally how many spam words are contained in them. The idea is that most normal, everyday messages won't contain words such as winning or cash, but many spam messages do. And, as with most of our feature engineering, we'll end up with a numerical feature that is what we need for our models.

The comprehension we use here to count the intersecting words between an SMS message and the external spam word list is long. But what it does is simple, loop through and create a list of words for each SMS message, then intersect it with the spam list and count occurrences (Listing 11-7).

---

[5]https://nlp.stanford.edu/IR-book/html/htmledition/stemming-and-lemmatization-1.html

***Listing 11-7.*** Create Counts of Spam Words

```
sms_df["external_spam_word_count"] = [len([(x) for x in sent.split() if x
in spam_list]) for sent in sms_df["sms_clean"].values]
```

And is it going to help our model? (Listing 11-8 and Figure 11-7)

***Listing 11-8.*** Does It Help Predicting Spam?

```
sms_df[["outcome", "external_spam_word_count"]].groupby('outcome').describe()
```

| | **external_spam_word_count** | | | | | | | |
| | **count** | **mean** | **std** | **min** | **25%** | **50%** | **75%** | **max** |
| **outcome** | | | | | | | | |
| **ham** | 4518.0 | 1.283311 | 1.482789 | 0.0 | 0.0 | 1.0 | 2.0 | 14.0 |
| **spam** | 653.0 | 3.762634 | 1.962145 | 0.0 | 2.0 | 4.0 | 5.0 | 11.0 |

***Figure 11-7.*** *Spam word counts by "**outcome**" variable*

Yes, it will! If you look at the mean count of ham vs. spam words, spam has more than twice the amount of external spam words than ham does.

# Feature Extraction with Sklearn's TfidfVectorizer

At this point we have gathered enough quantitative features. Let's run the TFIDF vectorizer, which is like one massive feature engineering calculation of every word in our dataset against every other (Listing 11-9 and Figure 11-8).

***Listing 11-9.*** Vectorizing the Data

```
vectorizer = TfidfVectorizer()
vectors = vectorizer.fit_transform(sms_df['sms_clean'])
vectorized_df = pd.DataFrame(vectors.toarray())
vectorized_df.head()
```

| | 0 | 1 | 2 | 3 | 4 | 5 | 6 | 7 | 8 | 9 | ... | 9536 | 9537 | 9538 | 9539 | 9540 | 9541 | 9542 | 9543 | 9544 | 9545 |
|---|---|---|---|---|---|---|---|---|---|---|---|---|---|---|---|---|---|---|---|---|---|
| 0 | 0.0 | 0.0 | 0.0 | 0.0 | 0.0 | 0.0 | 0.0 | 0.0 | 0.0 | 0.0 | ... | 0.0 | 0.0 | 0.0 | 0.0 | 0.0 | 0.0 | 0.0 | 0.0 | 0.0 | 0.0 |
| 1 | 0.0 | 0.0 | 0.0 | 0.0 | 0.0 | 0.0 | 0.0 | 0.0 | 0.0 | 0.0 | ... | 0.0 | 0.0 | 0.0 | 0.0 | 0.0 | 0.0 | 0.0 | 0.0 | 0.0 | 0.0 |
| 2 | 0.0 | 0.0 | 0.0 | 0.0 | 0.0 | 0.0 | 0.0 | 0.0 | 0.0 | 0.0 | ... | 0.0 | 0.0 | 0.0 | 0.0 | 0.0 | 0.0 | 0.0 | 0.0 | 0.0 | 0.0 |
| 3 | 0.0 | 0.0 | 0.0 | 0.0 | 0.0 | 0.0 | 0.0 | 0.0 | 0.0 | 0.0 | ... | 0.0 | 0.0 | 0.0 | 0.0 | 0.0 | 0.0 | 0.0 | 0.0 | 0.0 | 0.0 |
| 4 | 0.0 | 0.0 | 0.0 | 0.0 | 0.0 | 0.0 | 0.0 | 0.0 | 0.0 | 0.0 | ... | 0.0 | 0.0 | 0.0 | 0.0 | 0.0 | 0.0 | 0.0 | 0.0 | 0.0 | 0.0 |

5 rows × 9546 columns

*Figure 11-8.* *TfidfVectorizer output*

Wow! "**TfidfVectorizer**" created 9,546 new features and did it pretty quickly! There are a few things to remember here: "**TfidfVectorizer**" is a powerful feature engineering tool, as it considers not only every word but the entire context as well. It results in a very sparse matrix, so this may be a problem for very large corpuses, but it will work just fine in our case. If that is a problem, it can also limit the number of words it will consider in its vectors by feeding it a limited set of vocabulary words.

Now we just need to join these new features to the previous features we created earlier.

# Preparing the Outcome Variable

With all supervised models, we need to be clear on what we are trying to predict. Here we want to predict whether a message is ham or not ham. So, ham messages need to be labeled as 1, while spam messages need to be labeled as 0. This can easily be done using the Pandas "**Categorical()**" function (Listing 11-10).

*Listing 11-10.* Removing Duplicate Rows

**Input**:

```
print(all_df[outcome].head())
all_df[outcome] = pd.Categorical(all_df[outcome],
categories=["spam","ham"])
all_df[outcome] = all_df[outcome].cat.codes
print(all_df[outcome].head())
```

**Output**:

```
0       ham
1       ham
2       spam
3       ham
4       ham
Name: outcome, dtype: object
0       1
1       1
2       0
3       1
4       1
Name: outcome, dtype: int8
```

By printing a before and after transformation of our outcome feature, we confirm that hams use the digit "**1**" and spams use "**0**."

# Modeling with Sklearn's RandomForestClassifier

We have now collected all the quantitative features needed to start running our random forest classifier. Random forest, as its name implies, will create many sets of random feature-trees and train and predict using those trees against the outcome variable. It then will bring all those predictions back together (i.e., ensemble them back), with the assumption that it will catch many more nuances than a single model would. In essence, a random forest isn't really "**a**" single model but a collection of many, with differing views and understanding of the data.

The "**sklearn.ensemble**" library has an efficient and easy-to-use random forest classifier, aptly named RandomForestClassifier. Here we ask it to run two parallel jobs, as most computers today have at least two CPUs and the model will run much faster, and 100 decision tree classifiers (the official documentation lists plenty more powerful options to explore[6]). See listing 11-11.

---

[6]http://scikit-learn.org/stable/modules/generated/sklearn.ensemble. RandomForestClassifier.html

***Listing 11-11.*** Running the Random Classifier Model

```
rf_model = RandomForestClassifier(n_jobs=2, random_state=0, n_estimators=100)
rf_model.fit(X_train, y_train)
```

Once we have the "**rf_model**" trained, we can run predictions on the test set to measure the model's performance. The model offers predictions in both class and probability formats (Listings 11-12 and 11-13).

***Listing 11-12.*** Predicting Spam vs. Ham

**Input**:

```
prediction_classes = rf_model.predict(X_test)
```

**Output**:

```
[1 1 1 ... 1 1 1]
```

***Listing 11-13.*** Getting Probabilities for Spam vs. Ham

**Input**:

```
prediction_probas = rf_model.predict_proba(X_test)
```

**Output**:

```
[[0.08 0.92]
 [0.14 0.86]
 [0.   1.  ]
 ...
 [0.01 0.99]
 [0.   1.  ]
 [0.   1.  ]]
```

# Measuring the Model's Performance

The sklearn.metrics library has a large amount of functions to help us measure how well a model is performing. We'll start with the "**classification_report**" (Listing 11-14).

***Listing 11-14.*** Getting Model Metrics

**Input**:

```
sklearn.metrics import classification_report
print(classification_report(y_test, prediction_classes))
```

**Output**:

|         | precision | recall | f1-score | support |
|---------|-----------|--------|----------|---------|
| 0       | 0.92      | 0.40   | 0.56     | 302     |
| 1       | 0.92      | 0.99   | 0.96     | 2108    |
| avg / total | 0.92  | 0.92   | 0.91     | 2410    |

We see that our model does a great job predicting ham messages (second row) and does really well on "**precision**" (how well you did among what you labeled) for spam messages. The model struggles a bit on "**recall**" (how well you did among the full test set) and "**f1-score**" (score between 1 and 0 based on precision and recall) for spam messages. This is to be expected because the data is skewed, and we don't have as many spam messages to train on.

The "**confusion matrix**" is another powerful tool to visualize how well a model performs. It is related to the precision and recall but uses a different terminology: true positive, false positive, true negative, and false negative (Figure 11-9).

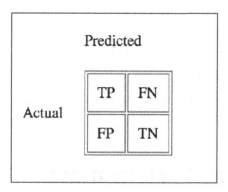

***Figure 11-9.***  *How to read a confusion matrix*

The sklearn.metrics "**confusion_matrix**" function takes the ground truth outcome labels, the predicted classes, and the label order. It is important to state you want the positive label first (i.e., is ham). It makes interpreting the confusion matrix much easier (Listing 11-15).

***Listing 11-15.*** Getting Confusion Matrix Metrics

**Input:**

```
from sklearn.metrics import confusion_matrix
cm = confusion_matrix(y_test, prediction_classes, [1,0])
print('Total length of test set: %i' % len(y_test))
print('total hams in test set: %i' % sum(y_test==1))
print('total spams in test set: %i' % sum(y_test==0))cm
```

**Output:**

```
Total length of test set: 2410
total hams in test set: 2108
total spams in test set: 302

array([[2097,   11],
       [ 180,  122]])
```

In the upper left corner of the array, we see that the model succeeded in correctly predicting 2,097 ham messages out of 2,108 (2,097 + 11) and succeeded in predicting 122 spam messages out of 302 (180 + 122). Those are the "**TP**" and "**TN**." TP means true positive where it's succeeded in predicting a message is ham (ham = 1), and TN means true negative where it succeeded in predicting that a message isn't ham (ham = 0).

Digging deeper, "**FN**" is when the model labels a message a spam when it is in fact ham (ham = 0 but really ham =1). In the array, the ham row is the top one, and the model predicted 11 messages as ham when they were in fact spam messages. In the context of ham vs. spam, this number is the one that hurts a lot, as it's when a user's personal messages get dumped into the spam folder.

"**FP**" is when the model labels a message as ham when it is in fact spam (setting ham = 1 when in reality ham = 0). In the first cell of the ham array, array at the bottom, the model predicted 180 messages as ham when they were in fact spam.

The concept of the confusion matrix is very important, as a model will give you additional flexibility beyond the base accuracy at a probability threshold of 0.5, where anything below is 0 and anything above is 1. By playing around with this threshold, you can squeeze more use out of it depending on whether more FN or more FP is better for your business needs than a middle-of-the-road threshold. After all, this is what our web application dashboard is all about.

The ROC chart is a great tool to understand the value of a binary classifier and how we can vary the threshold for different effects (Figure 11-10).

Let's consider the ROC curve for our model. Overall, its doing really well with an area under the curve (AUC) score of 0.93. The AUC score ranges between 0.5 and 1, where 0.5

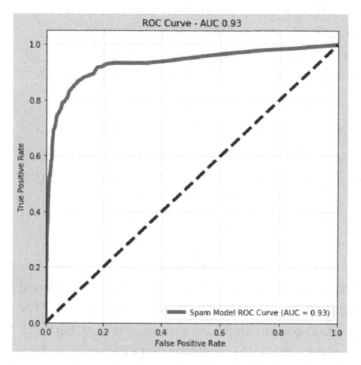

***Figure 11-10.***  *The ROC curve of our ham vs. spam model*

is random and 1 is perfect. So, an ideal AUC of 1 would make the green line in the chart go from the bottom left at 0 straight up to 1, then straight across from left to right (in other words, it covers the entire upper left triangle).

The AUC represents an area where the score is constant but, by sliding up or down the outer edge of the curve, you can play with the "**true positive rate**" and "**false**

**positive rate**." This may seem like a strange concept that you have flexibility with a model without affecting its AUC score. At a high level, part of the reason for this flexibility is that it really isn't a "**flexibility**" as much as a "**compromise**." The higher up the curve you go (higher up the green line), the more your predicted positive outcomes are accurate but the fewer of them you end up with, as those you've traveled over switch from positive to negative outcomes.

This goes back to our earlier example that an emergency room with limited resources may rather have higher precision predictions and fewer of them, thus sliding up the green AUC line. While in our case with ham vs. spam, an email user wouldn't tolerate having good messages disappear in the spam box, thus sliding down the AUC line would be preferable (knowing that more negative messages would be relabeled as positive, therefore accepting that more real spam messages would end up in the inbox).

# Interacting with the Model's Threshold

Let's take a look at how our dashboard will illustrate the flexibility of the probability threshold. If we take a standard probability cutoff of 0.5 (i.e., in our case anything above 0.5 is ham and below is spam). See Listing 11-16 and Figure 11-11.

***Listing 11-16.*** Confusion Matrix with 0.5 Cutoff

```
prediction_tmp = [1 if x >= 0.5 else 0 for x in prediction_probas[:, 1]]
```

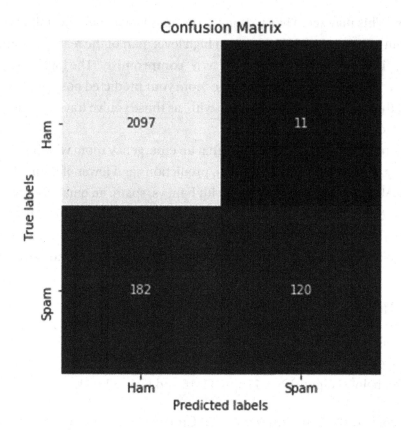

***Figure 11-11.*** *A Confusion Matrix with a Simple 0.5 Cutoff*

We get the following graphical confusion matrix, and we read that the model correctly identified 2,097 as ham and 120 as spam but mislabeled 11 ham messages as spam–not good! Now if we lower our threshold to 0.3, let's see how many ham messages get labeled as spam. See Listing 11-17 and Figure 11-12.

***Listing 11-17.*** Confusion Matrix with 0.3 Cutoff

```
prediction_tmp = [1 if x >= 0.3 else 0 for x in prediction_probas[:, 1]]
```

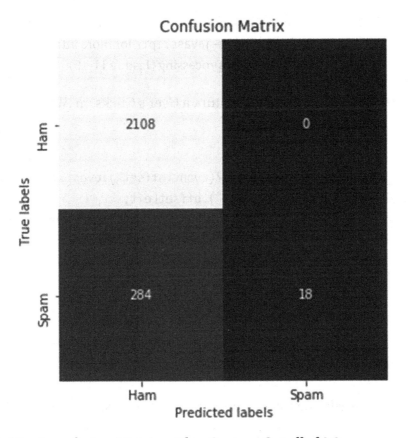

***Figure 11-12.*** *A confusion Matrix with a Custom Cutoff of 0.3*

With a threshold cutoff of 0.3, the model didn't mislabel any ham messages as spam! But the cost is that we went from only mislabeling 182 spam messages as ham up to 284 (i.e., advertisements making it into the user's inbox). And the spam folder only received ten correctly labeled spam messages instead of 120. There you have it; *that is the cost.* In situations like ham vs. spam, that cost is trivial because nobody wants to lose any personal messages and they are willing to tolerate a lot of spam in order to achieve that goal.

# Interacting with Web Graphics

A nice feature we are going to implement here is to allow the users to click on the AUC image to experiment with the model's threshold. This is a very intuitive way of getting your users to interact with the web page and the concepts surrounding this chapter. Capturing a user's click event is easily done using JavaScript and capturing "**event offsets**" (code based

on Emanuele Feronato post;, see http://www.emanueleferonato.com/2006/09/02/
click-image-and-get-coordinates-with-javascript/ for more information on this
approach) and passing it back to Flask for processing (Listing 11-18).

***Listing 11-18.*** JavaScript Code to Capture a User's Clicks on AUC Chart

```
function point_it(event)
{
        cur_x_coord = event.offsetX?(event.offsetX):event.pageX-document.
        getElementById("pointer_div").offsetLeft;
        cur_y_coord = event.offsetY?(event.offsetY):event.pageY-document.
        getElementById("pointer_div").offsetTop;

        <!-- send coordinates back to Flask application -->
        fetchdata(cur_x_coord, cur_y_coord)
}
```

And this is translated in Flask to a new threshold using a series of if/then statements
(Listing 11-19).

***Listing 11-19.*** Translating User Clicks into Cutoff Thresholds

```
x_image_coord = int(request.args.get('new_x_coord'))
y_image_coord = int(request.args.get('new_y_coord'))

new_thres = 0.0
# translate coordinates to threshold
if (y_image_coord >= 360 and y_image_coord < 390):
    new_thres = 0.1
elif (y_image_coord >= 340 and y_image_coord < 360):
    new_thres = 0.2
elif (y_image_coord >= 290 and y_image_coord < 340):
    new_thres = 0.3
elif (y_image_coord >= 260 and y_image_coord < 290):
    new_thres = 0.4
elif (y_image_coord >= 220 and y_image_coord < 260):
    new_thres = 0.5
```

```
elif (y_image_coord >= 185 and y_image_coord < 220):
    new_thres = 0.6
elif (y_image_coord >= 150 and y_image_coord < 185):
    new_thres = 0.7
elif (y_image_coord >= 115 and y_image_coord < 150):
    new_thres = 0.8
elif (y_image_coord >= 75 and y_image_coord < 115):
    new_thres = 0.9
elif (y_image_coord < 75):
    new_thres = 1
```

# Building Our Web Application—Local Flask Version

We thought through our model and dashboard concept, so now it is time to build it. Let's start by building a local Flask version.

Let's download the files for Chapter 11 onto your local machine. Once you have downloaded and unzipped everything, open a command line window, and change the drive into the "**web-application**" folder. Your "**web-application**" folder should contain the following files as shown in Listing 11-20.

***Listing 11-20.***  Web aApplication Files

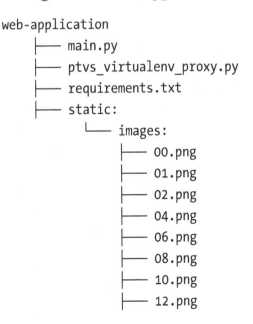

```
web-application
      ├── main.py
      ├── ptvs_virtualenv_proxy.py
      ├── requirements.txt
      ├── static:
            └── images:
                  ├── 00.png
                  ├── 01.png
                  ├── 02.png
                  ├── 04.png
                  ├── 06.png
                  ├── 08.png
                  ├── 10.png
                  ├── 12.png
```

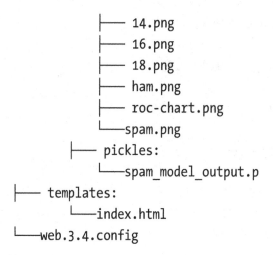

```
├─── 14.png
├─── 16.png
├─── 18.png
├─── ham.png
├─── roc-chart.png
└───spam.png
├─── pickles:
└───spam_model_output.p
├─── templates:
└───index.html
└───web.3.4.config
```

Start a virtual environment (see Listing 11-21).

***Listing 11-21.*** Starting Up the Virtual Environment

```
$ python3 -m venv hamspamenv
$ source hamspamenv/bin/activate
```

Then install all the required Python libraries by running the "**pip3 install -r**" command (Listing 11-22).

***Listing 11-22.*** Installing Required Libraries

```
$ pip3 install -r requirements.txt
```

And run the web application on your local machine (Listing 11-23 and Figure 11-13).

***Listing 11-23.*** Taking the Web Application for a Spin

```
$ python3 main.py
```

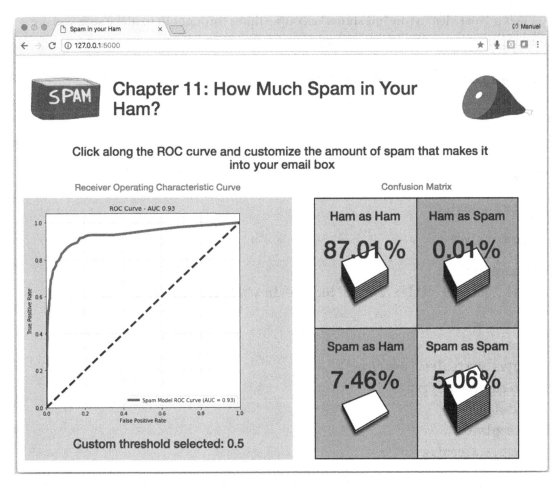

*Figure 11-13.* *Local version of our web application*

# Deploying to Microsoft Azure

It's time to deploy our web application to the cloud. We'll do a very brief fly-by, as we've seen these steps a bunch of times already.

## Git for Azure

Initialize a Git session (Listing 11-24).

*Listing 11-24.* Initializing Git

```
$ git init
```

It is a great idea to run "**git status**" a couple times throughout to make sure you are tracking the correct files (Listing 11-25).

***Listing 11-25.*** Running "**git status**"

**Input**:

```
$ git status
```

**Output**:

```
On branch master

No commits yet

Untracked files:
  (use "git add <file>..." to include in what will be committed)

    hamspamenv/
    main.py
    ptvs_virtualenv_proxy.py
    requirements.txt
    static/
    templates/
    web.3.4.config
```

Add all the web-application files from the "**web-application**" file using the "**git add**," command and check "**git status**" again (Listing 11-26).

***Listing 11-26.*** Adding to Git

**Input**:

```
$ git add .
$ git status
```

**Output**:

```
Changes to be committed:
  (use "git rm --cached <file>..." to unstage)
```

```
new file:    hamspamenv/lib/python3.6/site-packages/werkzeug/urls.py
new file:    hamspamenv/lib/python3.6/site-packages/werkzeug/useragents.py
new file:    hamspamenv/lib/python3.6/site-packages/werkzeug/utils.py
new file:    hamspamenv/lib/python3.6/site-packages/werkzeug/websocket.py
new file:    hamspamenv/lib/python3.6/site-packages/werkzeug/wrappers.py
new file:    hamspamenv/lib/python3.6/site-packages/werkzeug/wsgi.py
```

...

You may have noticed that we have added a lot of files to our "**git add.**" command. As per instructions from "**git status**," it tells us how to remove files that we don't want to commit to Git with the "**rm**" command. Let's remove all files and folder from the virtual environment "**hamspamenv**" that aren't needed for the project (Listing 11-27).

*Listing 11-27.* Removing "**hamspamenv**" from Git

**Input**:

```
$ git rm -r --cached hamspamenv
$ git status
```

**Output**:

```
Changes to be committed:
  (use "git rm --cached <file>..." to unstage)

    new file:    main.py
    new file:    ptvs_virtualenv_proxy.py
    new file:    requirements.txt
    new file:    static/images/00.png
    new file:    static/images/01.png
    new file:    static/images/02.png
    new file:    static/images/04.png
    new file:    static/images/06.png
    new file:    static/images/08.png
    new file:    static/images/10.png
    new file:    static/images/12.png
    new file:    static/images/14.png
    new file:    static/images/16.png
    new file:    static/images/18.png
```

```
    new file:    static/images/ham.png
    new file:    static/images/roc-chart.png
    new file:    static/images/spam.png
    new file:    static/pickles/spam_model_output.p
    new file:    templates/index.html
    new file:    web.3.4.config
```

```
Untracked files:
  (use "git add <file>..." to include in what will be committed)

    hamspamenv/
```

We now have only the files we need. So, do a local "**git commit**" and add a comment that makes sense in case you need to revisit past actions in the future (Listing 11-28).

***Listing 11-28.***  Git Commit

**Input**:

```
$ git commit -am 'hamspamenv deployment on Azure'
```

**Output**:

```
[master (root-commit) 8c03a49] hamspamenv deployment on Azure
 20 files changed, 450 insertions(+)
 create mode 100644 main.py
 create mode 100644 ptvs_virtualenv_proxy.py
 create mode 100644 requirements.txt
 create mode 100644 static/images/00.png
 create mode 100644 static/images/01.png
 create mode 100644 static/images/02.png
 create mode 100644 static/images/04.png
 create mode 100644 static/images/06.png
 create mode 100644 static/images/08.png
 create mode 100644 static/images/10.png
 create mode 100644 static/images/12.png
 create mode 100644 static/images/14.png
 create mode 100644 static/images/16.png
 create mode 100644 static/images/18.png
 create mode 100644 static/images/ham.png
```

```
create mode 100644 static/images/roc-chart.png
create mode 100644 static/images/spam.png
create mode 100644 static/pickles/spam_model_output.p
create mode 100644 templates/index.html
create mode 100644 web.3.4.config
```

For more information on the Git Deployment to Azure App Service, see https://docs.microsoft.com/en-us/azure/app-service/app-service-deploy-local-git.

# The azure-cli Command Line Interface Tool

## Step 1: Logging In

Create an "**az**" session (Listing 11-29 and Figure 11-14).

***Listing 11-29.*** Logging into "**az**"

```
$ az login
```

```
[manuels-MacBook-Pro-2:web-application manuel$ az login
 To sign in, use a web browser to open the page https://microsoft.com/devicelogin
 and enter the code B54YSXKF2 to authenticate.
```

***Figure 11-14.*** *Logging into Azure from azure-cli*

Follow the instructions, point a browser to the given URL address, and enter the code accordingly (Figure 11-15).

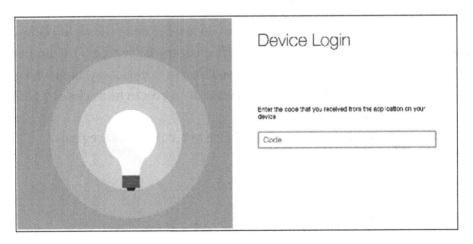

***Figure 11-15.*** *Authenticating session*

If all goes well (i.e., you have an Azure account in good standing), it will connect the azure-cli terminal to the cloud server. Also, once you are authorized, you can safely close the browser window.

Make sure your command-line tool is pointing to this chapter's "**web-application**" folder.

# Step 2: Create Credentials for Your Deployment User

This user will have appropriate rights for FTP and local Git use. Here I set the user-name to "**flaskuserXX**" and password to "**flask123**". You should only have to do this once; then you can reuse the same account. In case it gives you trouble, simply create a different user name (or add a number at the end of the user name and keep incrementing it like I do; Listing 11-30).

*Listing 11-30.* Setting Deployment User

```
$ az webapp deployment user set --user-name flaskuser30 --password flask123
```

As you proceed through each "**azure-cli**" step, you will get back JSON replies confirming your settings. In the case of the "**az webapp deployment**," most should have a null value and no error messages. If you have an error message, then you have a permission issue that needs to be addressed ("**conflict**" means that name is already taken so try another, and "**bad requests**" means the password is too weak).

# Step 3: Create Your Resource Group

This is going to be your logical container. Here you need to enter the region closest to your location (see https://azure.microsoft.com/en-us/regions/). Going with "**West US**" for this example isn't a big deal even if you're worlds away, but it will make a difference in a production setting where you want the server to be as close as possible to your viewership for best performance.

Here I set the name to https://azure.microsoft.com/en-us/regions/ myResourceGroup (Listing 11-31).

*Listing 11-31.* Creating Group

```
$ az group create --name myResourceGroup --location "West US"
```

## Step 4: Create Your Azure App Service Plan

Here I set the name to "**myAppServicePlan**" and select a free instance (sku) (Listing 11-32).

*Listing 11-32.* Creating Service Plan

```
$ az appservice plan create --name myAppServicePlan --resource-group
myResourceGroup --sku FREE
```

## Step 5: Create Your Web App

Your "**webapp**" name needs to be unique, and make sure your "**resource-group**" and "**plan**" names are the same as what you set in the earlier steps. In this case I am going with "**amunateguihamspam.**" For a full list of supported runtimes, run the "**list-runtimes**" command (Listing 11-33).

*Listing 11-33.* Supported Runtimes

```
$ az webapp list-runtimes
```

To create the web application, use the "**create**" command (Listing 11-34).

*Listing 11-34.* Creating the Webapp

```
$ az webapp create --resource-group myResourceGroup --plan myAppServicePlan
--name amunateguihamspam --runtime "python|3.4" --deployment-local-git
```

The output of "**az webapp create**" will contain an important piece of information that you will need for subsequent steps. Look for the line "**deploymentLocalGitUrl**" (Figure 11-16).

```
(hamspamenv) manuels-MacBook-Pro-3:web-application manuelamunategui$ az web
app create --resource-group myResourceGroup --plan myAppServicePlan --name
amunateguihamspam --runtime "python|3.4" --deployment-local-git
Local git is configured with url of 'https://flaskuser30@amunateguihamspam.
scm.azurewebsites.net/amunateguihamspam.git'
{
  "availabilityState": "Normal",
  "clientAffinityEnabled": true,
  "clientCertEnabled": false,
  "cloningInfo": null,
  "containerSize": 0,
  "dailyMemoryTimeQuota": 0,
  "defaultHostName": "amunateguihamspam.azurewebsites.net",
  "deploymentLocalGitUrl": "https://flaskuser30@amunateguihamspam.scm.azure
websites.net/amunateguihamspam.git",
  "enabled": true,
  "enabledHostNames": [
```

*Figure 11-16.* *Output of "**az webapp create**"; note your deployment Git URL*

Start "**Git**" if you haven't already (and install it if you never used it before at `https://git-scm.com/book/en/v2/Getting-Started-Installing-Git`).

## Step 6: Push Git Code to Azure

Now that you have a placeholder web site, you need to push out your Git code to Azure (Listing 11-35).

*Listing 11-35.* Adding Remote User

```
# if git remote is say already exits, run 'git remote remove azure'
$ git remote add azure "https://flaskuser30@amunateguihamspam.scm.
azurewebsites.net/amunateguihamspam.git"
```

Finally, push it out to Azure (Listing 11-36).

*Listing 11-36.* Push It Out (enter the "**webapp deployment user**" password when prompted)

```
$ git push azure master
```

It will prompt you for your "**webapp deployment user**" password you set up earlier. This may take a while, as we have to upload a bunch of corollary files like images and dataset. If all goes well, you should be able to enjoy the fruits of your labor. Open a web browser and enter your new URL that is made of your "**webapp**" name followed by "**.azurewebsites.net**" (http://amunateguihamspam.azurewebsites.net).

On the other hand, if the azure-cli returns error messages, you will have to address them (see the "**Troubleshooting**" section). Anytime you update your code and want to redeploy it, see Listing 11-37.

***Listing 11-37.*** Committing and Pushing Out

```
$ git commit -am "updated output"
$ git push azure master
```

You can also manage your application directly on Azure's web dashboard. Log into Azure and go to App Services (Figure 11-17).

***Figure 11-17.*** *Managing your application directly in the Microsoft Azure dashboard*

# Important Cleanup!

This is a critical step; you should never leave an application running in the cloud that you don't need, as it does incur charges (or use up your free credits if you are on the trial program). If you don't need it anymore, take it down (Listing 11-38).

***Listing 11-38.*** Tear-Down Time (you will be asked to confirm this action)

```
$ az group delete --name myResourceGroup
```

Or delete it using Azure's web dashboard under "**App Services.**" And finally, deactivate your virtual environment (Listing 11-39).

**Listing 11-39.**  End Your Virtual Session if You Didn't Do So Earlier

```
$ deactivate hamspamenv
```

# Troubleshooting

It can get convoluted to debug web application errors. One thing to do is to turn on logging through Azure's dashboard (Figure 11-18).

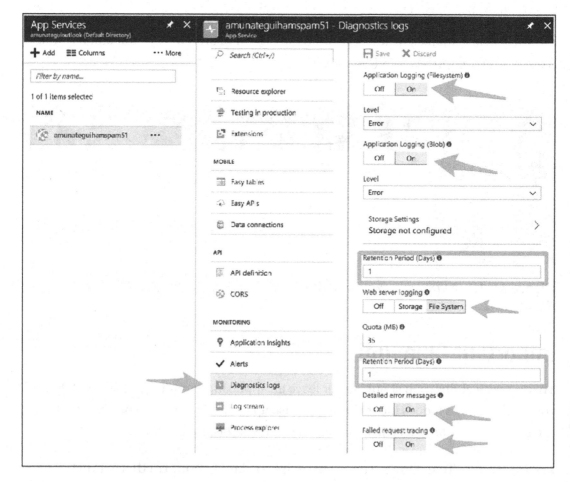

**Figure 11-18.**  Turning on Azure's Diagnostics logs

Then you turn the logging stream on to start capturing activity (Figure 11-19).

*Figure 11-19.* *Capturing log information*

You can also check your file structure using the handy Console tool built into the Azure dashboard (Figure 11-20).

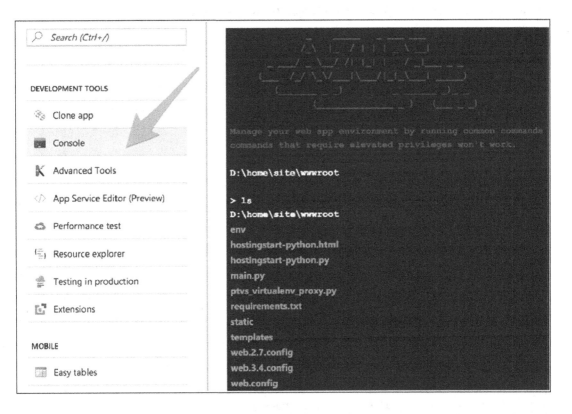

*Figure 11-20.* *Azure's built-in command line tool*

You can also access the tail of the log in your command window (Listing 11-40).

***Listing 11-40.*** Access the Log

```
$ az webapp log tail --resource-group myResourceGroup --name
amunateguihamspam
```

You can even check if your **"requirement.txt"** file works by calling the "**env\scripts\ pip**" function (Listing 11-41).

***Listing 11-41.*** Checking That You Can Install Your Python Libraries

```
$ env\scripts\pip install -r requirements.txt
```

# Conclusion and Additional Resources

Azure is one of the top three web hosting platforms currently available and thus a great choice for enterprise and machine learning solutions.

For additional information on the Azure deployment process, see the detailed and clear Azure document "**Create a Python web ap in Azure.**"[7]

---

[7]https://docs.microsoft.com/en-us/azure/app-service/
app-service-web-get-started-python

# CHAPTER 12

# Case Study Part 3: Enriching Content with Fundamental Financial Information

Predicting the stock market with fundamental financial data aggregation on PythonAnywhere.

We're going to keep adding features to our "**Pair Trading Booth**" web application (Figure 12-1).

© Manuel Amunategui, Mehdi Roopaei 2018
M. Amunategui and M. Roopaei, *Monetizing Machine Learning*, https://doi.org/10.1007/978-1-4842-3873-8_12

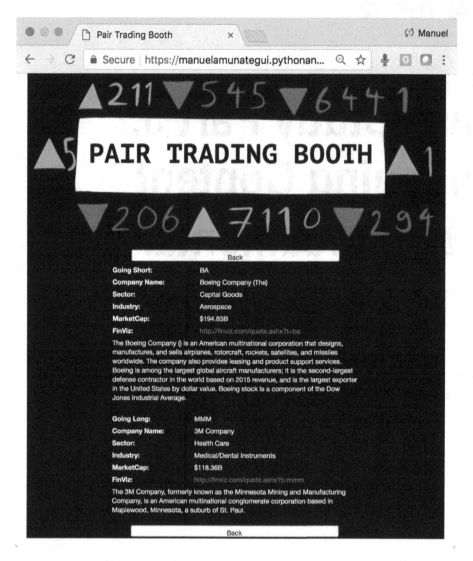

***Figure 12-1.*** *The final web application for this chapter*

So far, we told our visitors about the best pair trade to make, showed them the related financial charts, and now we're going to give them critical fundamental details about the companies behind the stocks mentioned.

- Full name of the company

- Short introduction of the company from Wikipedia.com

- Market capitalization

- Market sector

- Market industry

- Dynamic link to financial site: Finviz.com

---

**Note**    Download the files for Chapter 12 by going to `www.apress.com/9781484238721` and clicking the source code button. Open Jupyter notebook "**chapter12.ipynb**" to follow along with this chapter's content.

---

# Accessing Listed Stocks Company Lists

A stock symbol is short and vague and can mean different things if you drop it into a web browser for Internet searches. We need to tie it to its full company name to guarantee its uniqueness. The Nasdaq website[1] offers a great series of CSV files for us to use that matches the symbol to additional corollary information including the full company name. Point your browser to

`https://www.nasdaq.com/screening/company-list.aspx`.

Download all three files to your local machine. Make sure to rename them in the following format, otherwise they will all be called "**companylist.csv**":

- companylist_NASDAQ.csv

- companylist_AMEX.csv

- companylist_NYSE.csv

Let's find matches between our ten stock symbols and their location in the downloaded files. This is easily done by using the list of the ten symbols we are interested in and looping through each symbol in the company lists (Listing 12-1). Because these stocks are very well known, you will find a match for each one of them (this may not be the case for smaller cap companies).

---

[1]`https://www.nasdaq.com/`

***Listing 12-1.*** Checking Stock Symbols in "**companylist**"

**Input**:

```
stock_symbols = ['BA','GS','UNH','MMM','HD','AAPL','MCD','IBM','CAT','TRV']

print('Symbols found in the Nasdaq list:')
list(set(stock_symbols) & set(list(stock_company_info_nasdaq['Symbol'])))
```

**Output**:

```
Symbols found in the Nasdaq list:
['AAPL']
```

We found out that the "**companylist_NASDAQ.csv**" list contains one symbol, "**AAPL**". We can now pull that row out of the company list "**CSV**" file and save it. We proceed in the same manner for the other symbols in the other two company lists (Listing 12-2 and Figure 12-2).

***Listing 12-2.*** Querying "**companylist**"

```
stock_company_info_nasdaq[stock_company_info_nasdaq['Symbol'] == 'AAPL']
```

| | Symbol | Name | LastSale | MarketCap | IPOyear | Sector | industry |
|---|---|---|---|---|---|---|---|
| **196** | AAPL | Apple Inc. | 183.83 | $932.76B | 1980.0 | Technology | Computer Manufacturing |

***Figure 12-2.*** *The extra intelligence we extract from the Nasdaq company list*

We now know that "**AAPL**" equates to "**Apple Inc.**" and we can also get the market cap, IPO year, sector, and industry for that stock symbol. All this information is of great use to our users. Also, we now have the exact spelling of the company name, which we can use to pull additional information about this company from www.wikipedia.org. As mentioned before, some of these symbol names are too simple and won't necessarily return the correct information from a web search, but if we combine a symbol name with the actual company name, we have a much better chance of pulling exactly what we're looking for. (Keep in mind you may still find edge cases where you will pull something unrelated.)

# Pulling Company Information with the Wikipedia API

Wikipedia has a great and easy to use API in Python that we will leverage to add to depth to our application. We will pull the introductory paragraph for each company that we are recommending (the introductory paragraph is simply the first paragraph returned; Listing 12-3).

***Listing 12-3.*** Wikipedia Query

**Input**:

```
import wikipedia
description = wikipedia.page("Apple Inc.").content
description = description.split('\n')[0]
description
```

**Output**:

```
"Apple Inc. is an American multinational technology company headquartered
in Cupertino, California, that designs, develops, and sells consumer
electronics, computer software, and online services. The company's hardware
products include the iPhone smartphone, the iPad tablet computer, the
Mac personal computer, the iPod portable media player, the Apple Watch
smartwatch, the Apple TV digital media player, and the HomePod smart
speaker. Apple's software includes the macOS and iOS operating systems,
the iTunes media player, the Safari web browser, and the iLife and iWork
creativity and productivity suites, as well as professional applications
like Final Cut Pro, Logic Pro, and Xcode. Its online services include
the iTunes Store, the iOS App Store and Mac App Store, Apple Music, and
iCloud."
```

## Building a Dynamic FinViz Link

FinViz.com is a treasure trove of financial fundamental data. We are not going to scrape from them; instead we're going to build dynamic links so that our users can opt to go there for the additional information. This ensures we're not stealing information from others. It is my recommendation to always open a link in a new page; this ensures that the user still has an easy way for getting back to your property (Listing 12-4 and Figure 12-3).

***Listing 12-4.*** Finviz Link Making

**Input:**

```
predictions_df = pd.DataFrame(all_user_predicted_ratings, columns = movies_
df.index)
symbol = 'AAPL'
url = r'http://finviz.com/quote.ashx?t={}'.format(symbol.lower())
url
```

**Output:**

```
http://finviz.com/quote.ashx?t=aapl
```

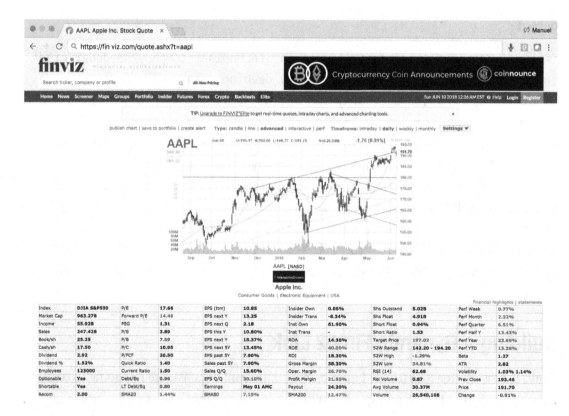

***Figure 12-3.***  *Finviz output link for Apple, Inc*

# Exploring Fundamentals

Go ahead and download the files for this chapter into a folder called "**chapter-12**." Open up the Jupyter notebook to follow along.

Let's abstract our three fundamental offerings into three clean and simple-to-use functions that can be easily integrated into our web application. We'll start with the "**GetCorollaryCompanyInfo()**" function. This function will pull the company name, the sector, the industry, and the market capitalization of both the long and short stock symbols in our trade (Listing 12-5).

*Listing 12-5.* Abstracting by Creating the "**GetCorollaryCompanyInfor()**" Function

```
def GetCorollaryCompanyInfo(symbol):
    CompanyName = "No company name"
    Sector = "No sector"
    Industry = "No industry"
    MarketCap = "No market cap"

    if (symbol in list(stock_company_info_nasdaq['Symbol'])):
        data_row = stock_company_info_nasdaq[stock_company_info_
        nasdaq['Symbol'] == symbol]
        CompanyName = data_row['Name'].values[0]
        Sector = data_row['Sector'].values[0]
        Industry = data_row['industry'].values[0]
        MarketCap = data_row['MarketCap'].values[0]

    elif (symbol in list(stock_company_info_amex['Symbol'])):
        data_row = stock_company_info_amex[stock_company_info_
        amex['Symbol'] == symbol]
        CompanyName = data_row['Name'].values[0]
        Sector = data_row['Sector'].values[0]
        Industry = data_row['industry'].values[0]
        MarketCap = data_row['MarketCap'].values[0]
```

```
elif (symbol in list(stock_company_info_nyse['Symbol'])):
    data_row = stock_company_info_nyse[stock_company_info_
    amex['Symbol'] == symbol]
    CompanyName = data_row['Name'].values[0]
    Sector = data_row['Sector'].values[0]
    Industry = data_row['industry'].values[0]
    MarketCap = data_row['MarketCap'].values[0]

return (CompanyName, Sector, Industry, MarketCap)
```

We'll also build a function to handle the pulling of Wikipedia information using the company we got out of the "**GetCorollaryCompanyInfo()**" function. This function will return the first paragraph of the entry found (Listing 12-6).

***Listing 12-6.*** Abstracting by Creating the "**GetWikipediaIntro()**" Function

```
def GetWikipediaIntro(symbol):
    description = wikipedia.page("Apple Inc.").content
    return(description.split('\n')[0])
```

Finally, we'll build a function to create a link to the Finviz.com financial website. This function doesn't do much but append the stock symbol to the end of the link (Listing 12-7).

***Listing 12-7.*** Abstracting by Creating the "**GetFinVizLink()**" function

```
def GetFinVizLink(symbol):
    return(r'http://finviz.com/quote.ashx?t={}'.format(symbol.lower()))
```

# Designing a Web Application

Go ahead and download the code for this chapter if you haven't already done so; open a command line window and change the drive to the "**web-application**" folder. Your "**web-application**" folder should contain the following files as shown in Listing 12-8.

***Listing 12-8.*** Web Application Files

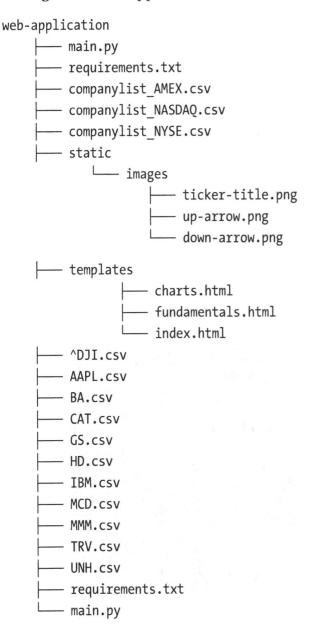

```
web-application
    ├── main.py
    ├── requirements.txt
    ├── companylist_AMEX.csv
    ├── companylist_NASDAQ.csv
    ├── companylist_NYSE.csv
    ├── static
    │       └── images
    │               ├── ticker-title.png
    │               ├── up-arrow.png
    │               └── down-arrow.png
    ├── templates
    │           ├── charts.html
    │           ├── fundamentals.html
    │           └── index.html
    ├── ^DJI.csv
    ├── AAPL.csv
    ├── BA.csv
    ├── CAT.csv
    ├── GS.csv
    ├── HD.csv
    ├── IBM.csv
    ├── MCD.csv
    ├── MMM.csv
    ├── TRV.csv
    ├── UNH.csv
    ├── requirements.txt
    └── main.py
```

As usual, we'll start a virtual environment to segregate our Python library installs and create the "**requirements.txt**" file if needed (Listing 12-9).

***Listing 12-9.*** Starting the Virtual Environment

```
$ python3 -m venv pairtrading
$ source pairtrading/bin/activate
```

Then install all the required Python libraries by running the "**pip install -r**" command (Listing 12-10).

***Listing 12-10.*** Installing Requirements and Taking the Web Application for a Local Spin

```
$ pip3 install -r requirements.txt
$ python3 main.py
```

Run the web application, as per usual, and make sure it works. Also, try the various options on the page to make sure everything works as advertised, especially the "**Access Fundamentals**" button (Figure 12-4).

***Figure 12-4.*** *The local version of the pair-trading application*

Deactivate out of your virtual environment when finished (Listing 12-11).

***Listing 12-11.*** Deactivating Virtual Environment

```
deactivate pairtrading
```

# Uploading Web Application to PythonAnywhere

Let's upload our updated code to PythonAnywhere. Log in to your PythonAnywhere account and find the folder "**pair-trading-booth**" that we created previously. Click the "**Files**" link in the top menu bar and enter the "**pair-trading-booth**" directory (Figure 12-5).

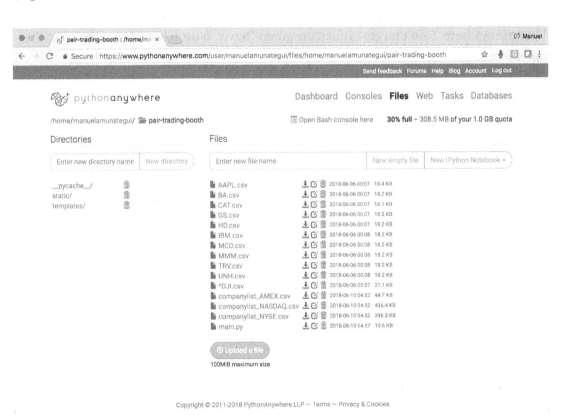

*Figure 12-5. Our pair-trading application on PythonAnywhere*

All the stock financial CSV files needed should already be there (if not, run through Chapter 5 again). You will need to upload the Nasdaq, Amex, and NYSE company files under the main "**pair-trading-booth**" directory, along with the ten-stock CSV files already there.

- companylist_AMEX.csv

- companylist_NASDAQ.csv

- companylist_NYSE.csv

385

You will also need to update the "**main.py**" and "**index.html**" files and upload the new "**fundamentals.html**" file to display our stock and derived charts (or create it as a new file on PythonAnywhere and copy/paste the code into it). The best way to proceed is to simply open those files in a local editor and copy and paste the content into PythonAnywhere.

For example, let's update "**main.py**," open the file in your local editor and open the file in PythonAnywhere, then copy and paste the new version into your "**main.py**" file on PythonAnywhere. Don't forget to click the green "**Save**" button before moving on to the other files (Figure 12-6).

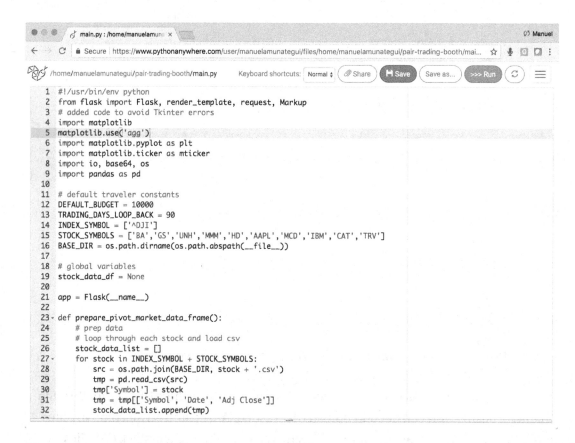

*Figure 12-6.* *Updating the "**main.py**" code base on PythonAnywhere to handle the creation of dynamic charts*

Proceed in the same way in the templates folder for file "**index.html**" and also create a new HTML file called "**fundamentals.html**" (Figure 12-7).

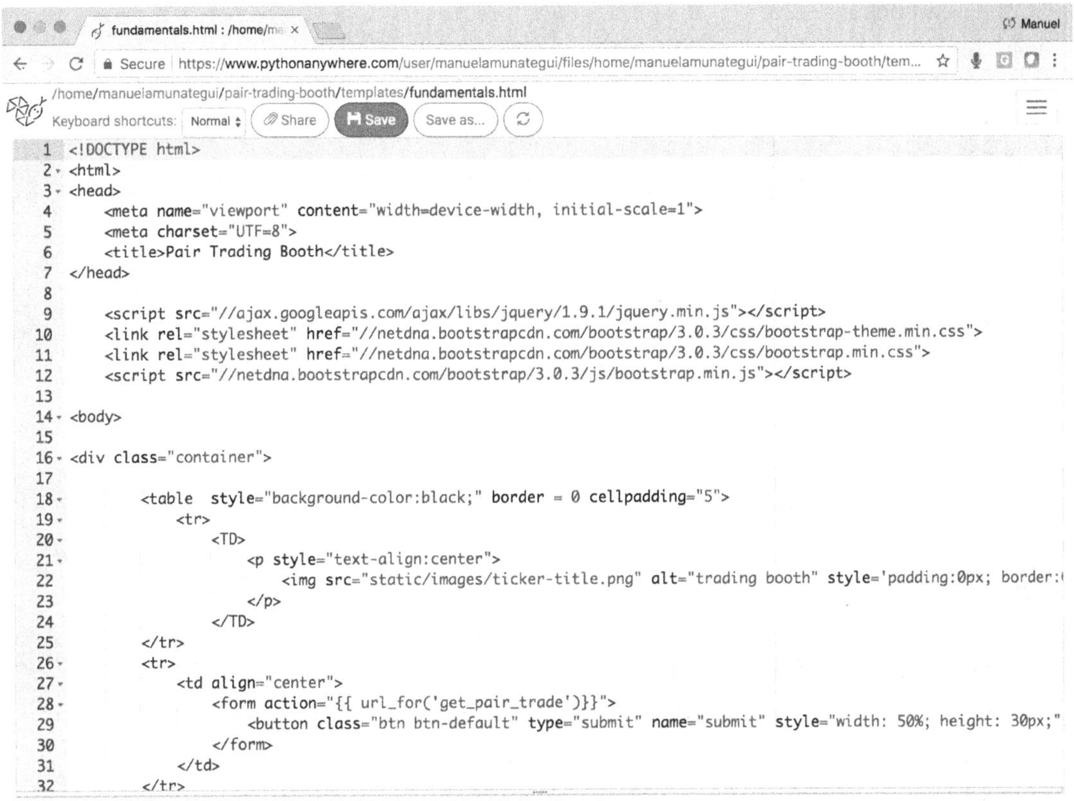

*Figure 12-7.  Creating a new file called "**fundamentals.html**" on PythonAnywhere to handle the creation of dynamic charts*

Next you need to "**pip3**" install Wikipedia as it isn't included in the base Python 3 build on PythonAnywhere (Figure 12-8).

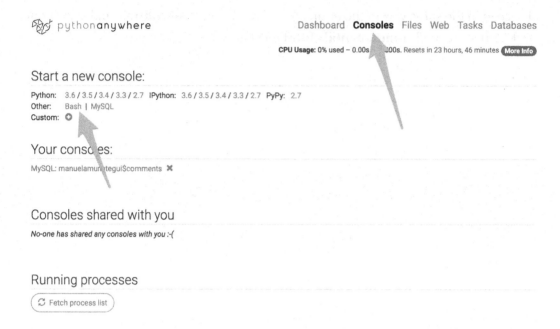

*Figure 12-8.* *Opening a bash console to pip install libraries not included in the original Python build*

Once the bash console is open you are ready to pip install any needed libraries. Go ahead and install the Wikipedia library with the following command (you need to add two dashes and user to override permission denied messages; Listing 12-12 and Figure 12-9).

*Listing 12-12.* Installing Requirements

```
$ pip3 install wikipedia --user
```

*Figure 12-9.* *Installing Wikipedia library on PythonAnywere*

Close your bash console by using the "**exit**" command. You can also kill the console completely if you don't need it anymore by going back to the console page in the PythonAnywhere dashboard and clicking the "**x**" under the bash console you opened (Figure 12-10).

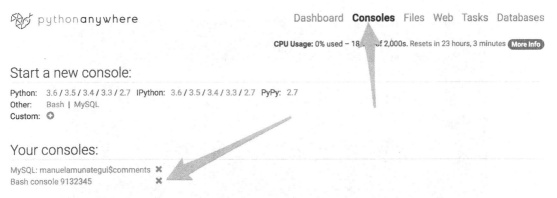

*Figure 12-10.* *Closing the bash control if you don't need it anymore*

Now you are ready to refresh your web service and fire up the web application. Click the "**Web**" menu tab and hit the big, green button to reload the application. That is very easy to do; click on the "**Consoles**" button and open a bash console (Figure 12-11).

*Figure 12-11.* *Hit the "**Reload**" button to update your web server*

If you enter the URL of your PythonAnywhere site into your browser, you should see the new "**Pair Trading Booth**" site in all its glory. Go ahead and take it through its paces (Figure 12-12).

*Figure 12-12.* *The new "**Pair Trading Booth**" site enhanced with charts*

# Conclusion

In this chapter, we took a third pass at the "**Pair Trading Booth**" web application and enhanced it with collateral fundamental information. We joined the current data with additional external data from Nasdaq. We plugged into the Wikipedia API to extract a high-level description of each company. We also created dynamic links for even more information if the user chooses.

# CHAPTER 13

# Google Analytics

Advanced intelligence for free.

Let's look at a simple tool to better understand how our users interact with our web applications. This is a huge boon to web application developers. Building our own analytic tracker would require adding a lot of custom Flask code to every page to track users, along with a database to save those interactions and an analytical engine to make sense of it. That's a lot of work! Instead, with Google Analytics, all we have to do is add a JavaScript snippet of code at the top of each page. That's it; add a few generic lines to every page, no editing required, and Google Analytics will handle everything else.

Google Analytics will tell us where users came from, how much time they spend on the site and on each page, the paths they take, etc. This is a must tool to not only better understand users, but also to refactor and create new content. There is a free version and a costly premium version—we'll focus on the free one here. The free version gives you plenty of insight for small web applications like the one we're building here.

---

**Note** There are no downloads for this chapter.

---

## Create a Google Analytics Account

Navigate to Google Analytics to create a free account at `https://analytics.google.com/analytics/web/provision`.

This will show a simple graphic of the process of tracking a web page and the "**Sign up**" button (Figure 13-1).

© Manuel Amunategui, Mehdi Roopaei 2018
M. Amunategui and M. Roopaei, *Monetizing Machine Learning*, https://doi.org/10.1007/978-1-4842-3873-8_13

**Figure 13-1.** *The Google Analytics process and sign up*

Where it asks for a website name, enter you PythonAnywhere.com account (the Pair Trading one; don't worry you can track up to 100 accounts). It will ask you a few basic questions and you can go with the defaults (Figure 13-2).

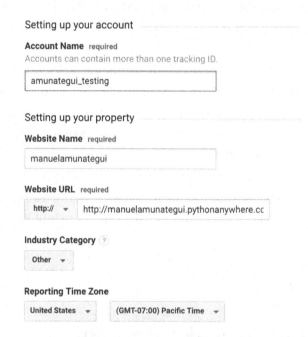

**Figure 13-2.** *My answers on Google Analytics*

Finally, click the blue button at the bottom of the page "**Get Tracking ID**" and accept the terms of service.

# JavaScript Tracker

The first page you will see once you log into your Google Analytics dashboard is the "**Admin**" tab with the key snippet of JavaScript needed to track a web page (Listing 13-1). This is the heart of the tracking system. You just need to add your API key where it says "**<<ADD-YOUR-GOOGLE-ANALYTICS-TRACKING-ID>>**" and drop this on all your pages.

***Listing 13-1.*** The JavaScript Tracking Snippet

```
<!-- Google Analytics -->
<script>
(function(i,s,o,g,r,a,m){i['GoogleAnalyticsObject']=r;i[r]=i[r]||function(){
(i[r].q=i[r].q||[]).push(arguments)},i[r].l=1*new Date();a=s.createElement(o),
m=s.getElementsByTagName(o)[0];a.async=1;a.src=g;m.parentNode.insertBefore(a,m)
})(window,document,'script','https://www.google-analytics.com/analytics.
js','ga');

ga('create', '<<ADD-YOUR-GOOGLE-ANALYTICS-TRACKING-ID>>', 'auto');
ga('send', 'pageview');
</script>
<!-- End Google Analytics -->
```

Copy it and drop it in the "**<head>**" section of any website you own and want to track. The head of every HTML template for the Pair Trading Booth should look like Listing 13-2 (make sure you enter yours, as this one will collect traffic analytics for my account).

***Listing 13-2.*** This is Mine

```
<head>
  <meta name="viewport" content="width=device-width, initial-scale=1">
  <meta charset="UTF=8">
  <title>Pair Trading Booth</title>

    <!-- Google Analytics -->
    <script>
```

```
(function(i,s,o,g,r,a,m){i['GoogleAnalyticsObject']=r;i[r]=i[r]||
function(){
(i[r].q=i[r].q||[]).push(arguments)},i[r].l=1*new Date();a=s.create
Element(o),
m=s.getElementsByTagName(o)[0];a.async=1;a.src=g;m.parentNode.insert
Before(a,m)
})(window,document,'script','https://www.google-analytics.com/
analytics.js','ga');

ga('create', 'UA-118908159-1', 'auto');
ga('send', 'pageview');
</script>
<!-- End Google Analytics -->
</script>
</head>
```

# Reading Your Analytics Report

After adding your Google Analytics tracking code to your site, save it and propagate the
**"Pair Trading Booth."** Point a browser to your site, view the source to make sure the new
code is there, then refresh your Google Analytics Dashboard. You should see that one
active user is on the site–that is you (Figure 13-3).

***Figure 13-3.*** *Google Analytics home showing 1 active user*

If you click on the "**REAL-TIME**" tab in the left-hand pane, you will even get to see where that active user is located. I was in Spain when I took the screenshot–pretty cool, right? (Figure 13-4)

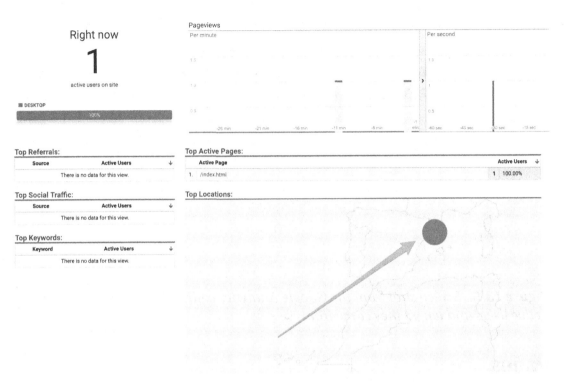

***Figure 13-4.***  *My real-time, active user–me in Spain!*

Obviously, you will have to run your web application for many days and have actual traffic to start looking for interesting patterns. You will also need to add your tracking code for the "**Pair Trading Booth**" on all HTML template pages to start seeing who goes where and for how long.

# Traffic Sources

Once you have collected traffic patterns, you can find out where you users came from. There is a great training course from the Google Analytics training team: `https://analytics.google.com/analytics/academy/` (screen shot from the demo account provided by the Google Analytics training team; Figure 13-5).

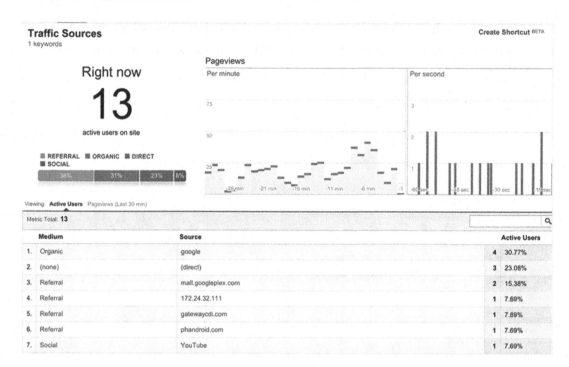

***Figure 13-5.*** *Screen shot from the Google Analytics course that shows multiple active users and where they came from*

# Pages

Once you have collected traffic patterns, you can find out when users visited a particular page, the total hit counts, and how much time they spent (from the demo account provided in the Google Analytics training; Figure 13-6).

**Figure 13-6.** *Page view information*

# Conclusion and Additional Resources

As a quick recap, you need the Google Analytics JavaScript tracker on all HTML pages you want to track. Google Analytics will collect a wide range of behaviors such as traffic source, language, browser type, etc. It considers one session to be any activity followed by 30 minutes of inactivity. You can customize all sorts of aspects and the information you access is always anonymous. This is a highly recommended tool for anyone who wants to understand the value of a particular page in comparison witho others, to get ideas of what your users like and don't like, and to gauge interest and upsell potentials.

For additional information on customizing your Google Analytics data, I highly recommend Google's Google Analytics Academy free course at `https://analytics.google.com/analytics/academy/`.

# A/B Testing on PythonAnywhere and MySQL

This is an ambitious chapter, so we'll limit the scope in order to distill the essence of this rich topic without going overboard. We'll start by building a simple MySQL database and table to track whether a visitor liked or didn't like the art work on the landing page. Because this is A/B Testing, we're going to create two landing pages and switch them randomly when users visit the site (Figure 14-1).

*Figure 14-1.* *The final web application for this chapter*

© Manuel Amunategui, Mehdi Roopaei 2018
M. Amunategui and M. Roopaei, *Monetizing Machine Learning*, https://doi.org/10.1007/978-1-4842-3873-8_14

> **Note**    Download the files for Chapter 14 by going to `www.apress.com/`
> `9781484238721` and clicking the source code button. You will need to install
> MySQL on your local machine in order to follow along with the Jupyter notebook
> "**chapter14.ipynb**."

In analytics, A/B testing means using two different versions of something and measuring how people react to each. This is commonly used in websites to try out new designs, products, sales, etc. In our case, we're going to expose our visitors to two different versions of our landing page. The versions are going to be assigned randomly and the visitor will be offered the opportunity to give the page a thumbs-up if they liked it. In the background, we're going to be tracking this traffic and whether or not a user gives a thumbs-up. If the user doesn't give the thumbs-up, we'll assume that it was a down vote.

This is an important topic and can yield valuable knowledge about your business and your users. There is a famous anecdote where Marisa Meyer, while at Google, ran an A/B test to determine which shade of blue, out of 40, the users preferred.[1] Obviously, one can go overboard with these types of tests.

# A/B Testing

The goal of A/B testing is to expose different products to the public and measure their reactions. In our web application, we're going to show a web page with two different images: an angry face and a friendly one. We will add a simple label to the page asking the visitor to give the image a thumbs-up if they liked it. In the background we're going to count each visit and count each thumbs-up. To keep things simple, we'll count an initial visit as a thumbs-down and update it to a thumbs-up if the visitor clicks the voting button. See Figures 14-2 and 14-3 for a version of each image.

---

[1]https://iterativepath.wordpress.com/2012/10/29/testing-40-shades-of-blue-ab-testing/

***Figure 14-2.*** *Image one*

***Figure 14-3.*** *Image two*

## Tracking Users

There are various ways of tracking anonymous visitors. Popular methods include the usage of cookies and databases. Each has its advantage and purpose. A cookie offers the advantage of tracking a user over longer periods regardless of whether they closed their browser or turned their computer off. A web page can easily check the visitor's computer for previous cookies and compare it with their database to determine if this is a repeat visitor or not.

We won't need to use cookies, as we will only consider single visits. Instead, we'll keep track of users using an HTML hidden tag and send that tag back using a post request. When a visitor first visits the page, we'll insert a row in the database with the page background image, a timestamp, and a unique identifier (a very long string that is unique to that user; the odds of creating two of the same are infinitesimal) referred to as a UUID. As mentioned, we assume that a first page visit is a thumbs-down and write it to the database. As we build the page, we insert a hidden HTML tag containing the UUID so that if the user interacts with the page by clicking the thumbs-up button, we'll pass the UUID back to the web server, so we can update the row previously entered in the database. This approach allows us to serve many visitors at the same time without worrying about who has what page. There are many ways you can tweak and improve this process depending on your needs. You can even pass that UUID from client to server and back as many times as you want and always know which session and user you are dealing with.

## UUID

The Universally Unique Identifier (UUID) is 128 bits long strong, and is guaranteed to be unique. We'll use the handy "**uuid**" Python library to automatically generate a guaranteed unique identifier (Listing 14-1).

*Listing 14-1.* The "**uuid**" Kibrary

**Input:**

```
import uuid
str(uuid.uuid4())
```

**Output:**

```
'e7b1b80e-1eca-43a7-90a3-f01927ace7c9'
```

In the "**uuid**" library, the "**uuid4()**" function generates a new random ID without relying on your computer's identifier, so it is unique and private. Check out the docs for additional UUID details and options at `https://docs.python.org/3/library/uuid.html`.

# MySQL

We're going to use the MySQL Community Server, which is a great and popular free database. It is perfect to support our A/B testing needs. It is an open-source relational database that can support a wide range of needs and is being used by big players including WordPress, and large media companies like Google and Facebook.

Go ahead and download the version of MySQL Community Server for your OS at `https://dev.mysql.com/downloads`. You will also find the installation instruction for your OS if you have any questions or issues. We won't use any front end, though there are quite a few of them available in case you want to use one (Figure 14-4).

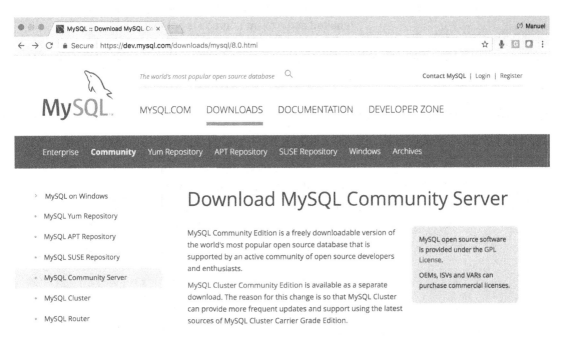

***Figure 14-4.*** *Find and download the correct version for your operating system*

You will be prompted with a series of questions including setting up root password and password encryption type (Figure 14-5).

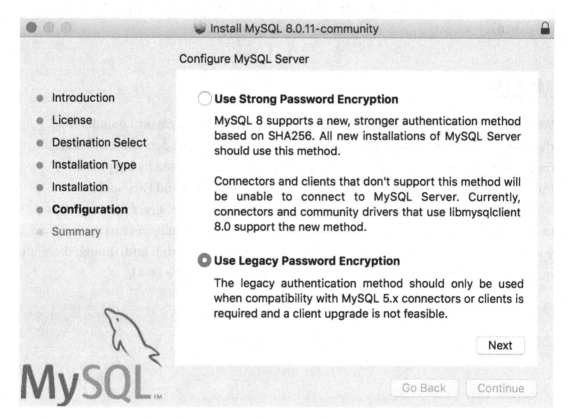

*Figure 14-5. Keeping it simple and using the legacy password system*

You can also start and stop your database through the control center for your operating system (this can also be done through the command line; Figure 14-6).

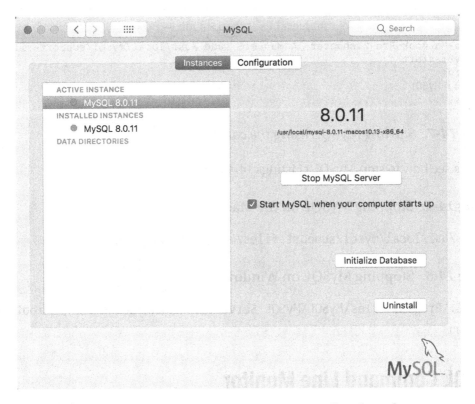

***Figure 14-6.*** *Setting MySQL server to start automatically when the computer starts*

# Command Line Controls

To start and stop MySQL (in most cases it should start automatically after your install it and restart your machine). Check out the docs for other operating systems, changes since this book was published, and additional commands at https://dev.mysql.com/doc/. Let's see how to start MySQL (Listings 14-2 and 14-3, and Figure 14-7).

***Listing 14-2.*** Starting MySQL on the Mac

```
$ sudo /usr/local/mysql/support-files/mysql.server start
```

***Listing 14-3.*** Starting MySQL on Windows

```
C:\> "C:\Program Files\MySQL\MySQL Server 8.0\bin\mysqld"
```

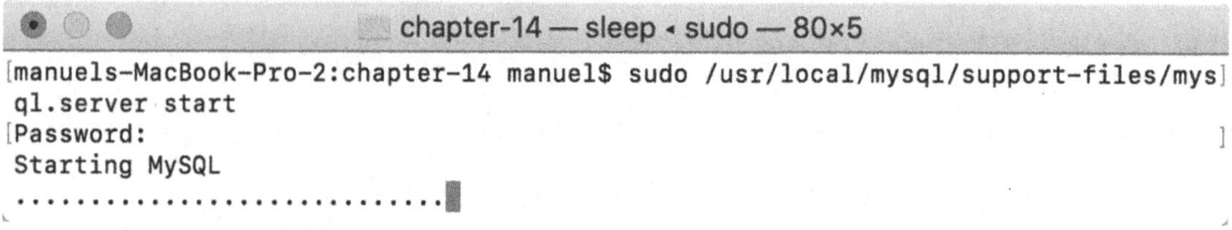

```
● ● ●                    chapter-14 — sleep ◂ sudo — 80×5
[manuels-MacBook-Pro-2:chapter-14 manuel$ sudo /usr/local/mysql/support-files/mys]
ql.server start
[Password:                                                                       ]
 Starting MySQL
 ...........................▮
```

***Figure 14-7.*** *Starting MySQL using the command line*

Let's see how to stop MySQL (Listings 14-4 and 14-5).

***Listing 14-4.*** Stopping MySQL on the Mac

```
$ sudo /usr/local/mysql/support-files/mysql.server stop
```

***Listing 14-5.*** Stopping MySQL on Windows

```
C:\> "C:\Program Files\MySQL\MySQL Server 8.0\bin\mysqladmin" -u root
shutdown
```

# MySQL Command Line Monitor

The command-line monitor is a handy tool that allows you to manage users and permissions, create databases and tables, and much more (see the docs for other operating systems and additional commands at https://dev.mysql.com/doc/).

To enter the monitor, change the drive to your MySQL directory or export a path, then enter the "**mysql -u root -p**" command and you will be prompted for your password that you created during the installation process (Listing 14-6 and Figure 14-8).

***Listing 14-6.*** Code Input

```
$ export PATH=$PATH:/usr/local/mysql/bin
$ mysql -u root -p
```

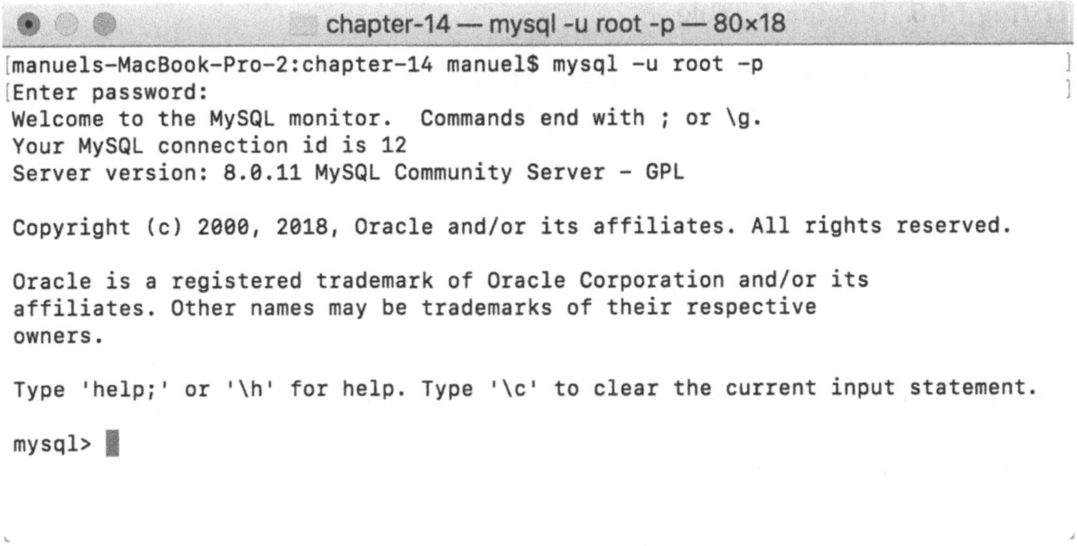

```
●  ●  ●                chapter-14 — mysql -u root -p — 80×18

[manuels-MacBook-Pro-2:chapter-14 manuel$ mysql —u root —p          ]
[Enter password:                                                    ]
Welcome to the MySQL monitor.  Commands end with ; or \g.
Your MySQL connection id is 12
Server version: 8.0.11 MySQL Community Server — GPL

Copyright (c) 2000, 2018, Oracle and/or its affiliates. All rights reserved.

Oracle is a registered trademark of Oracle Corporation and/or its
affiliates. Other names may be trademarks of their respective
owners.

Type 'help;' or '\h' for help. Type '\c' to clear the current input statement.

mysql> ▌
```

***Figure 14-8.*** *Running the MySQL Command Line Monitor*

You will know that you entered the monitor once your prompt changes to "**mysql>**." Let's create a user, a database, and a table for our A/B testing.

# Creating a Database

Let's create a database named "**ABTesting**" (Listing 14-7).

***Listing 14-7.*** Creating a Database

```
mysql> CREATE DATABASE ABTesting;
```

# Creating a Table

Let's create a new table using the "**CREATE TABLE**" statement. Whenever you are creating a new table, it is a good idea to drop it first, otherwise you will get an error (but make sure that you really do want to drop it as you will lose all data contained therein). We will create a table called "**tblFrontPageOptions**" that will have a unique identifier field called "**uuid**," a Boolean flag called "**liked**" to hold whether or not the user clicked the thumbs up, a page_id to mark whether this was an "**A**" or "**B**" page, and an automated timestamp field (Listing 14-8).

***Listing 14-8.*** Creating a Table

```
mysql> DROP TABLE ABTesting.tblFrontPageOptions;
mysql> CREATE TABLE ABTesting.tblFrontPageOptions (
        uuid VARCHAR(40) NOT NULL,
    liked BOOLEAN NOT NULL DEFAULT 0,
        pageid INT NOT NULL,
        time_stamp  TIMESTAMP NOT NULL DEFAULT CURRENT_TIMESTAMP);
```

You can easily test that your table is working by inserting some data into it using an "**INSERT INTO**" statement (Listing 14-9).

***Listing 14-9.*** Inserting Data

```
mysql> INSERT INTO ABTesting.tblFrontPageOptions (uuid, liked, pageid)
VALUES(9999, 1, 2);
```

To check that the data did indeed make it into the table, we use a "**SELECT \***" statement (Listing 14-10).

***Listing 14-10.*** Querying Data

**Input**:

```
mysql> SELECT * FROM ABTesting.tblFrontPageOptions;
```

**Output**:

```
+------+-------+---------+---------------------+
| uuid | liked | page_id | time_stamp          |
+------+-------+---------+---------------------+
| 9999 |     1 |       2 | 2018-05-19 14:28:44 |
+------+-------+---------+---------------------+
1 row in set (0.00 sec)
```

We're looking good; the table now has a new row in it. If you want to start with a clean state, you can drop and re-create the table with the previous code. There are plenty of great primers on SQL syntax on the Internet, but a great place to start is the w3schools at https://www.w3schools.com/sql. Exit out of the "**mysql>**" prompt and open up the Jupyter notebook for the chapter to practice inserting data into our table and reading it out through the "**mysql.connector**" Python library.

# Creating A Database User

We are going to create a user dedicated to our A/B testing web application. It is a bad idea to use your root password in your Flask code. Our user will be called "**webuser**" and its password will be "**thesecre**" (Listing 14-11).

***Listing 14-11.*** Creating a user

```
mysql> CREATE USER 'webuser'@'localhost' IDENTIFIED BY 'thesecret';
```

Next, we will grant this user all privileges, and once you are more comfortable with MySQL (no, not my SQL, the MySQL product... you know they're probably joking like that all day long over at the MySQL headquarters...), you can tone this down to just read/write permissions for specific tables). See Listing 14-12.

***Listing 14-12.*** Granting Rights

```
mysql> GRANT ALL PRIVILEGES ON ABTesting.* TO 'webuser'@'localhost' WITH
GRANT OPTION;
```

Finally, you can check that the "**webuser**" user was successfully added with the following handy command (Listing 14-13).

***Listing 14-13.*** Checking Users

```
mysql> SELECT User FROM mysql.user;

+------------------+
| User             |
+------------------+
| mysql.infoschema |
| mysql.session    |
| mysql.sys        |
| root             |
| webuser          |
+------------------+
5 rows in set (0.00 sec)
```

You can exit out of the MySQL command line tool by simply entering the "**exit**" command.

# Python Library: mysql.connector

Most of the interaction with our database will be done through Python and Flask using the handy "**mysql.connector**" library. Here you can refer to the corresponding Jupyter notebook for Chapter 14 to follow along (please install the "**requirements_jupyter.txt**" file to get the libraries needed). Keep in mind that in this chapter we will not create a local Flask version, so get familiar with the commands using the notebook, then we'll jump directly to the cloud.

We will program three types of functions using the "**SELECT**," "**INSERT**," and "**UPDATE**" SQL functions. If you are not familiar with these classic SQL functions, check out the great primer from w3schools (I know, I keep pushing that site; it's that good and I swear that I have no relations with them whatsoever) at https://www.w3schools.com/sql.

## SELECT SQL Statement

"**SELECT**" is the most common SQL command and is used to read data from a table. This is easily done using the "**mysql.connector**" library. You first create a connection to the database by calling the "**connect()**" function and passing permissioned credentials for a database. The connection returns a cursor to communicate and send orders to the database. This is done using a query string holding the "**SELECT**" statement. We will use this approach for all our SQL statements. The difference with this statement versus "**INSERT**" and "**UPDATE**" is that we are expecting to receive data back from the database. After executing our query string through the "**execute()**" function, we can access the returned data through a loop. Each loop represents one row of data. Notice that an open cursor and connection are both closed at the end of the call, as we don't want to hold onto resources longer than we need to (Listing 14-14).

*Listing 14-14.* "**select**" Statement with Cursor

**Input:**

```
cnx = mysql.connector.connect(user='webuser', password='thesecret',
database='ABTesting')
cursor = cnx.cursor()
query = "SELECT * FROM ABTesting.tblFrontPageOptions"
cursor.execute(query)
```

```
for (uuid, liked, pageid, time_stamp) in cursor:
        print("uuid: {} liked:{} pageid: {} on {:%m/%d/%Y %H:%M:%S}".format(
        uuid, liked, pageid, time_stamp))
cursor.close()
cnx.close()
```

**Output** (example, your output will only contain what's in the table so far):

uuid: 704a44d0-29f4-4a2d-bc6a-fe679017f7e9 liked:1 pageid: 2 on 05/19/2018 17:03:29

# INSERT SQL Statement

The "**INSERT**" statement allows us to insert data into the ABTesting table of our database. The first time a visitor hits the web page, we insert a row containing the following fields, a "**UUID**," a "**liked**" flag turned to false, the "**pageid**" representing which of the two background images the user is viewing, and a "**timestamp**." As done in the "**SELECT**" statement, we first open a connection to the database, then create a cursor and pass it our query statement. In this case, our query statement isn't a "**SELECT**" but an "**INSERT**." We also use the handy "**%s**" statement as a variable placeholder to then be filled by whatever value is held in the "**args**" tuple. Here we are inserting a new unique ID, with "**liked**" set to false (or thumbs down), and the page ID viewed.

Also, whenever you are inserting or updating a table, don't forget to call the "**commit()**" function to commit your inserts before closing the connection. If you don't commit, your changes will get ignored (Listing 14-15).

*Listing 14-15.* SQL "**INSERT**" Statement with Cursor

```
cnx = mysql.connector.connect(user='webuser', password='thesecret',
database='ABTesting')
cursor = cnx.cursor()
query = "INSERT INTO ABTesting.tblFrontPageOptions (uuid, liked, pageid)
VALUES (%s, %s, %s);"
args = ("704a44d0-29f4-4a2d-bc6a-fe679017f7e9", 0, 1)
cursor.execute(query, args)
cursor.close()
cnx.commit()
cnx.close()
```

## UPDATE SQL Statement

The "**UPDATE**" statement is similar to the "**INSERT**" statement, but instead of adding a new row at the end of the table, you are updating an existing row. In order to update a specific row, you have to be able to find the correct row before updating it. In this case, as we have the handy "**UUID**" that is guaranteed unique, we can easily find that specific row and not have to worry about updating another in error. In order to properly build the "**UPDATE**" statement, we have to pass it two values, the "**UUID**" of "**704a44d0-29f4-4a2d-bc6a-fe679017f7e9**" and the "**liked**" flag set to true (Listing 14-16).

***Listing 14-16.*** SQL "**update**" Statement with Cursor

```
cnx = mysql.connector.connect(user='webuser', password='thesecret',
database='ABTesting')
cursor = cnx.cursor()
query = "UPDATE ABTesting.tblFrontPageOptions SET liked = %s WHERE uuid = %s;"
args = (1, "704a44d0-29f4-4a2d-bc6a-fe679017f7e9")
cursor.execute(query, args)
cursor.close()
cnx.commit()
cnx.close()
```

Again, don't forget to call the "**commit()**" function to commit your changes before closing the connection (if you don't, your changes will get ignored).

## Abstracting the Code into Handy Functions

We need to abstract all our SQL code into simple to use functions. We start by creating two global variables to hold the MySQL user account and password. This enables us to only have to set it once and not worry about it during subsequent SQL calls. It also comes in handy whenever you need to change user accounts (Listing 14-17).

***Listing 14-17.*** Abstracting Account Data

```
mysql_account = 'webuser'
mysql_password = 'thesecret'
mysql_database = 'ABTesting'
mysql_host = 'localhost'
```

We also can abstract the Uuid-generating code to keep things clean and simple (Listing 14-18).

***Listing 14-18.*** Abstracting "**GetUUID()**" function

```
def GetUUID():
    return (str(uuid.uuid4()))
```

Next, we create a function to insert new visits into the database. This function will get a new UUID from "**GetUUID()**," set the "**liked**" to false as we assume all new visits don't like or don't want to interact with the site, the "**pageid**" representing the image that was randomly selected for them, and the timestamp that is automatically generated by MySQL (Listing 14-19).

***Listing 14-19.*** Abstracting "**InsertInitialVisit()**" Function

```
def InsertInitialVisit(uuid_, pageid):
    try:
        cnx = mysql.connector.connect(user=mysql_account, password=mysql_
        password, database=mysql_database, host=mysql_host)
        cursor = cnx.cursor()
        query = "INSERT INTO ABTesting.tblFrontPageOptions (uuid, liked,
        pageid) VALUES (%s,%s,%s);"
        args = (uuid_, 0, pageid)
        cursor.execute(query, args)
        cursor.close()
        cnx.commit()
        cnx.close()
    except mysql.connector.Error as err:
        app.logger.error("Something went wrong: {}".format(err))
```

When a user interacts with the page and clicks the thumbs-up button, Flask uses the "**UpdateVisitWithLike()**" function to update the row using the unique identifier for the session and turns the "**liked**" flag to true (Listing 14-20).

***Listing 14-20.*** Abstracting "**UpdateVisitWithLike()**" Function

```python
def UpdateVisitWithLike(uuid_):
    try:
        cnx = mysql.connector.connect(user=mysql_account, password=mysql_
        password, database=mysql_database, host=mysql_host)
        cursor = cnx.cursor()
        query = "UPDATE ABTesting.tblFrontPageOptions SET liked = %s WHERE
        uuid = %s;"
        args = (1, uuid_)
        cursor.execute(query, args)
        cursor.close()
        cnx.commit()
        cnx.close()
    except mysql.connector.Error as err:
        app.logger.error("Something went wrong: {}".format(err))
```

Finally, we create the administrative dashboard to view how the A/B testing is going by offering total visit counts, total thumbs up and down, and how many thumbs up for each image. We also offer a log view where we dump all the content from the ABTesting table (Listing 14-21).

***Listing 14-21.*** Abstracting "**GetVoteResults()**" Function

```python
def GetVoteResults():
    results = "
    total_votes = 0
    total_up_votes = 0
    total_up_votes_page_1 = 0
    total_up_votes_page_2 = 0
    try:
        cnx = mysql.connector.connect(user=mysql_account, password=mysql_
        password, database=mysql_database, host=mysql_host)
        cursor = cnx.cursor()
        query = "SELECT * FROM  ABTesting.tblFrontPageOptions"
        cursor.execute(query)

        for (uuid_, liked, pageid, time_stamp) in cursor:
```

```
            total_votes += 1
            if liked==1 and pageid==1:
                total_up_votes_page_1 += 1
            if liked==1 and pageid==2:
                total_up_votes_page_2 += 1
            if liked == 1:
                total_up_votes += 1
            results += ("uuid: {} liked:{} pageid: {} on {:%m/%d/%Y
            %H:%M:%S}".format(uuid_, liked, pageid, time_stamp)) + "<br />"
        cursor.close()
        cnx.close()
    except mysql.connector.Error as err:
        app.logger.error("Something went wrong: {}".format(err))

    return (results, total_votes, total_up_votes, total_up_votes_page_1,
    total_up_votes_page_2)
```

# Designing a Web Application

Let's download the files for Chapter 14 and unzip them on your local machine if you haven't already done so. Your "**web-application**" folder should contain the following files as shown in Listing 14-22.

***Listing 14-22.*** Web Application Files

```
web-application
        ├── main.py
        ├── static
                    └──images
                            ├── background1.jpg
                            ├── background2.jpg
        └── templates
                    ├── admin.html
                    └── index.html
```

# Running a Local Version

Sorry folks, there's no local version this time. Instead, we'll use the PythonAnywhere wizard to create a MySQL instance in the cloud and upload all the needed data, file by file, as we've done previously.

# Setting Up MySQL on PythonAnywhere

It is really easy to set up MySQL on PythonAnywhere using the built-in wizard. Click the "**Databases**" link in the upper right hand of the dashboard and proceed through the setup just like we did earlier on the local version (Figure 14-9).

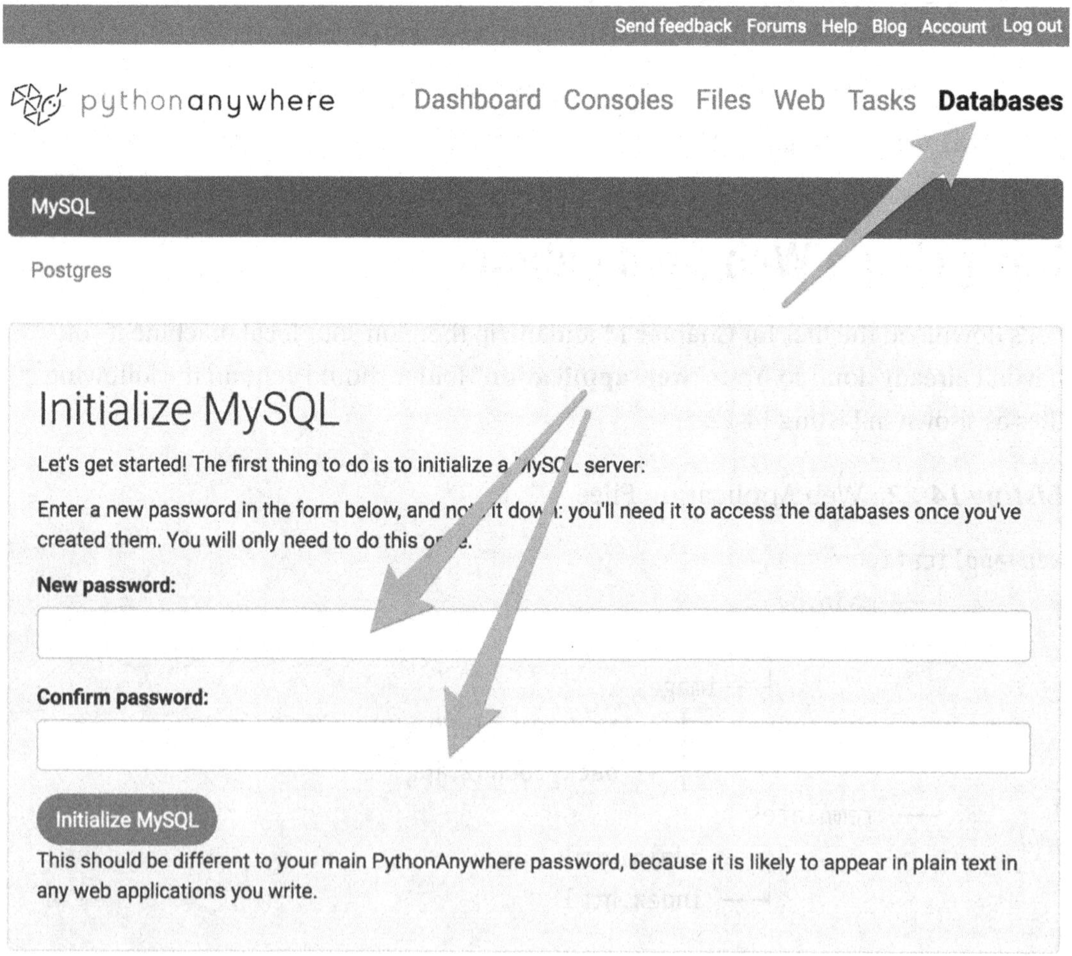

*Figure 14-9.* *Setting up MySQL on PythonAnywhere*

After you initialize MySQL, you will be able to create a database and get into the MySQL console to create the "**tblFrontPageOptions**." There are two caveats you will have to contend with. First, PythonAnywhere appends your account name in front of the database name. In my case, database "**ABTesting**" becomes "**amunateguioutloo$ABTesting**." This isn't a big deal, but we will have to update any code that talks to the database. The second issue is that the user it creates for you is the one you will have to add to your script, as it won't let you create additional users using the "**CREATE USER**" command (Figure 14-10).

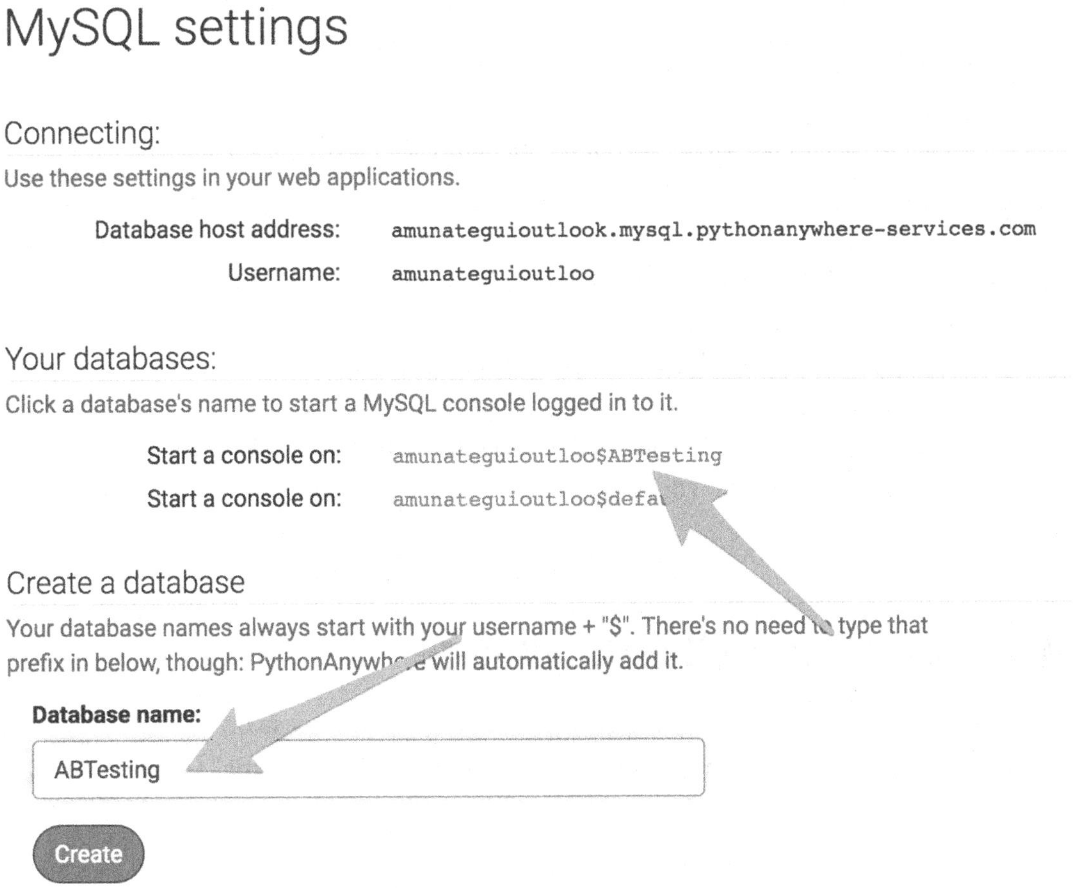

# MySQL settings

## Connecting:
Use these settings in your web applications.

|  |  |
|---|---|
| Database host address: | `amunateguioutlook.mysql.pythonanywhere-services.com` |
| Username: | `amunateguioutloo` |

## Your databases:
Click a database's name to start a MySQL console logged in to it.

|  |  |
|---|---|
| Start a console on: | `amunateguioutloo$ABTesting` |
| Start a console on: | `amunateguioutloo$defa` |

## Create a database
Your database names always start with your username + "$". There's no need to type that prefix in below, though: PythonAnywhere will automatically add it.

**Database name:**

ABTesting

Create

*Figure 14-10. Creating the ABTesting database and clicking the console link to set things up*

Click the console link for the "**...$ABTesting**" database and create the "**tblFrontPageOptions**" table and "**webuser**" account. Make sure to update the database to reflect your database name (Listing 14-23).

***Listing 14-23.*** Create Table Command and PythonAnywhere Confirming That the Table "**tblFrontPageOptions**" was Successfully Created

**Input**:

```
CREATE TABLE amunateguioutloo$ABTesting.tblFrontPageOptions (
      uuid VARCHAR(40) NOT NULL,
   liked BOOLEAN NOT NULL DEFAULT 0,
      pageid INT NOT NULL,
      time_stamp  TIMESTAMP NOT NULL DEFAULT CURRENT_TIMESTAMP);
```

**Output**:

```
mysql> CREATE TABLE amunateguioutloo$ABTesting.tblFrontPageOptions (
    -> uuid VARCHAR(40) NOT NULL,
    ->    liked BOOLEAN NOT NULL DEFAULT 0,
    -> pageid INT NOT NULL,
    -> time_stamp  TIMESTAMP NOT NULL DEFAULT CURRENT_TIMESTAMP);
Query OK, 0 rows affected (0.03 sec)
```

That is all we need to do in the console; everything else will be done through Python.

# A/B Testing on PythonAnywhere

Let's upload and update all the code to work with our new database on PythonAnywhere. Under "**Files**" create a new folder called "**ABTesting**" (Figure 14-11).

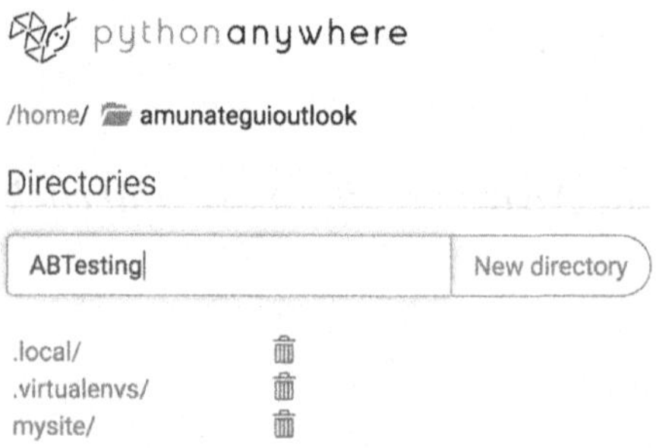

***Figure 14-11.***   *Creating the new folder to host our ABTesting site*

Next, we need to upload all the files, one-by-one, up to the site just like we did with the other PythonAnywhere projects we already did (Figures 14-12 through 14-14).

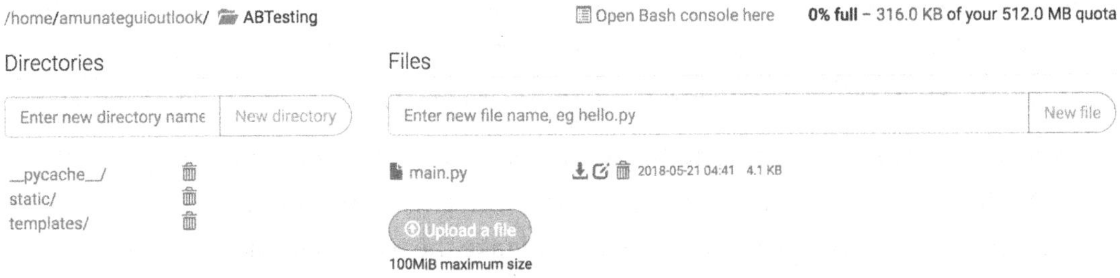

**Figure 14-12.** *Upload "**main.py**" under the "**ABTesting**" folder*

**Figure 14-13.** *Upload both images under "**ABTesting/static/**"*

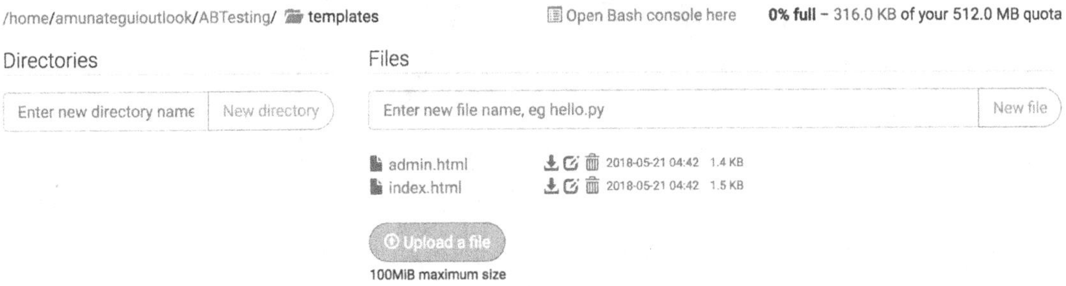

**Figure 14-14.** *Upload both HTML files under "**ABTesting/templates/**"*

Once you have uploaded all files, you need to go into "**main.py**" and update the database account and all table references. You will need to update the following variables with the ones assigned to you by PythonAnywhere. Click "**Databases**" in the upper right corner of your PythonAnywhere dashboard to access the variables (Listing 14-24).

***Listing 14-24.*** Assigned account data

```
mysql_account='<<ENTER-YOUR-DATABASE-USERNAME>>'
mysql_password='thesecret'
mysql_database='<<ENTER-YOUR-DATABASE-USERNAME>>$ABTesting'
mysql_host="<<ENTER-YOUR-DATABASE-USERNAME>>.mysql.pythonanywhere-services.com"
```

Make sure to replace in the "**main.py**" code all the "**<<ENTER-YOUR-DATABASE-USERNAME>>**" with your database user name, otherwise it will not work.

Hit the big green button on the web tab and take the web application for a spin. Go ahead and vote away (then check the administrative page; Figure 14-15).

***Figure 14-15.*** *The "**Do You Like Me?**" web application running on PythonAnywhere*

# A/B Testing Results Dashboard

In order to view the results of our A/B testing operation, we need to create a dashboard. Though this isn't essential, and you could just as well query the results directly through MySQL using SQL statements, a dashboard will allow anybody to look at the results throughout your testing without needing SQL knowledge or querying permissions to the ABTesting table (Figure 14-16).

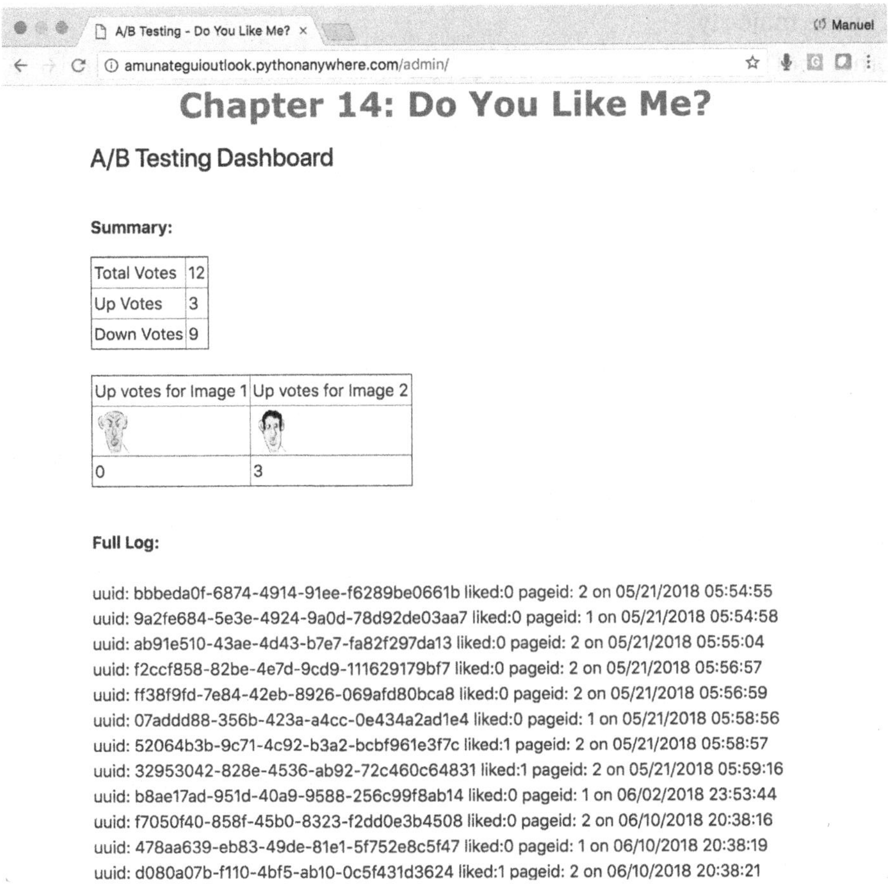

*Figure 14-16.* *A simple dashboard with the latest results of our A/B test*

We will keep things simple here and offer the total votes, the total up and down votes, the up votes per image, and the full log of all participants.

## Conclusion

A/B testing is one of the popular tools to better understand your users. It is also a loaded science with many ways to approach it. Here, we made the assumption that any new visit doesn't like the site, thus defaults with a thumbs-down. This doesn't necessarily mean they thought the page was bad, as it could also mean they didn't have time to read the question. So, in this scenario, I would look closer at the number of up votes per image rather than worry about the down votes; in either case you can extract which image was favored by the majority.

Another tool we introduced here is MySQL; it is a great open-source and free relational database that is widely used and supported.

# CHAPTER 15

# From Visitor to Subscriber

A look at some simple authentication schemes.

In this chapter, we're going to briefly look at different ways to handle subscribers. We'll look at a simple login mechanisms but quickly move on to plugins. The gist of this book is to quickly get your ideas up and running, so having to build your own subscriber mechanisms goes against the book's core philosophy. If you have the time, knowledge, and/or staff to do it, then you'll probably save some money but it is not an easy task. Whenever you are dealing with other people's personal and financial data, a whole new layer of responsibility is required. I prefer and recommend pushing this out to those that do it well and lease it out in the form of plugins. They are easy to use and allow you to focus on the important stuff–your business ideas!

We'll look at different ways of getting payments from visitors using a very simple site model.

- Text-based authentication (a concept to extend using a database for a home rolled solution–not recommended)

- Memberful–simple subscription or product purchase; unlocks access to videos in sites like vimeo

- Paypal donations

- Stripe payments

In the next chapter, we'll look at an example of a more robust style of paywall for subscribers using our pair-trading web application along with Memberful.com.

© Manuel Amunategui, Mehdi Roopaei 2018
M. Amunategui and M. Roopaei, *Monetizing Machine Learning*, https://doi.org/10.1007/978-1-4842-3873-8_15

---

**Note**    Download the files for Chapter 15 by going to `www.apress.com/9781484238721` and clicking the source code button. There is no Jupyter notebook for this chapter, but there are a series of HTML and Flask files to experiment with.

---

# Text-Based Authentication

One way of monetizing an online presence is to convert visitors into subscribers. If your content is exclusive and/or is frequently updated, then visitors may be willing to pay to access this on a regular basis and at a deeper level. This can be done in different ways and at different levels. At a high level, you need to separate your free content from your paid content by employing an authentication process to restrict access to certain areas.

The simplest approach is to hard-code a universal account/password into Flask Directly, or use a text file to handle multiple accounts.

---

**Warning**    This approach is only suitable for demos and/or short-term projects where security isn't an issue. You should never use such an approach to store anything private, valuable, and certainly nothing having anything remotely to do with money—hold on to those ideas until next chapter.

---

## Flask-HTTPAuth—Hard-Coded Account

We'll start with the base Flask-HTTPAuth example from the documentation[1] (Listing 15-1).

***Listing 15-1.*** Simple Authentication

```
from flask import Flask
from flask_httpauth import HTTPBasicAuth

app = Flask(__name__)
auth = HTTPBasicAuth()
```

---

[1]https://flask-httpauth.readthedocs.io/en/latest/

```python
users = {
    "john": "hello",
    "susan": "bye"
}

@auth.get_password
def get_pw(username):
    if username in users:
        return users.get(username)
    return None

@app.route('/')
@auth.login_required
def index():
    return "Hello, %s!" % auth.username()

if __name__ == '__main__':
    app.run()
```

It doesn't get much simpler than this. Save the code into a Python script (or download the files for this chapter and run the Flask script "**authentication-simple.py**"). If you are missing Python libraries, pip3 install via the associated "**authentication_ requirements.txt**" file as we've done in previous chapters).

Enter either account "**john**" with password "**hello**" or "**susan**" with password "**bye**" (Listing 15-2 and Figure 15-1).

***Listing 15-2.*** Running Local Flask Script "**authentication-simple.py**"

```
$ python3 authentication-simple.py
```

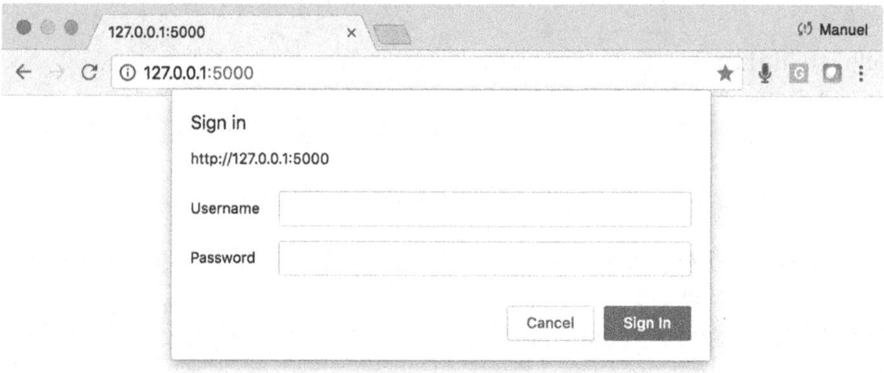

*Figure 15-1.* *Username and password required to proceed*

# Digest Authentication Example

In order to get a session authenticated (i.e., offer the user the ability to move between pages within your domain without having to sign in on each page), you need to use a form of authenticated cookie. Once the user for a session is authenticated, you simple pass the "**@auth.login_required**" before any Flask function and it will only let the session proceed if the visitor is authenticated; otherwise it will pop-up the login box. The code can be found under script "**authentication-digest.py**" (Listing 15-3).

*Listing 15-3.* Digest Authentication

```
from flask import Flask
from flask_httpauth import HTTPDigestAuth

app = Flask(__name__)
app.config['SECRET_KEY'] = 'secret key here'
auth = HTTPDigestAuth()

users = {
    "john": "hello",
    "susan": "bye"
}

@auth.get_password
def get_pw(username):
    if username in users:
```

```
        return users.get(username)
    return None

@app.route('/')
@auth.login_required
def index():
    return "Hello, %s!" % auth.username()

@app.route('/paywall')
@auth.login_required
def paywall():
    return "%s, you are on page 2!" % auth.username()

if __name__ == '__main__':
    app.run()
```

Give it a whirl and run Flask script "**authentication-digest.py**" and enter account "**john**" with password "**hello**" or "**susan**" with password "**bye**" (Listing 15-4).

***Listing 15-4.*** Digest Authentication

```
$ python3 authentication-digest.py
```

Once you are authenticated, add to the end of the URL, "/**paywall**." This will show you that you are now using and authenticated session where a subscriber only has to login once (Figure 15-2).

susan, you are on page 2!

***Figure 15-2.*** *Navigating through the site while authenticated*

# Digest Authentication Example with an External Text File

This is very much the same concept as before, but now we read the data from an external text file instead of a dictionary inside the Flask script. This will allow an administrator (or you) to add and remove names and passwords without having to affect the source or restart the web server, as the file is ready at each authentication. Keep in mind the username/passwords are written in the text file without quotes or comas, and one per line. The code can be found under script "**authentication-digest-external.py**" (Listing 15-5).

***Listing 15-5.*** Digest Authentication with External File

```python
from flask import Flask
from flask_httpauth import HTTPDigestAuth

app = Flask(__name__)
app.config['SECRET_KEY'] = 'secret key here'
auth = HTTPDigestAuth()

@auth.get_password
def get_pw(username):
        for user in open("users-file.txt","r").readlines():
                if username in user:
                        user={user.split(':')[0]:user.split(':')[1].rstrip()}
                        return user.get(username)
        return None

@app.route('/')
@auth.login_required
def index():
    return "Hello, %s!" % auth.username()

@app.route('/paywall')
@auth.login_required
def paywall():
    return "%s, you are on page 2!" % auth.username()

if __name__ == '__main__':
    app.run()
```

Give it a whirl and run Flask script "**authentication-digest-external.py**" and enter either account "**john**" with password "**hello**" or "**susan**" with password "**bye**" (Listing 15-6).

***Listing 15-6.*** Digest Authentication with External File

```
$ python3 authentication-digest-external.py
```

Once you are authenticated, add to the end of the URL, "/**paywall**." This will show you that you are now using an authenticated session where a subscriber only has to log in once (Figure 15-3)

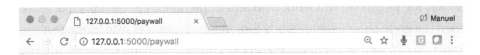

## susan, you are on page 2!

***Figure 15-3.*** *Navigating through the site while authenticated from text file*

---

**Note**   The code seen so far in this chapter should be used carefully, as it doesn't use security features to adequately store user credentials. This should only be used for prototypes, one-offs, or internal presentations on secure intranets. The next section will show a much more robust and recommended approach.

---

# Simple Subscription Plugin Systems

Using a professional and externally managed plugin is the approach I recommend when building commercial-grade paywalls or subscription-only pages. Let the professionals deal with encryption, security, storing sensitive information, credit-card payment, etc., so you can focus on building great content and services! This is the way to go for us **"weekend warriors."**

# Memberful

Memberful is the plugin we will work with and implement. I personally like Memberful[2] and think it is a great choice for anybody looking for an easy way to manage a paywall section of a website. It offers credit card payment through Stripe,[3] offers user-management features, and is discreet. Memberful has a series of educational videos to help you better understand how things work on their end, as this could become an important tool toward your web monetization goals.

Let's look at a simple example of purchasing something from Memberful. Here we will only look at buying items; we'll worry about subscriptions and paywalls in the next chapter. To set up a product for purchase, you simply go to your Memberful dashboard and set up an item for sale there and they will give you a simple URL to put on your page. When a visitor goes through the purchasing process, they will see a pop-up box appear inside your site. This is the beauty of Memberful: your visitors never feel like they're leaving the site to do the purchase. Big A+.

So, go ahead and sign up for a free account at Memberful. You will not need to use a credit card in this chapter (Figure 15-4).

---

[2]https://memberful.com/
[3]http://stripe.com/

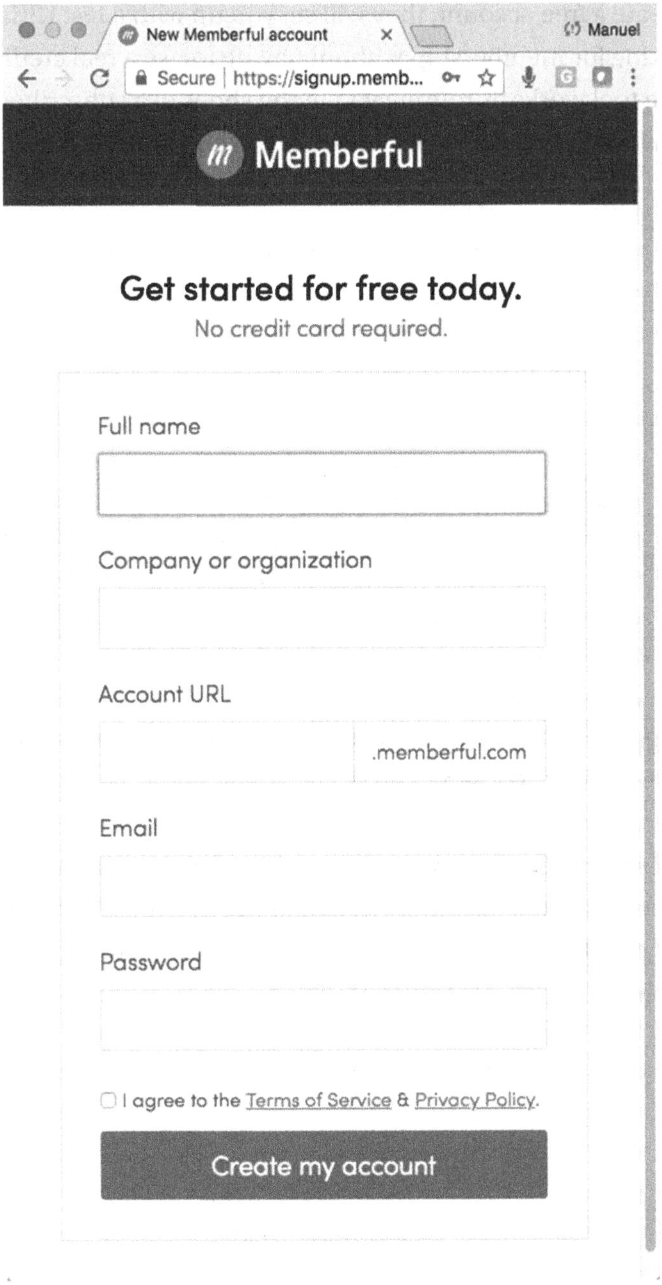

**Figure 15-4.** *Signing up for a free Memberful account–no credit card required in this chapter*

When you create a free account, they will set you up with a test URL to simulate the process of asking for membership without having to use a real credit card. This will show you how the tool works by putting you in the shoes of a subscriber. Sign up for an account and, when it asks you whether you are using WordPress, select "**I'm using something else**" (Figure 15-5).

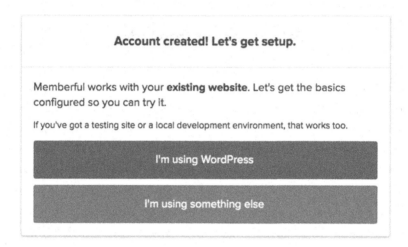

***Figure 15-5.*** *Sign up for an account and select "**I'm using something else**" when asked if you are using WordPress*

Next, you will see a page with some code and a video. I would highly recommend watching the video, as it explains things well. Enter you PythonAnywhere website address and copy the HTML JavaScript code. The JavaScript creates the popup window, so the user never leaves your site (Figure 15-6).

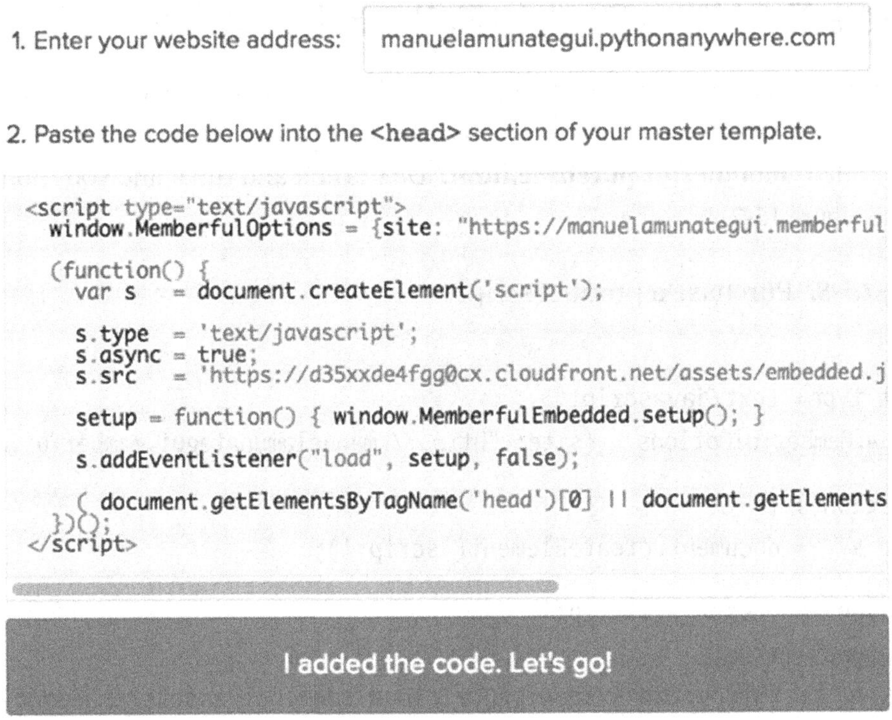

1. Enter your website address:   manuelamunategui.pythonanywhere.com

2. Paste the code below into the **<head>** section of your master template.

```
<script type="text/javascript">
    window.MemberfulOptions = {site: "https://manuelamunategui.memberful

    (function() {
        var s  = document.createElement('script');

        s.type  = 'text/javascript';
        s.async = true;
        s.src   = 'https://d35xxde4fgg0cx.cloudfront.net/assets/embedded.j

        setup = function() { window.MemberfulEmbedded.setup(); }

        s.addEventListener("load", setup, false);

        ( document.getElementsByTagName('head')[0] || document.getElements
    })();
</script>
```

I added the code. Let's go!

**Figure 15-6.** *Enter you PythonAnywhere URL and copy the HTML code*

In return, you will get a "**Purchase link HTML code**" (Listing 15-7).

**Listing 15-7.** Fake Product Purchase Link

```
<a href="https://<<ADD-YOUR-ACCOUNT>>.memberful.com/
checkout?plan=30287">Buy Sample Plan for $25/month.</a>
```

Let's create a very simple test web page (no Flask required) to house our purchase link. Once done, click on it and make that fake purchase. You can use any of the fake test credit card numbers listed in the help docs, or simply use "**4242 4242 4242 4242.**"

# Create a Real Web Page to Sell a Fake Product

Build the simple HTML shown in Listing 15-8 and make sure to replace the purchase link with yours (i.e., with a valid account). You can find the base script in the directory for this chapter called "**memberful-purchase.html**." Do the edit and run it and you should see a page like Figure 15-7.

***Listing 15-8.*** Purchase a product script

```html
<html>
<script type="text/javascript">
  window.MemberfulOptions = {site: "https://manuelamunategui.memberful.com"};

  (function() {
    var s   = document.createElement('script');

    s.type  = 'text/javascript';
    s.async = true;
    s.src   = 'https://d35xxde4fgg0cx.cloudfront.net/assets/embedded.js';

    setup = function() { window.MemberfulEmbedded.setup(); }

    s.addEventListener("load", setup, false);

    ( document.getElementsByTagName('head')[0] || document.getElementsBy
    TagName('body')[0] ).appendChild( s );
  })();
</script>

<body>
<h1>Membership</h1>
<p><a href="https://<<ADD-YOUR-ACCOUNT>>.memberful.com/
checkout?plan=30287">Buy Sample Plan for $25/month.</a></p>
</body>
</html>
```

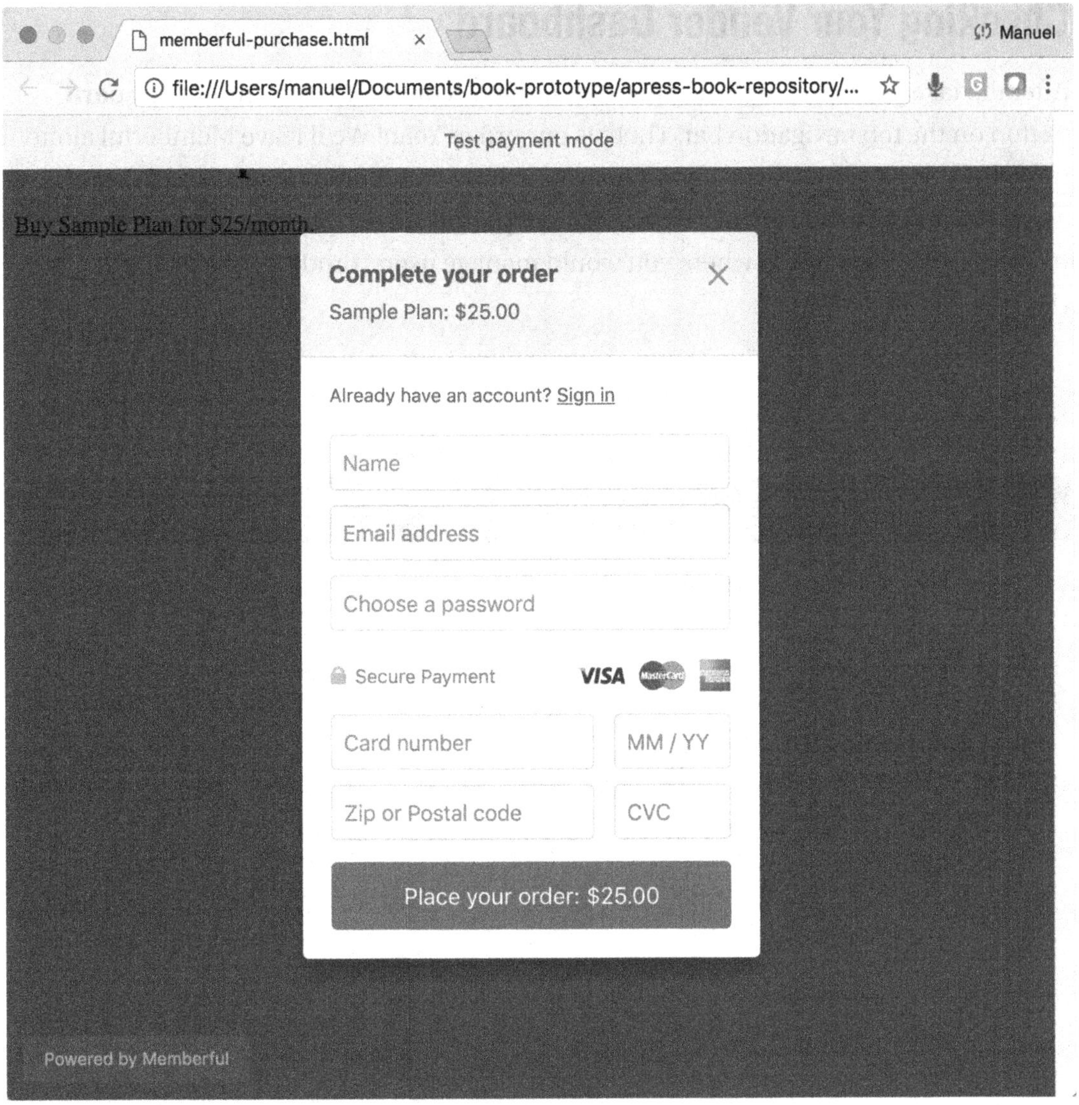

*Figure 15-7.* *Memberful pop-up on your site; go ahead make the order using the fake credit card number*

# Checking Your Vendor Dashboard

After the fake purchase, log into your Memberful account and click the "**Dashboard**" button on the top navigation bar. There is our order! Yeah! We'll leave Memberful alone until the next chapter. Obviously, if you were selling real products, you would create many of these purchasing links with your own product descriptions and pictures. It is also in the Dashboard where you would manage users, products, refunds, etc. (Figures 15-8 and 15-9).

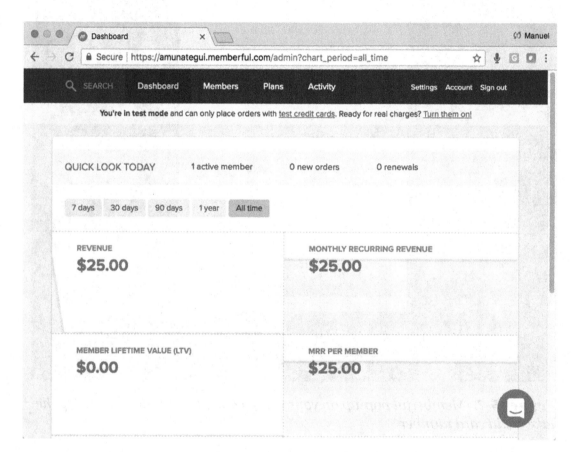

*Figure 15-8.*  *Viewing order activities*

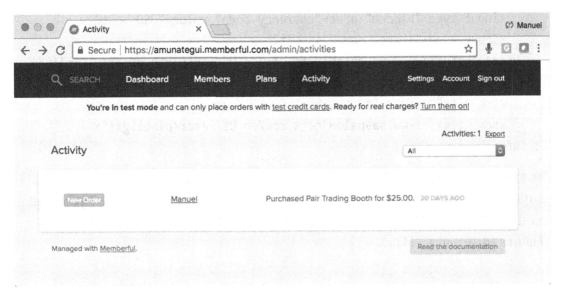

**Figure 15-9.** *Managing orders*

# Taking Donations with PayPal

Setting up a donation button from PayPal is one of the easiest things you can do to raise funds for personal and nonprofit efforts (read the disclaimers or find another type of PayPal option www.paypal.com/buttons). Obviously, you need an account in good standing, and all you need to do is drop the HTML forms code into your web page and it will take care of everything else–it will even display the button for you. This is a painless option where you don't even need Flask, as this payment option just requires HTML– nothing else (Listing 15-9).

***Listing 15-9.*** Paypal Donation Code

```
<form action="https://www.paypal.com/cgi-bin/webscr" method="post">
      <input type="hidden" name="business" value="amunategui@gmail.com">
      <input type="hidden" name="cmd" value="_donations">
      <input type="hidden" name="item_name" value="Donate to support these
      great blog posts!">
      <input type="hidden" name="item_number" value="Support">
```

```
        <input type="hidden" name="currency_code" value="USD">
        <input type="image" name="submit"
        src="https://www.paypalobjects.com/en_US/i/btn/btn_donate_LG.gif"
        alt="Donate">
        <img alt="" width="1" height="1"
        src="https://www.paypalobjects.com/en_US/i/scr/pixel.gif">
</form>
```

Once you have the code inside your web page, you will see the yellow "**Donate**" button. When you click on it using a valid PayPal account, it will take you to PayPal and ask the donator a series of questions to get the financial donation completed (Figures 15-10 and 15-11).

*Figure 15-10.* *PayPal donation button on your site*

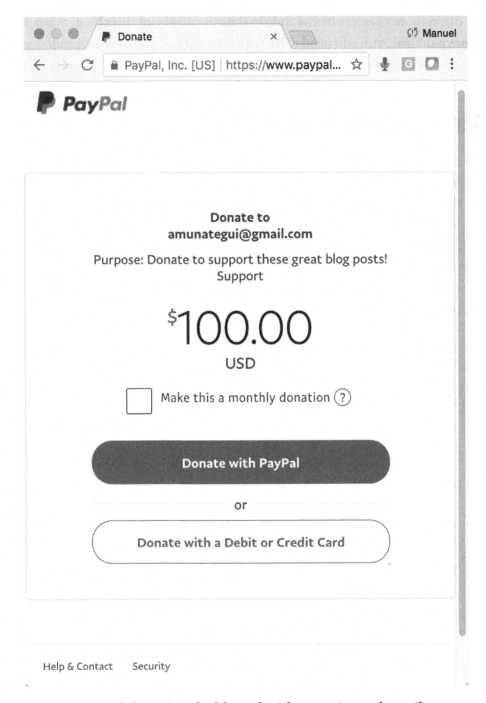

*Figure 15-11.* *PayPal donation dashboard with your site and email information*

You can view an example of this code in the downloads for this chapter under the name "**paypal.html**," you will need to have an account and get your own code if you want to use this type of method for your own fundraising needs.

# Making a Purchase with Stripe

Stripe is a simple, powerful, and widely used payment platform with a lot of Flask support. It is also widely trusted, which is important if you want visitors to give you money. We will follow a simple example from the official docs (`https://stripe.com/docs/checkout/flask`).

First sign up for a free account and it won't require any payment information if you only want to test it out using the developer tools.

Sign up for a free Stripe account, navigate to the "**Developers**" section and click on "**API Keys**" (Figure 15-12).

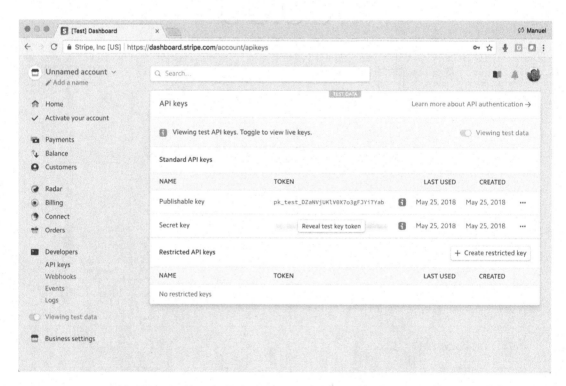

***Figure 15-12.***   *The developer section showing the publishable and secret keys*

The code to get a test purchase is very simple and we will follow along with their example (`https://stripe.com/docs/checkout/flask`). They recommend taking the "**Publishable key**" and "**Secret key**" and creating OS variables out of them so you don't hardcode them (a great practice). Use the "**export**" command in your terminal on the MAC, the control panel on Windows, or add the variables directly into Flask (Listing 15-10).

***Listing 15-10.*** Exporting Your API Keys

```
$ export SECRET_KEY="<<YOUR-SECRET-KEY>>"
$ export PUBLISHABLE_KEY="<<YOUR-PUBLISHABLE-KEY>>"
```

You also need to pip3 install Stripe and Flask (Listing 15-11).

***Listing 15-11.*** Installing Needed Libraries

```
$ sudo pip3 install --upgrade stripe
$ sudo pip3 install flask
```

The "**main.py**," our Flask controller, imports Stripe, sets the secret and publishable keys, and offers two pages–the "**index.html**" page where you would put your items for sale, and the "**charge.html**" where you process the purchase for items using the "**stripe. Charge.create()**" function–and offers a confirmation page. This is fairly straightforward and therein lies its effectiveness (Listing 15-12).

***Listing 15-12.*** A Look at "**Main.py**"

```
import os
from flask import Flask, render_template, request
import stripe

stripe_keys = {
  'secret_key':  os.environ['SECRET_KEY'],
  'publishable_key': os.environ['PUBLISHABLE_KEY']
}
```

```python
stripe.api_key = stripe_keys['secret_key']

app = Flask(__name__)

@app.route('/')
def index():
    return render_template('index.html', key=stripe_keys['publishable_key'])

@app.route('/charge', methods=['POST'])
def charge():
    # Amount in cents
    amount = 500

    customer = stripe.Customer.create(
        email='customer@example.com',
        source=request.form['stripeToken']
    )

    charge = stripe.Charge.create(
        customer=customer.id,
        amount=amount,
        currency='usd',
        description='Flask Charge'
    )

    return render_template('charge.html', amount=amount)

if __name__ == '__main__':
    app.run(debug=True)
```

The templates also use a neat trick of using a layout file ("**layout.html**"). This allows you to create a skeleton HTML page that you can reuse throughout your site. For example, you only need to create branding and drop-down links once, and have every page inherit it.

You then leverage the Jinja2 tags "**{% block content %}{% endblock %}**" in the HTML to ingest new code (Listing 15-13).

***Listing 15-13.***  Using a Template HTML Page–"**layout.html**"

```
<!DOCTYPE html>
<html>
<head>
  <title>Stripe</title>
  <style type="text/css" media="screen">
    form article label {
      display: block;
      margin: 5px;
    }

    form .submit {
      margin: 15px 0;
    }
  </style>
</head>
<body>
  {% block content %}{% endblock %}
</body>
</html>
```

And any code that wants to be housed in the layout file uses the Jinja2 tag
"**{% extends "layout.html" %}**" (Listing 15-14).

***Listing 15-14.***  Jinja2 Tag for "**layout.html**" and Variable "**key**"

```
{% extends "layout.html" %}
{% block content %}
  <form action="/charge" method="post">
    <article>
      <label>
        <span>Amount is $5.00</span>
      </label>
    </article>

    <script src="https://checkout.stripe.com/checkout.js" class="stripe-
    button"
```

445

```
            data-key="{{ key }}"
            data-description="A Flask Charge"
            data-amount="500"
            data-locale="auto"></script>
  </form>
{% endblock %}
```

Create a free account on Stripe.com and export your API keys as shown previously. Then run the sample code (you can find it in the downloads for this chapter under formerly named "**stripe-payments**"). Go ahead and run the code locally (Listing 15-15 and Figure 15-13).

**Listing 15-15.**  Running the Stripe Flask Sample

```
$ python3 main.py
```

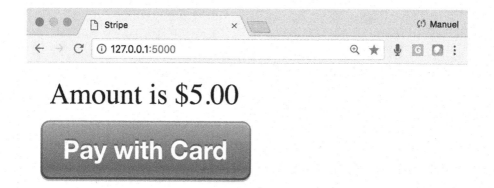

***Figure 15-13.***  *Your own Stripe.com purchase button*

If you go through the sample purchase (you should be able to enter any fake credit card number) and then log into your Stripe.com dashboard, you should see the order (Figure 15-14).

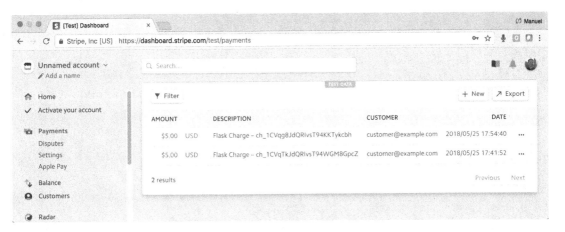

***Figure 15-14.*** *Your own Stripe.com purchase button*

# Conclusion

This chapter provided a very brief introduction to some of the authentication, donation, and purchase plugins that can be used with your web application. I will reiterate that any of the "**roll-your-on**" solutions presented here are not for any serious use and certainly not for anything remotely commercial. In Chapters 16 and 17, we will look at a real solution you can tailor for your paywall and subscription needs.

# CHAPTER 16

# Case Study Part 4: Building a Subscription Paywall with Memberful

Let's finalize our case study with a subscription-based paywall using Memberful and credit card payments on PythonAnywhere.

In this chapter, we will hide all trading content behind a paid-only subscription paywall (Figure 16-1). This is our last project, let's make it a special one! Let's extend our pair-trading site with a real paywall subscription system using Memberful.com. This is more involved than what we've seen in the previous chapter. Earlier we saw how to sell products or charge for subscriptions. These are one-step processes. Instead, we want a way of allowing paying customers to navigate the entire site, including content behind the paywall, and only have to log in once during each session.

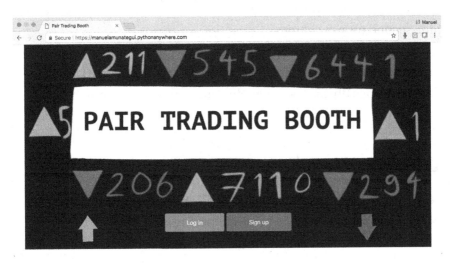

***Figure 16-1.*** *The final web application for this chapter*

© Manuel Amunategui, Mehdi Roopaei 2018
M. Amunategui and M. Roopaei, *Monetizing Machine Learning*, https://doi.org/10.1007/978-1-4842-3873-8_16

We're assuming here that the pretend trading advice is valuable enough to get people to pay for it. These are the tools needed to pull that off. Also, because this chapter is a paywall and it requires tying up a specific address to Memberful, we will not be able to run this locally. The Pair Trading Booth code we are going to use is the same as in the previous sections; all we will do here is add an additional landing page and the Memberful paywall features. In simple terms, no Jupyter or local Flask code is used in this chapter.

---

**Note**   Download the files for Chapter 16 by going to `www.apress.com/9781484238721` and clicking the source code button.

---

# Upgrading Your Memberful and PythonAnywhere Pay Accounts

In order to build this paywall, you will need to use paid accounts. You need to upgrade your Memberful.com account to the "**Pro plan**," which requires a valid credit card number and costs $25 a month. This is required so that you can access the API and webhooks, which are disabled on the free account. You will also need to upgrade your PythonAnywhere account to the lowest paid level–at the time of writing it is referred to as "**Hacker $5/month**." This is required because the authentication needs to use custom ports and it is only allowed on paid accounts. I recommend turning it on for a few days to try it out, and if you don't think it is useful, downgrade back to free accounts before incurring the second month charges.

In a nutshell, we want to offer members the ability to buy subscriptions using Memberful (so we don't have to deal with any user or payment data), have them log into their account only once during a session, then allow them to peruse the site freely until they log out.

## Upgrading Memberful

Let's upgrade our account in order to get a handle on the OAuth state and custom ports. This will allow our application to let visitors log in and access the pages behind the paywall. Go to the Memberful website, then to "**Account**" and "**Plans and billing**" (Figure 16-2).

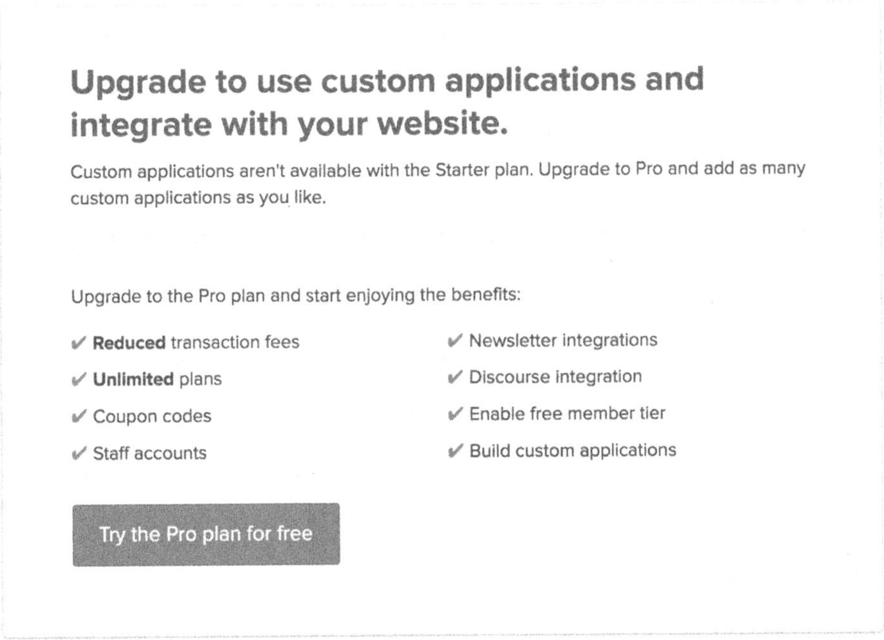

**Upgrade to use custom applications and integrate with your website.**

Custom applications aren't available with the Starter plan. Upgrade to Pro and add as many custom applications as you like.

Upgrade to the Pro plan and start enjoying the benefits:

- ✔ **Reduced** transaction fees          ✔ Newsletter integrations
- ✔ **Unlimited** plans                    ✔ Discourse integration
- ✔ Coupon codes                           ✔ Enable free member tier
- ✔ Staff accounts                         ✔ Build custom applications

Try the Pro plan for free

*Figure 16-2.*  *Accessing the "Pro plan"*

Go ahead and upgrade to the "**Pro plan**." In the "**Application Details**" pane, check all the checkboxes and add a landing membership page. This is critical, as it is where Memberful will redirect visitors once they've signed up or signed in (Listings 16-1 and 16-2, and Figure 16-3).

*Listing 16-1.*  You Application Name in Memberful is Your PyhonAnywhere.com account. It Should Look Like

```
http://<<YOUR PYTHON ANYWHERE SITE>>.pythonanywhere.com/
```

*Listing 16-2.*  Your "**OAuth Redirect URL**" Should Look Like

```
https://<<YOUR PYTHON ANYWHERE SITE>>.pythonanywhere.com/member
```

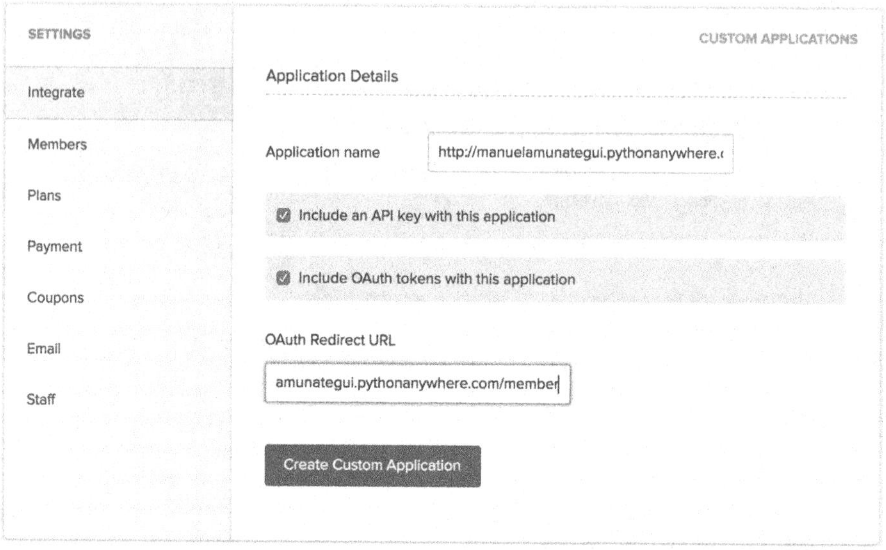

***Figure 16-3.*** *Settings for the paywall of our web application*

Next copy your OAuth "**Identifier**" and "**Secret**" keys. These are the keys we will use in the Flask portion of the web application (Figure 16-4).

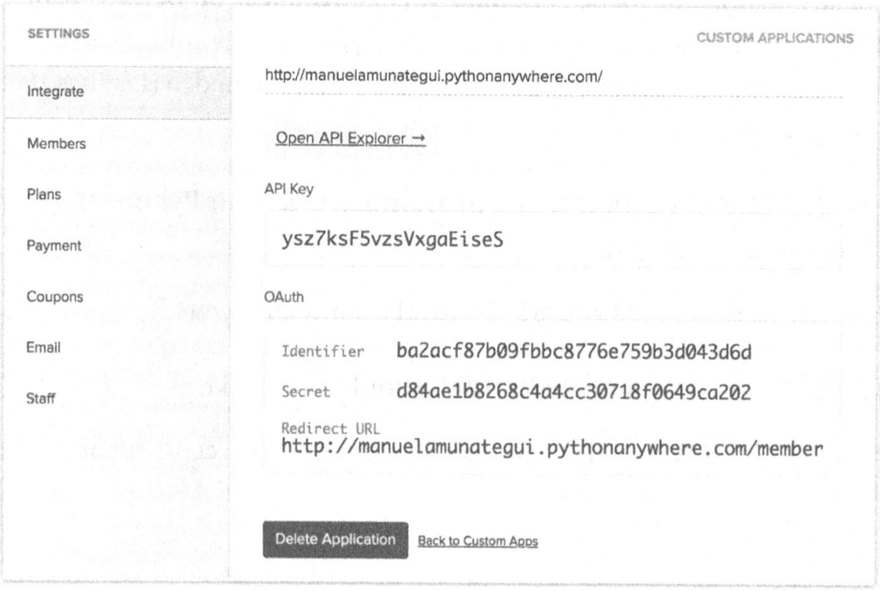

***Figure 16-4.*** *What your page should look like after creating a custom application*

Under the "**Custom Apps**" tab, set your "**Login application**" to the site in the drop down (it should already contain your PythonAnywhere account you added earlier in the "**Custom Application**" automatic login application; Figure 16-5).

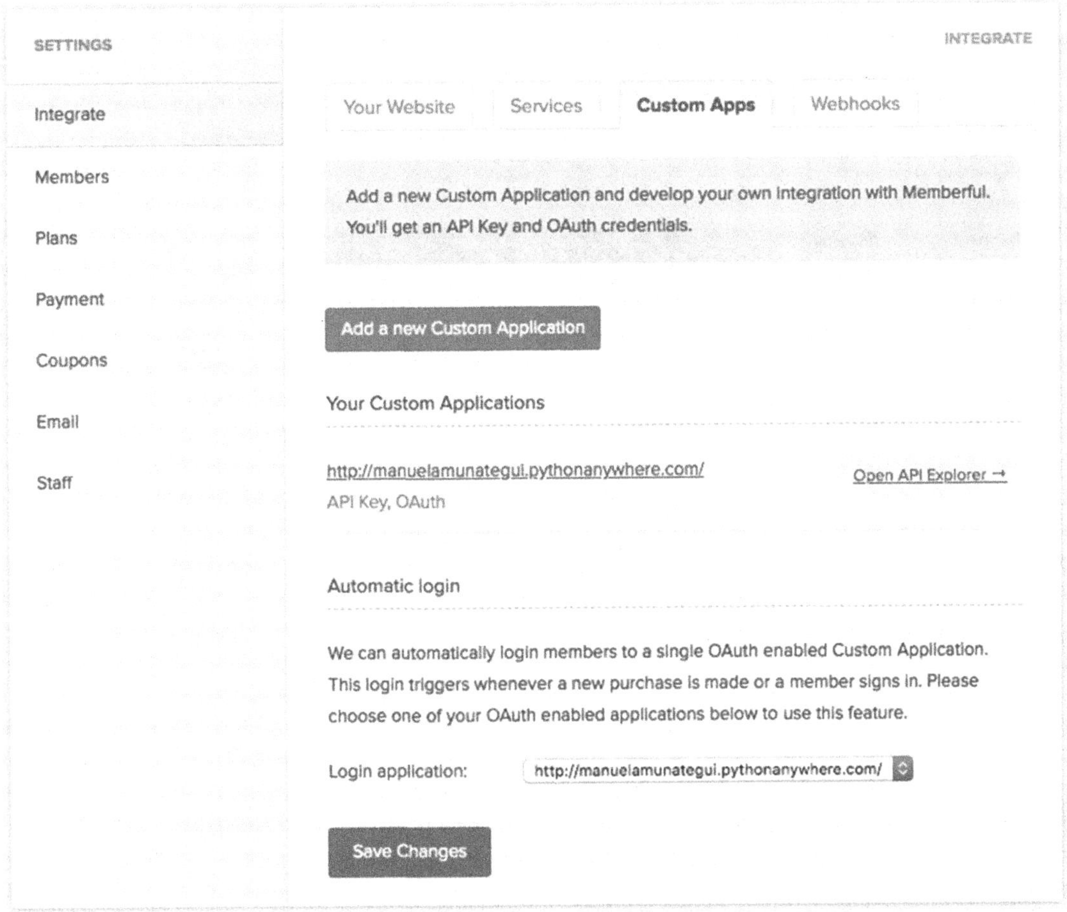

*Figure 16-5.* *Select your "**Login application**" to reflect your PythonAnywhere account*

# Upgrading PythonAnywhere

Under your account tab in PythonAnywhere, opt for the first paid plan (i.e., cheapest), the "**Hacker**" plan (Figure 16-6).

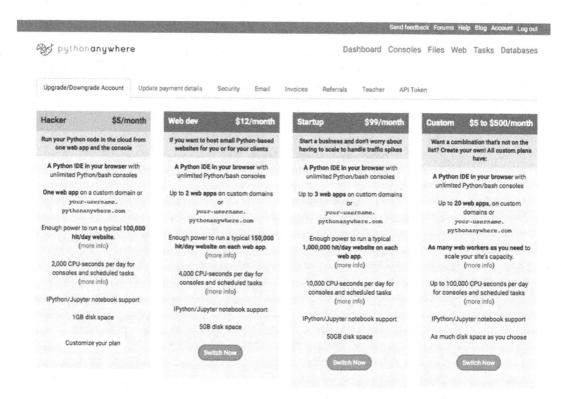

***Figure 16-6.***  *Choices of paid PythonAnywhere accounts; the cheapest paid account will allow you to follow along with this chapter's project*

# Pip Install Flask-SSLify

We need to install "**Flask-SSLify**" because it isn't part of the Python 3 build on PythonAnywhere. Flask-SSLify will force all pages to use "**HTTPS**" and give you enhanced security. You already used "**pip3 install**" earlier with Wikipedia's API. Click on the "**Consoles**" link at the top of the PythonAnywhere account and access a bash console (Figure 16-7).

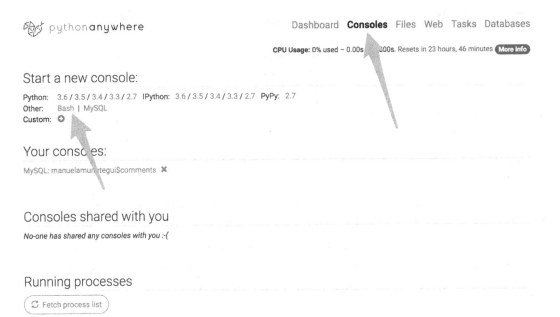

***Figure 16-7.*** *Opening a bash console to pip install libraries not included in the original Python build*

Once the bash console is open, you are ready to pip install needed libraries. Go ahead and install the Flask-SSLify library with the following command (you need to add two dashes and "**user**" to override permission denied messages; Listing 16-3).

***Listing 16-3.*** Installing Additional Libraries

```
$ pip3 install Flask-SSLify --user
```

# Memberful Authentication

Memberful supports the OAuth 2.0 protocol for authentication but because it requires you to tie your application directly to your Memberful account, creating a paywall is much easier than if you rolled your own. For more information, please refer to the official Memberful docs at https://memberful.com/help/integrate/advanced/memberful-api/.

# Two-Step Process and Flask Session Mechanism

The Flask session is a data holding object that can maintain state between pages using cookies. When you think about it, there are times when you want to remember what a user is doing, what is in their shopping cart, or who they are even if they jump from page to page. You could pass data back and forth using form variables, but that would get messy and confusing quickly (and not very secure). The Flask session remedies that by storing data on the client's computer using cookies. For this to work, cookies have to be enabled.

The system we will create to authenticate a user is easy to implement thanks to Memberful's powerful membership functions and integrated UI. When a visitor enters a page that is behind the paywall, our Flask application will check that the user is indeed a paying customer. We only need to check that once during a session, then rely on the session variable to remember that s/he has a right to see pages behind the paywall.

## Authentication Step 1

The first step is to get the authentication "**code**" from the Memberful site. This code is a temporary key that allows you to ask for a user's information. The code is automatically returned whenever a user clicks on the web application's Memberful authentication "**sign_in**" link (Listing 16-4).

***Listing 16-4.*** Authentication Sign-in Link

```
<a href="https://<<YOUR-ACCOUNT>>.memberful.com/auth/sign_in">Log in</a>
```

Once Memberful recognizes the querying URL as a valid Memberful client site, it returns a code attached to the redirect URL (Listing 16-5).

***Listing 16-5.*** Authentication Code

```
https://manuelamunategui.pythonanywhere.com/member/?code=483294e65b5dd2e65
862e3c1ba454dee&redirect_to=http%3A%2F%2Fmanuelamunategui.pythonanywhere.
com%2Fmember%2F
```

Our application reads the code "**code=483294e65b5dd2e65862e3c1ba454dee**" as a "**GET**" variable and holds on to it to build the second authentication step.

# Authentication Step 2

Our custom "**IsSubscriberLoggedIn()**" function does most of the authentication work. It first builds the "**access_token_req**" dictionary that holds the temporary "**code**" from the previous step, the application's Memberful Id and secret keys, and the redirect URL (all these can be found in your Memberful account under "**Settings**" ➤ "**Integrate**" ➤ "**Custom Apps**").

It then passes the access token dictionary as a "**POST**" to the "**oauth/token**" URL (Listing 16-6).

***Listing 16-6.*** The auth/token URL

```
https://<<YOUR-ACCOUNT>>.memberful.com/oauth/token
```

If all is correct, it will return the "**access_token**"; this is the real key that will unlock our visitor's personal data, like whether or not they are an actual member of our web application and whether or not their subscription is currently active (Listing 16-7).

***Listing 16-7.*** This Data Can Be Queried Using the Following URL

```
https://<<YOUR-ACCOUNT>>.memberful.com/api/graphql/member?access_
token=999999999
```

Everything in step 2 of the authentication process is done via form "**POST**" so as not to show sensitive information as a readable URL. This type of work is easily abstracted in our function "**IsSubscriberLoggedIn()**," which will enable our web application to easily check each visitor's member status (Listing 16-8).

***Listing 16-8.*** The Function "**IsSubscriberLoggedIn()**"

```
def IsSubscriberLoggedIn(code):

        # build the access token dictionary
    access_token_req = {
        "code": code,
        "client_id": MEMBERFUL_KEY,
        "client_secret": MEMBERFUL_SECRET,
        "redirect_uri": redirect_uri,
        "grant_type": "authorization_code" }
```

```
# build the oauth/token to access visitor's data
content_length=len(urlencode(access_token_req))
access_token_req['content-length'] = str(content_length)
r = requests.post(MEMBERFUL_SITE + '/oauth/token', data=access_token_req)
data = json.loads(r.text)

# build the graphql query to query specific values needed
r = requests.get(MEMBERFUL_SITE + '/api/graphql/
member?access_token=' +              data['access_token'] +
'&query={%20currentMember%20{%20fullName%20subscriptions%20{%20
active%20expiresAt%20}%20}%20}')
```

An area worth mentioning is graphQL (see the "**Queries and Mutations**" section of the Memberful API docs for more details at `https://memberful.com/help/integrate/advanced/memberful-api/#queries-and-mutations`). This tool, after a successful authentication, allows for the querying of specific subscriber information (Listing 16-9).

***Listing 16-9.*** From the Official Help Docs

**Input**:

```
query {
  member(id: 1) {
    id
    fullName
    email
    subscriptions {
      id
      plan {
        id
        name
      }
    }
  }
}
```

**Output:**

```
{
  "data": {
    "member": {
      "id": "1",
      "fullName": "John Doe",
      "email": "john.doe@example.com",
      "subscriptions": [
        {
          "id": "1",
          "plan": {
            "id": "1",
            "name": "Monthly"
          }
        }
      ]
    }
  }
}
```

In our case, we are only interested with two pieces of information: what is the user's name and whether their subscription is active (Listing 16-10).

***Listing 16-10.*** We Append the Following Variables to Our "**GET**" URL

```
query={%20currentMember%20{%20fullName%20subscriptions%20{%20active%20
expiresAt%20}%20}%20}
```

This translates to: give us the full name of this member along with whether their subscription is active and when it expires. Here we only use whether or not the subscription is active, but you could easily extend this by checking the expiration date and reminding the member to renew soon (Listing 16-11).

***Listing 16-11.*** If We Run It and Peek At Our graphQL Response, We Get

```
{"data":{"currentMember":{"fullName":"Manuel", "subscriptions":[{"active":true,
"expiresAt":1529879538}]}}}
```

# Calling Memberful Functions

This is the beauty of Memberful, it is trivial to use and they do all the hard work for us by storing user data and managing payments, refunds, renewals, etc. There are four URLs offered by Memberful that can be embedded on your website.

**To sign in:**

```
https://<<YOUR-ACCOUNT>>.memberful.com/auth/sign_in
```

**To buy a subscription (plan number will vary):**

```
https://<<YOUR-ACCOUNT>>.memberful.com/checkout?plan=29504
```

**To log out:**

```
https://<<YOUR-ACCOUNT>>.memberful.com/auth/sign_out
```

**To manage your account:**

```
https://<<YOUR-ACCOUNT>>.memberful.com/account
```

By adding a simple JavaScript snippet at the beginning of each page along with login/signup/purchase links, you'll automatically inherit the customer pop-up management system. This is extremely powerful, as the visitor feels that it is all built inside our web application (Listing 16-12).

*Listing 16-12.* Memberful JavaScript Code to Manage the In-site Pop-ups

```
<script type="text/javascript">
  window.MemberfulOptions = {site: "https://amunategui.memberful.com"};

  (function() {
    var s    = document.createElement('script');

    s.type  = 'text/javascript';
    s.async = true;
    s.src   = 'https://d35xxde4fgg0cx.cloudfront.net/assets/embedded.js';

    setup = function() { window.MemberfulEmbedded.setup(); }
```

```
s.addEventListener("load", setup, false);

( document.getElementsByTagName('head')[0] || document.getElements
ByTagName('body')[0] ).appendChild( s );
})();
</script>
```

You can find the JavaScript snippet on the Memberful site by clicking on the settings button on the top right corner, and then the '**Your Website**' tab (Figure 16-8).

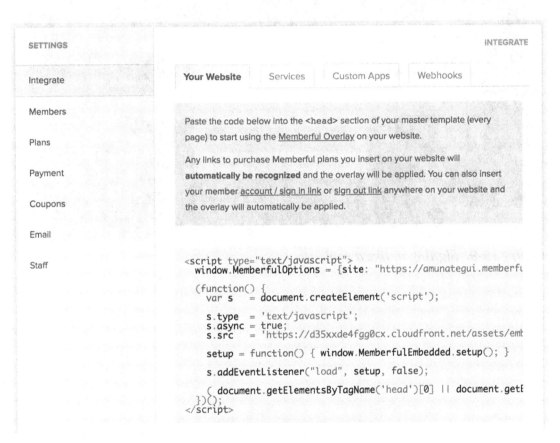

*Figure 16-8.*  *Accessing the "**MemberfulOptions**" JavaScript Snippet*

When the user accesses the web application and clicks on "**Log in**," they get that professional and integrated pop-up box that we all have come to expect on serious web sites (the fact that we aren't managing or storing any of the user data or financial data is our own little dirty secret; Figure 16-9).

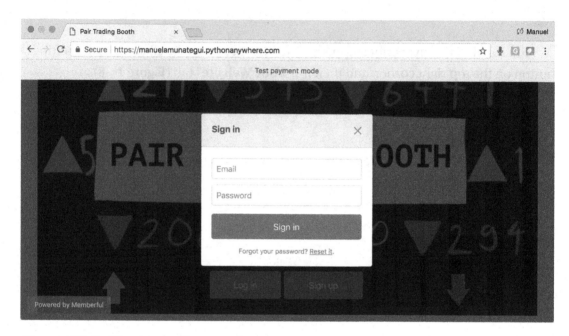

***Figure 16-9.*** *Signing in like a pro!*

# Designing a Subscription Plan on Memberful.com

In order to create a membership, a visitor has to come to the site and click on "**Sign up**." When they do so, they will see a subscription window. This can be customized in many ways, including different prices, tiers, and subscription lengths. We'll go with the defaults offered by the demo account (Figure 16-10).

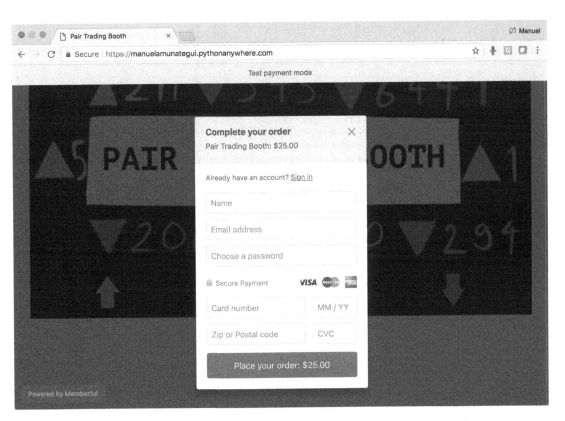

***Figure 16-10.*** *A new visitor creating a membership to the Pair Trading Booth site; yes!!!*

A membership plan can easily be created on the Memberful dashboard. Log in to your account and navigate to **Plans ➤ Sample Plan ➤ Plan** settings, copy the generated URL plan purchase link, and paste it in your sign-up button (Figure 16-11).

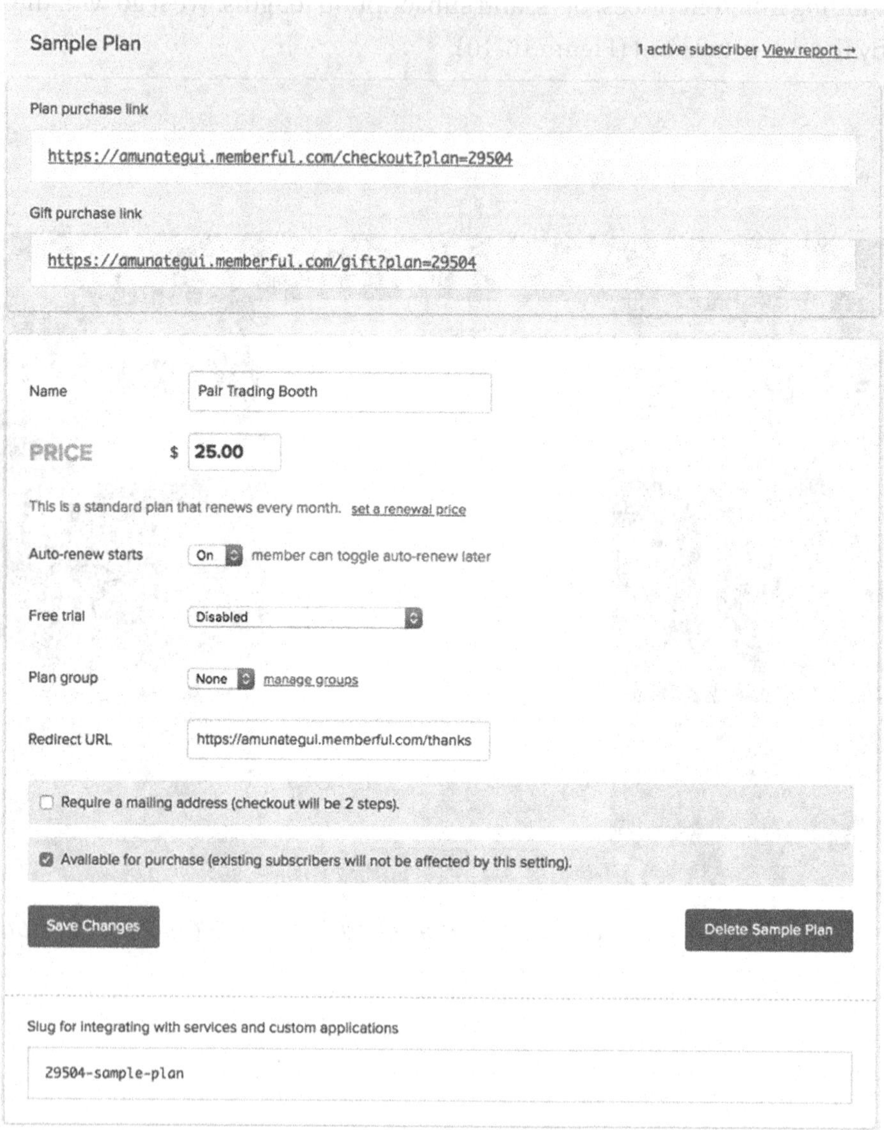

*Figure 16-11.* *Creating a sample plan; here we go with the defaults*

If you take the "**Plan purchase link**" and drop it into a browser, you will see what it offers (of course, we won't access it that way normally because we want it to appear as a pop-up inside our own web application). It is live, and you can create a test account using one of the fake credit card numbers supplied (Figure 16-12).

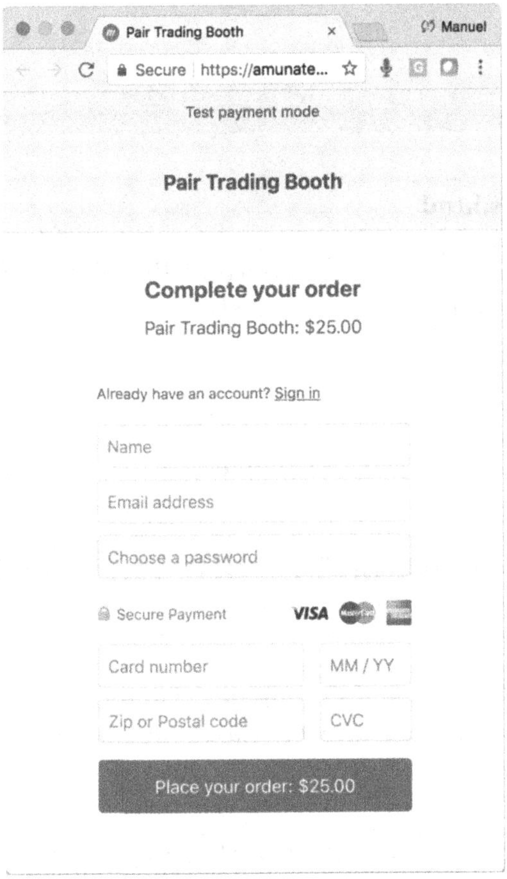

**Figure 16-12.**  *Don't forget to use a testing credit card number!*

Use any of the following fake credit card numbers (from the official docs at `https://memberful.com/help/general/using-test-credit-cards/`).

- **Visa**: 4242 4242 4242 4242

- **Mastercard**: 5555 5555 5555 4444

- **American Express**: 3782 822463 10005

# Uploading the Web Application to PythonAnywhere

As there is no local Flask version to run in this chapter, let's get it up to PythonAnywhere without further ado. If you have been following along with the previous "**Pair Trading Booth**" case studies, you will only need to update the following files:

- **main.py**
- **index.html**
- **charts.html**
- **fundamentals.html**

And you will need to add the new landing page that hides all trading information from all the nonmembers and those not logged in:

- **welcome.html**

Log into your PythonAnywhere account and replace the five files with the new versions found in the downloads for this chapter.

## Replacing Memberful and MySQL with Your Own Credentials

You will need to replace a few things in "**main.py**" before you can run the web application. There are three Memberful constants and one PythonAnywhere constant to set (Listing 16-13).

***Listing 16-13.*** Change the Following Constants in "**main.py**" on PythonAnywhere with Your Credentials

```
MEMBERFUL_KEY='<<ENTER-YOUR-MEMBERFUL-KEY-HERE>>'
MEMBERFUL_SECRET='<<ENTER-YOUR-MEMBERFUL-SECRET-HERE>>'
MEMBERFUL_SITE='<<ENTER-YOUR-MEMBERFUL-SITE-HERE>>'
PYTHONANYWHERE_SITE = '<<ENTER-YOUR-PYTHON-ANYWHERE-SITE-HERE>>'
```

Once everything is up, hit the big, green button to refresh the web application and take it for a spin. You will need to go through the sign-up process once with the fake credit card numbers; then you will be able to log in with those credentials (Figure 16-13).

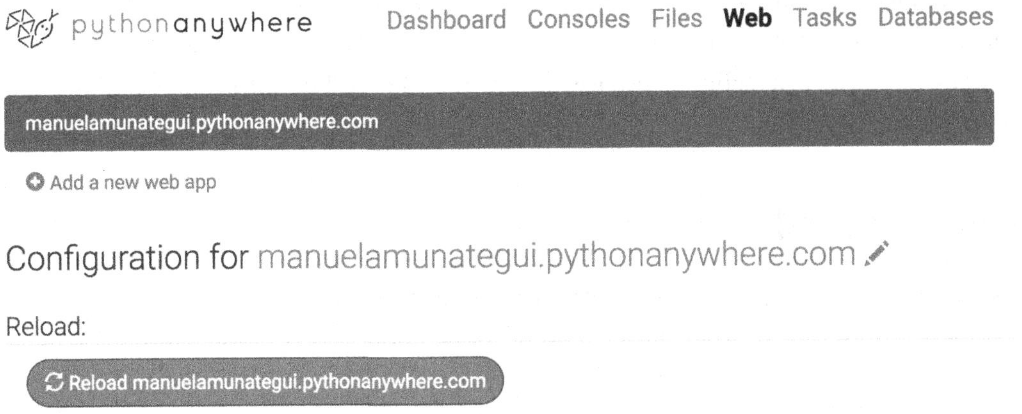

**Figure 16-13.** *Last step in the process before turning our paywall live*

# What's Going on Here?

Let's take a high-level look at some of the interesting elements going on in our paywall.

## main.py

This is the brains behind our web application; thus it will take the brunt of the Memberful additions. We need to add a handler to handle the landing page "**welcome. html**" where users go before logging in. Once they have logged in, they are directed to the "**/member/**" path where they can be directed through the following three authentication paths:

**Did the member request a logout?**

Check the "**request.args.get('action')**" variable and look for the "**logout**" value. If that is the case, log them out by clearing that member's session variable.

**Is the member already logged in?**

Check the session variable to see if there is a user name in it; if so, there's no need to authenticate again. Let them keep browsing behind the paywall.

**Did this visitor just arrive and is trying to log in?**

Get the "**request.args.get('code')**" and pass it to the "**IsSubscriberLoggedIn()**" function to make sure they have an active subscription. If they do, add their user name to the session object and they're good to browse behind the paywall.

The rest of the code in "**main.py**" is the same as the previous case studies. It has code to look for the most extreme stocks in our list of ten Dow 30 stocks, code to create dynamic price charts, and the ability to pull corollary information on the companies in play.

## welcome.html

The "**welcome.html**" page is the new landing page (Figure 16-14).

**Figure 16-14.**  *The new landing page*

It is a simple HTML page that follows the look and feel of the "**Pair Trading Booth**" web site. It offers two buttons: one to log in and the other to sign up.

## index.html

The "**index.html**" page gets two new buttons: one to log out and another to manage the user's account (Figure 16-15).

**Figure 16-15.** *Two new buttons: one to log out and one to access account info*

## Conclusion

There you have it: all the tools you need to create your own paywall to monetize your machine learning ideas. Being able to push out the management of subscribers and credit card payments in such an integrated manner is simply amazing. This is something that would have been a whole lot harder to achieve just a few years ago. You now can focus fully on your machine learning ideas, and let the membership pros, do the rest.

# CHAPTER 17

# Conclusion

The coverage in this book is ambitious and sacrifices had to be made, sections had to be omitted. What it may lack in technology introductions is hopefully made up for by quickly getting you up and running, and providing pointers on where to look for additional information. We only briefly covered databases and didn't cover custom domain names; thankfully plenty of others have written about that already.

I hope you found these chapters inspiring and that the gears in your head are spinning when thinking about all the things you could do with the Memberful paywall implementation.

Remember that "**compete agreement**"? Yes, the opposite of a non-compete agreement that I mentioned in the introduction of the book. So, its time to take anything you need from the book and take it on the road with you! We can't wait to see what you come up with!

## Turning It Off!

Let's quickly review how to turn off cloud instances as well as Memberful and PythonAnywhere accounts. Recall that in many cases you can turn instances off using command-line commands, but it is always a good idea (essential idea, really) to log into your account in the cloud and make sure everything is turned off. (Be warned: if you don't, you may get an ugly surprise at the end of the billing cycle.)

## Google Cloud (App Engine)

Navigate to your GCP account, to the "**App Engine**" dashboard, and to "**Versions**." Click your active version and stop it (Figure 17-1). If you have multiple versions, you can delete the old ones; you won't be able to delete the default one, but stopping it should be enough (if you really don't want any trace of it, just delete the entire project).

471

© Manuel Amunategui, Mehdi Roopaei 2018
M. Amunategui and M. Roopaei, *Monetizing Machine Learning*, https://doi.org/10.1007/978-1-4842-3873-8_17

**Figure 17-1.** *Stopping and/or deleting your App Engine version*

# Amazon Web Services (Beanstalk)

Log into your AWS account and make sure that your EC2 and Elastic Beanstalk accounts don't have any active services you didn't plan on having (Figures 17-2 and 17-3).

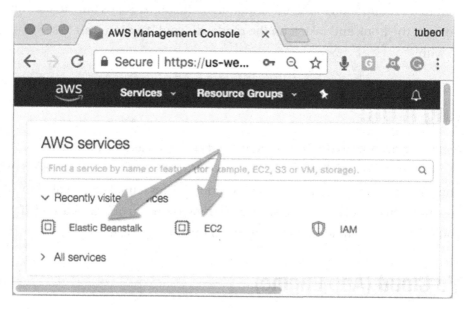

**Figure 17-2.** *Checking for any active and unwanted instances on the AWS dashboard*

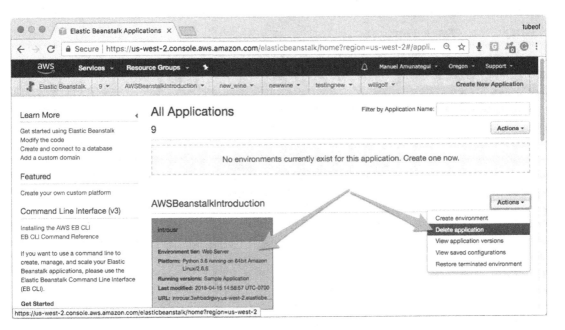

***Figure 17-3.*** *Locate the instance you want to terminate or delete, and select your choice using the "**Actions**" dropdown button*

In case you see an instance that seems to keep coming back to life after each time you "**Delete application**," check under EC2 "**Load Balancers**" and terminate those first, then go back and terminate the rogue instance again (Figure 17-4).

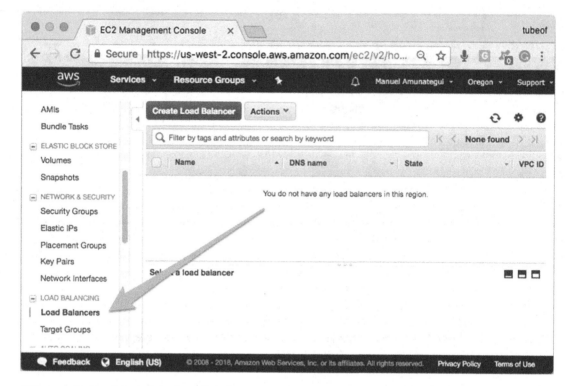

*Figure 17-4.* *"Load Balancers" can prevent an application from terminating; this can kick in if you inadvertently start multiple instances with the same name.*

## Microsoft Azure (AWS)

Log into the Azure Dashboard, enter "**All resources**" in the search bar, and delete everything you created (Figure 17-5).

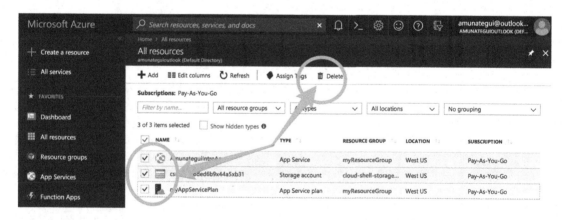

*Figure 17-5.* *Deleting unwanted resources under the "All resources" view*

# PythonAnywhere.com

If you have opted for a paid account and want to downgrade back to a free account, simply log into your PythonAnywhere dashboard and click the "**Account**" tab in the upper-right corner. This is where you can upgrade and downgrade your account depending on your needs. Click the "**Downgrade to a free account**" and your'e back into the free tier (Figure 17-6).

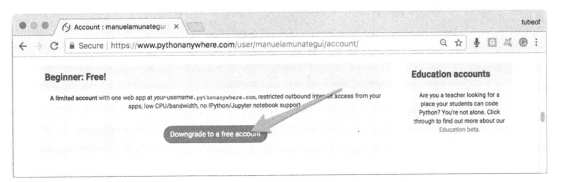

*Figure 17-6.*  *Downgrading to a free account on PythonAnywhere*

# Memberful.com

If you have gone through Chapter 16 and set up the paywall but would like to not incur additional charges, you can easily downgrade back into the free tier. Log into your Memberful account and click the "**Account**" button in the top-right corner, then choose the "**Plans and billing**" tab. On this page you will find an option to "**Downgrade to Starter**" link; click it and follow the instructions (Figure 17-7).

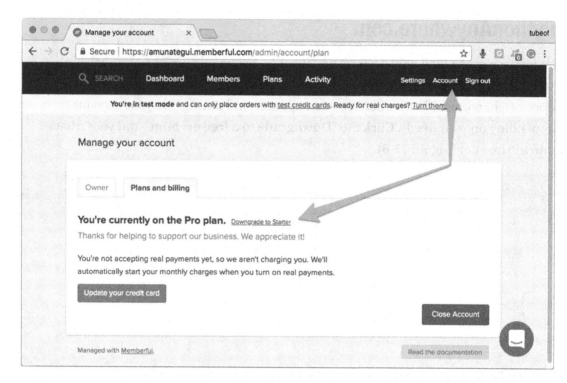

***Figure 17-7.*** *Downgrading to a starter account on Memberful*

That's it! And a huge thanks for reading this book!

# Index

477

Printed in the United States
By Bookmasters